New Generation of Europium and Terbium Activated Phosphors

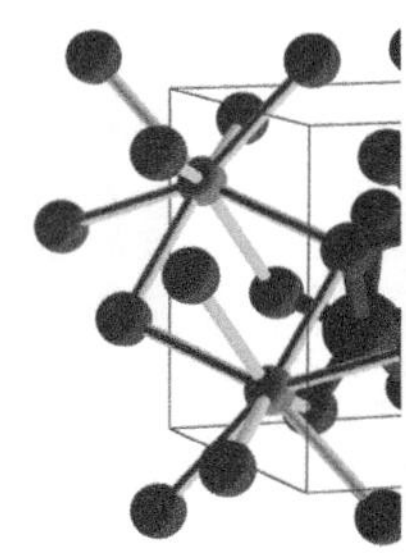

Mihail Nazarov
Do Young Noh

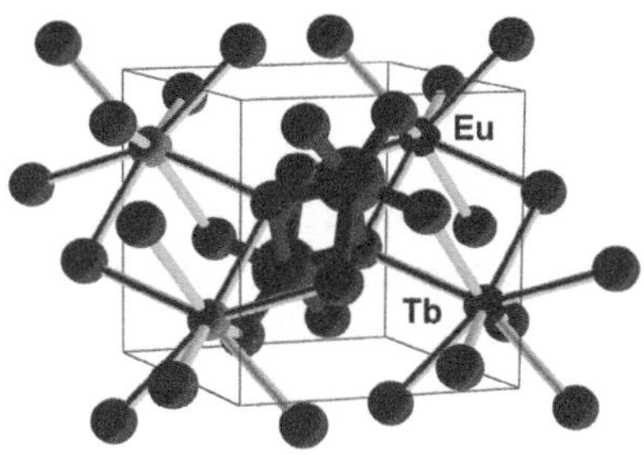

New Generation of Europium and Terbium Activated Phosphors

from syntheses to applications

PAN STANFORD PUBLISHING

Published by

Pan Stanford Publishing Pte. Ltd.
Penthouse Level, Suntec Tower 3
8 Temasek Boulevard
Singapore 038988

Email: editorial@panstanford.com
Web: www.panstanford.com

British Library Cataloguing-in-Publication Data
A catalogue record for this book is available from the British Library.

New Generation of Europium and Terbium Activated Phosphors: From Syntheses to Applications

ISBN 978-981-4310-77-2 (Hardcover)
ISBN 978-981-4364-05-8 (eBook)

Printed in the USA

Contents

Preface

Rare earth (RE) doped materials are of crucial importance to optoelectronics and are widely deployed in solid state lasers, solid state lighting, LED, etc. In recent decades, a large number of luminescent materials based on rare earth ions or rare earth host lattices have been invented. In many cases, rare earth phosphors dramatically improved the performance of the devices in which they were applied. Thus, it seems attractive to develop new phosphor materials with high efficiency and color saturation or modify well-known compositions by co-doping them with RE ions, in particular with Eu and Tb activators. The luminescence properties could be essentially varied and, therefore, some new applications could be proposed.

Since the 1950s, several classic books on phosphors and luminescence have been published, including *Luminescent Materials*, by Blasse and Grabmaier, *Phosphor Handbook*, edited by Shionoya, Yen and Yamamoto, and many other books. This book does not repeat these numerous publications. The authors did not aim revise the basic principles of luminescence. The main notions and definitions are given in the book only to better understand the proposed material, and an extensive bibliography is given for further reading. The authors do not aim to make a detailed analysis of all phosphors. It is impossible, because a huge amount of phosphors are known now, and every day their number increases. Only commercial phosphors used often and widespread are shown in the book, and more attention is attracted to new materials or well-known phosphors with improved properties. The old books became rapidly out of date and are not actual today. New phosphors appear, new methods of investigation are developed and new theories are created. All this stimulate the writing of new articles, reviews and books. The present book continues and adds to the series of books about luminescence and phosphors.

The goal of this book is to provide a connection between basic research and materials science through analysis of fundamental spectroscopic properties of Eu and Tb activated luminescent materials. In addition this book will serve as an updated reference for materials research by covering recently published articles, reviews and books on this topic. This book presents both the physics and the materials aspects of the field of solid state luminescence. The achievement of materials having purity and suitable morphology, as well as manufacturability, is very important for solid state luminescence. The book reviews key types of phosphors and solid state luminescence that are of current interest, including X-ray phosphors, display materials and light emitting materials and devices. This book concentrates on the luminescence and structural properties of the new generation of Eu and Tb activated phosphors, associated phenomena and related topics, from basic principles to the most recent discoveries. The book summarizes the present state of the art in this rapidly growing field. Recent developments in the areas of rare earth doped phosphors are discussed, and some new materials or well-known materials with improved properties that open up new possibilities are described. The areas of focus include X-ray phosphors, phosphors for light-emitting devices, emissive displays and fluorescent lamps. The book comprises theoretical and experimental analyses of various properties of phosphors, research methods and preparation techniques, and some promising applications.

The book consists of four parts.

The first part (Chapter 1) is a very general introduction to luminescence and luminescent materials.

The second part (Chapter 2) gives an overview of the synthesis of phosphors. This chapter reviews key types of synthesis methods that are of current interest, including solid state and wet-chemical methods.

In the Chapter 3 we show the measurements of phosphor properties. Some new and non-traditional methods of phosphor investigation as color cathodoluminescence or terahertz spectroscopy are presented and explained. Every method is demonstrated by real examples.

The main part of this book is Chapter 4. It consists of the original investigations and applications of the new generation of phosphors, activated by europium and terbium. The host materials have become the key problem in searching for novel phosphors. Vanadates, molybdates, tungstates, tantalates and niobates, known as self-activated compounds, have become interesting hosts for rare earth ions and, especially, for Eu^{3+} and Tb^{3+} activation. This is because they can transfer energy to activators successfully after being excited by the excitation source. Most of these hosts we consider in this book. It is shown that two activators can change the emission in all visible range and we can obtain the materials with predicted properties. High-efficiency red, green and yellow phosphors are of particular interest, though recent developments in more traditional materials with improved properties are highlighted. Much of the material is drawn from personal experience in synthesizing, studying, and applying modified and improved phosphors during the last two decades. Some of the data in this book were obtained during two years' work (M.Nazarov) in the Samsung Company, Suwon, Korea.

The development of new optoelectronic materials depends not only on materials engineering at a practical level, but also on a clear understanding of the properties of materials, and fundamental science behind these properties. The book therefore also includes such paragraphs as Stark levels and selection rules, electron-vibrational interaction in optical transitions, quantum efficiency, first-principles (*ab initio*) calculations, configurational coordinate diagram and others. One of the aims of this book is to describe the spectroscopic properties of some perspective phosphors activated with Eu and Tb to demonstrate their practical application. The theoretical analysis of $5d$–$4f$ transitions as well as transitions between the $4f$ levels in different compounds will be discussed. A short overview of the origin and the key characteristics of the broad phonon-assisted optical bands of the light absorption and emission is given. We discuss both the semiclassical and quantum mechanical approaches and deduce the key parameters of the bands, namely, Stokes shift, mirror low for absorption and emission, temperature dependence of the half-width, zero-phonon lines and vibronic satellites comprising a good balance of theoretical and experimental chapters. This book discusses the luminescent materials which have been, are or may be used in the application concerned. Their performance is discussed in terms of theoretical models presented in Chapter 4.

This book includes an extensive bibliography for further reading. Sufficient recent references are given to locate most of the literature on luminescence of modern phosphors.

The book is aimed at students, researchers and technologists engaged in the research, development and study of materials in electronics and photonics, and industrial scientists developing new materials and devices for the electronic, optoelectronic and communications industries. This book is interesting for people working in both an academic and an industrial environment, in physics, chemistry and material science. This book may also be used as a starting point to gain an understanding of various types and mechanisms of luminescence for students and professionals.

It is a pleasure to acknowledge the friendly advice of many scientists. Their helpful suggestions, corrections and criticisms have been a considerable aid in the preparation of this book. For helpful discussion on the theoretical aspects of absorption and luminescence, the authors are very grateful to Professor Boris Tsukerblat of Department of Chemistry, Ben-Gurion University of the Negev, Beer-Sheva, Israel.

Further appreciation is expressed to Dr. Elisabeth-Jeanne Popovici and her colleagues of the "Raluca Ripan" Institute for Research in Chemistry, Babes Bolyai University, Cluj-Napoca, Romania, for synthesis of many excellent $CaWO_4$ and $Y(Ta,Nb)O_4$ phosphors and useful discussions.

Mihail Nazarov

Chapter 1

PHOSPHORS AND LUMINESCENCE

A phosphor is a material that emits light as a result of it absorbing energy (Fig. 1.1). Luminescence is the term given to the emission of this light.

Hundreds of thousands of phosphors have been synthesized, each one having its own characteristic color of emission and period of time during which light is emitted after excitation ceases. Most phosphors are composed of a transparent microcrystalline host (or a matrix) and an activator, i.e., a small amount of intentionally added impurity atoms distributed in the host crystal. Therefore, the luminescence processes of a phosphor can be divided into two parts: the processes mainly related to the host, and those that occur around and within the activator. The interaction between the host and the activator is not explicitly discussed in this section; in this sense, the host is treated only as a medium for the activator. The interaction processes such as the transfer of the host excitation energy to the activator will be discussed for each phosphor in Chapter 4.

The activators are primarily responsible for the luminescence.

1.1 Luminescence Classification

Luminescence, whatever may be said about it in quantum physics, is well known to everyone. You meet luminescence every day in

New Generation of Europium and Terbium Activated Phosphors
Mihail Nazarov and Do Young Noh

www.panstanford.com

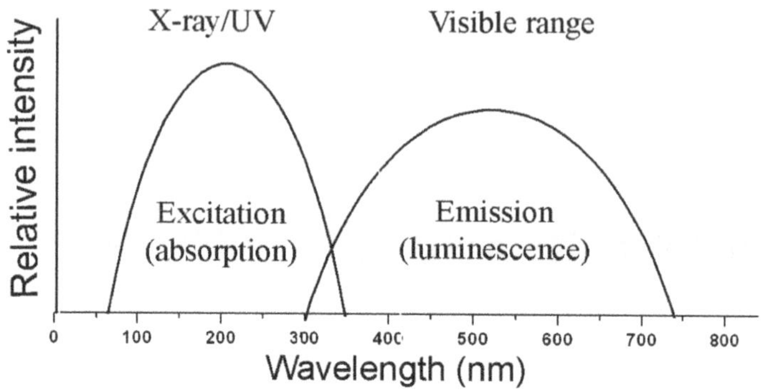

Figure 1.1 Diagram illustrating the energy absorbed by a phosphor (left), which is then transformed to emitted light of lower energy (right).

your office, laboratory, supermarket, on a street, and at home. You see luminescence when you switch on a fluorescent light, look at the screen of a computer or electron microscope, print on a paper, see bright traffic safety and advertising signs, deal with a document with an invisible marking, select an ideally white cotton, or relax in front of your TV set. In all these cases you see the emission of light. However, all is not luminescent that shines. When you heat a solid to a temperature in excess of about 600°C, it also shines and emits infrared radiation, but this is a thermal radiation. Luminescence (from the Latin *luminis* for light) is emission of light due to other causes than high temperature. Of course, during luminescence the temperature of a solid is raised; however, a luminescent material converts certain type of energy into electromagnetic radiation over and above thermal radiation. Besides, the decay time of luminescence (τ) exceeds a period of light oscillations of molecules ($T = 10^{-14}$–10^{-15} s).

1.1.1 *Types of Excitation*

A luminescent material, also called a phosphor, is a solid which converts the incoming radiation into visible light. Sometimes, such conversion can also be in other spectral regions, such as the ultraviolet or infrared, but usually it is a visible light (Fig. 1.1). Luminescence can occur as a result of many different kinds of excitation, excluding thermal radiation. In practice, most often we observe the next types of energy excitation and corresponding luminescence:

Electromagnetic (ultraviolet) radiation	→	Photoluminescence
X-ray	→	X-ray luminescence
Beam of energetic electrons	→	cathodoluminescence
Electric current or voltage	→	Electroluminescence
Mechanical energy	→	Triboluminescence
Energy of chemical reaction	→	Chemiluminescence
Chemical reactions in living things	→	Bioluminescence

Photoluminescence. Photoluminescence (PL) is the spontaneous emission of light from a material under optical excitation. Optically stimulated luminescence is excited by electromagnetic (often ultraviolet) radiation or by visible light or infrared. In this case ultraviolet (UV), vacuum ultraviolet (VUV), or infrared (IR) light is only a trigger for release of previously stored energy.

X-ray luminescence. X-ray luminescence (XL) can be defined as luminescence from materials which absorb X-rays and convert the absorbed energy efficiently into luminescence, in practice often ultraviolet or visible emission.

Cathodoluminescence. When excitation is done by electron beams generated at the electrical cathodes, the emission produced is called cathodoluminescence (CL). The screens of cathode ray tubes (CRT) and television tubes glow by this kind of emission. In cathode ray tubes zinc and cadmium sulphide phosphors often are used. Production of phosphors for TV screens is a very specialized technique, which requires a variety of colors and their appropriate persistence to smooth out the flicker of the scan.

Electroluminescence. Electroluminescence is luminescence caused by electric current or a strong electric field. Examples of electroluminescence are neon lights, the auroras and lightning flashes. This should not be mistaken for what occurs with the ordinary incandescent electric lights, in which electricity is used to produce heat, and it is the heat that in turn produces light. There is another type of electroluminescence known as injection luminescence. In this electron are injected from an external supply across a semiconductor p-n junction. On applying a direct current voltage across the junction, such that the electrons flow to the *p* region, luminescence is produced by the electron-hole recombination in that region. The light emitting diodes (LEDs), which are now commonly used as display devices in many scientific instruments, are based on this principle.

Triboluminescence. A large number of inorganic and organic materials subjected to mechanical stress, emit light, which is called triboluminescence or mechanoluminescence. It has been observed that all piezoelectric crystals exhibit triboluminescence. The spectra of triboluminescence light are similar to those of photoluminescence in many substances.

Chemiluminescence. It is the luminescence where the energy is supplied by chemical reactions. That glow in the dark plastic tubes sold in amusement parks is an example of chemiluminescence. Chemiluminescence is not a common accompaniment in chemical reactions, because the amount of energy released even in the exothermic reactions is not sufficient to cause electronic excitation, which needs a couple of eV energy. Secondly not all chemical molecules are capable of luminescence.

Bioluminescence. Bioluminescence is luminescence caused by chemical reactions in living things; it is a form of chemiluminescence. Fireflies, glow-worms, some bacteria and fungi and many sea creatures, both near surface and at great depths, are the striking examples of luminescence in living beings. The chemical reactions are the enzymic oxidations.

1.1.2 *Fluorescence and Phosphorescence*

In addition to the type of excitation, two other terms are used quite often to classify luminescent materials: *fluorescence*, from the Latin *fluor* = flow, and *phosphorescence*, from the Greek *phos* = light and *phoros* = carrying. Both can be related to the decay time (τ). For fluorescence the decay time is less 10 ms, usually $\tau \approx 10^{-8}$ to 10^{-9} s, and for phosphorescence $\tau > 0.1$ s, usually $\tau \approx 10^{-1}$ s to a few hours. Fluorescence is a spin-allowed transition ($\Delta S = 0$), which is very fast, whereas phosphorescence is a spin-forbidden transition ($\Delta S = 1$), which is very slow.

In modern usage, light emission from a substance during the time when it is exposed to exciting radiation is called fluorescence, while the afterglow if detectable by the human eye after the cessation of excitation is called phosphorescence. (In contrast to the nearly instantaneous "on-off" of fluorescence, phosphorescence involves a delayed emission of radiation following absorption. The delay may take as much as several minutes, but phosphorescence continues to appear after the energy source has been removed). However, it should be noted that these definitions are applied only to inorganic materials; for organic molecules, different terminology is used.

Typical application of inorganic fluors is in security documents (invisible marking on currency and stamps), fluorescent lamps and cathode ray display tubes. Organic fluors (called brighteners or optical whiteners) are used in the textile, paper and plastic industries to whiten materials. Phosphorescence is displayed by some inorganic crystalline solids, referred to as crystalline phosphors. As a rule, host crystal of a phosphor does not display any luminescence, but being doped with small amounts (10^{-2}–10^{-4}%) of some metals, called activators, the phosphor emits a light for a long time after the exciting emission is removed.

1.2 Luminescence Terminology

This section explains the terminology that is used to quantify how electrons make transitions between different atomic energy levels of the activators that lead to the phenomenon of luminescence. Starting with the principle quantum numbers, the most widely used electron coupling scheme, *L-S* or Russell-Saunders coupling, which is used to label the energy levels of a free activator ion, will be explained. An explanation of the rules that govern electronic transitions will be discussed next, as well how these transitions can be influenced by the environment around the activator. Finally radiative and non-radiative transitions in a phosphor will be explained using the configurational coordinate model. We hope that this minimized background in luminescence for beginners will help them to understand the application of these rules to new generation of phosphors considered in Chapter 4. One can find detailed information about spectroscopic properties of rare earths in optical materials in special books [1–3].

Table 1.1 Letter symbols assigned to each value of l.

l	0	1	2	3	4	5	6	7	8
Letter symbol	s	p	d	f	g	h	i	k	l

1.2.1 *Quantum Numbers, Term Symbols and L-S Coupling*

Every electron orbiting an atom can be assigned four unique quantum numbers, namely n, l, m_l, and m_s. Pauli's exclusion principle states that no two electrons orbiting a nucleus can have the same quantum numbers.

n, the principle quantum number, gives the number of the shells the electron is located in, and can be any positive integer number from 1 to infinity.

l, the angular momentum quantum number, can be any positive integer from $0 \leq l < n$. For example, when $n = 1$, l can only equal 0. When $n = 2$, $l = 0$ or 1. Thus, the number of atomic orbitals in each shell is determined by l. For each value of l, a symbol is assigned to distinguish the orbitals inside each shell, as shown in Table 1.1.

Therefore, for $n = 2$, $l = 0$ or 1, there is ($l = 0$) first orbital in the ($n = 1$) first shell ($l = 0, 1$) and 2*s* and 2*p* orbitals in the ($n = 2$) second shell.

m_l, the magnetic quantum number, represents the component of the angular momentum l along some reference axis (z). m_l can have positive and negative integer values, $-l \leq m_l \leq l$. For example, for $l = 2$, $m_l = -2, -1, 0, 1$, and 2.

m_s is the electron spin quantum number and it takes on only two values, ±½, and represents the spin angular momentum of the electron around its own axis. For the six 4*p* electrons, the four primary quantum numbers are represented in the table below:

Table 1.2 Representation of the six electrons in the 4*p* orbital.

n	l	m_l	m_s	
4	1	−1	+½, −½	4p, 6 electrons
4	1	0	+½, −½	
4	1	+1	+½, −½	

Table 1.3 Letter symbols assigned to each value of L for the L-S coupling scheme.

L	0	1	2	3	4	5	6	7	8
Term symbol	S	P	D	F	G	H	I	K	L

For hydrogen-like atoms (atoms or ions with only one electron in a shell), the energy of the electron is dependent only on the principle quantum number n. When there is more than one electron in a shell though, each electron still has a set of unique quantum numbers, but they repel each other through coulombic interactions. This leads to the splitting of the individual l orbitals, for instance making $2p$ higher in energy than $2s$, etc. To take into account the coulombic interactions, the total angular momentum of all electrons located above closed shells in the atom is summed:

$$L = \sum_i l_i$$

where L represents the term symbol. Like l, L takes on the same symbols but in capital letters, as seen in Table 1.3.

It is important to note here that L and l are not the same. Only when there is one electron located above a closed shell does L and l represent the same orbital. Because l is a vector quantity, L is determined by vector addition, and only those resultants that give integer values give valid L values. Like L, the spins can be added giving the spin quantum number S, which is not a term symbol. The number of wave functions that can represent the electron interactions is called the multiplicity and is given by $2S + 1$.

Examples of how the multiplicity can be calculated for one or more electrons is given in Ref. [4].

Optical line spectra of rare earth arise from transitions between levels of the $4f^N$ configuration. The positions of these levels arise from a combination of the Coulomb interaction between the electrons, the spin-orbit coupling and the crystalline electric field. The resultant splittings of the $4f^N$ configuration are shown schematically in Fig. 1.2. The electrostatic interaction yields terms ^{2S+1}L with separations of the order of 10^4 cm^{-1}. The spin-orbit interaction

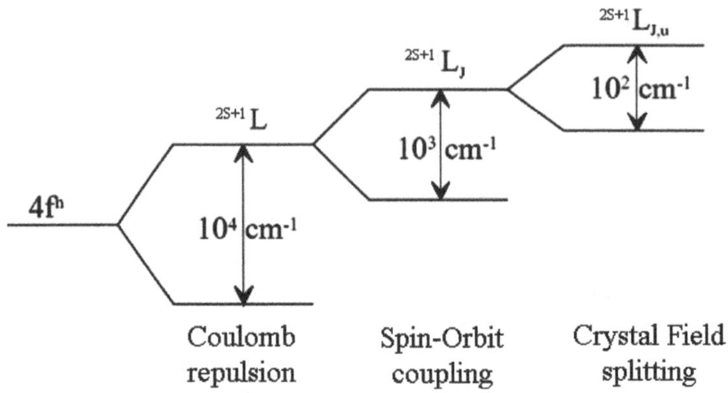

Figure 1.2 Schematic diagram of the splitting of rare earth energy levels due to the electrostatic, spin-orbit and crystal field interactions.

then splits these terms into J states with typical splitting of 10^3 cm^{-1}. Finally, the J degeneracy of the free ion states is partially or fully removed by the crystalline Stark field, yielding a Stark manifold usually extending over several hundred cm^{-1}.

Figure 1.3 shows how the six $4f$ electrons in the trivalent europium Eu^{3+} ion split into different levels. First, there are lowest two levels shown due to coulombic interactions with the level denoted by the term symbol ^{2S+1}L. After coulombic interactions, energy levels can be further split by the exchange interaction or spin correlation. When electrons occupy available energy levels, the electrons try to keep their spins the same, all +½ or −½. This is known as Hund's rule, and it states that for equivalent electrons having the same n and l values, the lowest energy state occurs when all the electrons exhibit the same spin. This interaction arises from the magnetic coupling between the magnetic moments associated with the spin and orbital motion of the electrons. For one electron above a closed shell, this effect is taken into account by adding l and m_s, both vector quantities, to give the total angular momentum, denoted by the vector $j \cdot j = l + m_s$ or $l - m_s$, since the spin of the electron can have ±½ values. For more than one electron, the total spin angular momentum (S) and the total orbital angular momentum (L) couple to give a total angular momentum of the ion (J), where $J = |L + S|$. The term symbol than can be written as $^{2S+1}L_J$. This is known as the L-S or Russell-Saunders coupling scheme [5] and can better approximate the true energy

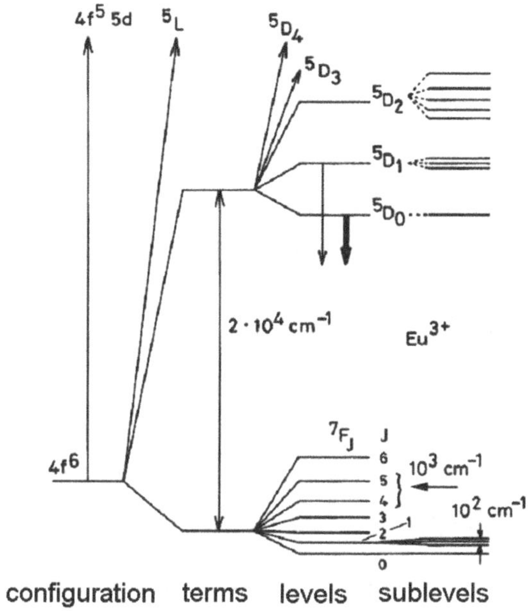

Figure 1.3 Spectroscopic terms and energy levels of Eu^{3+}.

states of the ion. The energy levels splitting due to the spin-orbit interaction is shown in Fig. 1.2 and is labeled according the *L-S* couple scheme and is less than the coulombic interaction. Hund's rules dictate how to arrange the terms from the lowest to highest energy. The rules are as follows: (1) the term with the maximum multiplicity lies lowest in energy, (2) for a given multiplicity the term with the largest value of *L* lies lowest in energy, and (3) for atoms with less than half-filled shells the level with the lowest value of *J* lies lowest in energy, while for shells more than half full the highest *J* lies at the lowest energy. Determination of the term symbols becomes quite complicated when there is more than one electron in the outer orbital. Some simple cases are treated in a cited book, for example Ref. [1]. The term symbol for the ground states of each atom can be determined with a simple calculation, of which the example for Eu^{3+} in Y_2O_3, shown in Fig. 1.3, is determined here. *ML* can take on the values 3, 2, 1, 0, −1, −2 for each individual electron, and the electrons will only occupy states with the same spin, $ms = +½$. This implies the highest value of *l* is 3, making *L* also equal to 3. *S* is then the addition of all the spins, $S = 6 \times ½ = 3$. *J* then is $L + S$, or 6, but can also take on the values 5,

4, 3, 2, 1, and 0. Thus, from Hund's first rule, the highest multiplicity has the lowest energy, $2S + 1 = 7$, then the highest value of L is also put at the lowest energy ($L = 3$), F. Since the $4f^6$ fills less than half the orbital, the lowest $J = 0$. Thus, the ground state of $4f^6$ is $^7F_0$. For more than half filled orbitals, the lowest level is assigned the highest J value (e.g., Tb^{3+} ground state is $^7F_6$). There are some 3003 combinations for $4f^6$, leading to following possible allowed states:

$$^1(S_4PD_6F_4G_8H_4I_7K_3L_4M_2N_2Q)^3\,(P_6D_5F_9G_7H_9I_6K_6L_3M_3NO)$$
$$^5(SPDF_2G_3H_2I_2KL)^7(F)$$

where the subscripts indicate the number of such terms and the superscript indicates the multiplicity (adapted from Ref. [1]). For example, $f^6 \rightarrow {}^1(S_4PD_6...)$ means there are four ^1S terms, one ^1P term, six ^1D, etc.

1.2.2 *Selection Rules*

For free ions, transitions are classified as either forbidden or allowed and generally only make transitions between adjacent states, as shown in Table 1.4. Transitions between states that have the same parity (wave functions with the same ever (u) or odd (g) symmetry), i.e. $f \rightarrow f$, are generally forbidden and if they occur, produce only very weak emission. This is known as Laporte's selection rule and basically means that electrons are not allowed to redistribute themselves within the same orbital and transitions only take place between even and odd wave functions. Allowed transitions are those that electrons transition between different orbitals or shells and the total angular momentum changes by 1, with the selection rules written as follows:

$$\Delta n = \pm 1, \pm 2, ..., \Delta l = \pm 1 \text{ and } \Delta j = 0, \pm 1$$

Table 1.4 Allowed and forbidden electron transitions types (free ion scheme).

Allowed Transitions	Forbidden Transitions
$s \leftrightarrow p$	$d \leftrightarrow d$
$f \leftrightarrow d$	$f \leftrightarrow f$

When a photon is emitted by a transitioning electron, the total momentum of the system must be conserved. A photon has an angular momentum or "spin" of 1, so when this photon is ejected, the total momentum of the system must also change by 1.

Electrons can make transitions that are termed electric dipole ($E1$), magnetic dipole ($M1$) and quadrupole ($E2$). $E1$ is allowed for the free ion while $M1$ and $E2$ are forbidden, and if they occur, are very weak. Each transition type is governed by certain selection rules. The selection rules in terms of the *L-S* coupling scheme for allowed $E1$ is

$$\Delta S = 0, \Delta L = 0, \pm 1 \text{ (except } 0 \rightarrow 0)$$

That is, the spin moment does not change and the transition occurs between at least two adjacent states. For *M1*, the rules are the same as $E1$ with the added rule $\Delta J = 0, \pm 1$. For $E2$, $\Delta S = 0$, $\Delta L = 0, \pm 1, \pm 2$.

When an activator ion is incorporated into a host lattice, the surrounding crystal field, or ligand field, can further split the energy levels of the activator ion. Each level can split into a number of sublevels, called the Stark effect, and as shown in Fig. 1.3 for crystal field splitting. The number of sublevels can be at most $(2J + 1)$ for J integer or $(J + ½)$ for J half-integer. The number of sublevels is determined by the symmetry of the ligand field that surrounds the activator ion.

We are now ready to consider how spectra can be interpreted in terms of energy transitions between these various levels.

1.3 Luminescent Materials

1.3.1 *The Past and Present of Phosphors*

Inorganic luminescent materials have been known since approximately the tenth century in China and Japan as well as since the end of the Middle Ages in Europe. The detailed history of luminescence can be found in one of the early books of Harvey [6] or in a special review dedicated to the 100 years of research and application of inorganic luminescent materials [7].

It is said that an alchemist, Vincentinus Casciarolo of Bologna, Italy, found a heavy crystalline stone with a gloss at the foot of a volcano and fired it in a charcoal oven intending to convert it to a noble metal. He obtained no metals but found that the sintered stone emitted red light in the dark after exposure to sunlight. This stone was called the "Bolognian stone" [8]. From the knowledge now known, the stone found appears to have been barite ($BaSO_4$), with the fired product being BaS, which is now known to be a host for phosphor materials.

At the end of the nineteenth century, a first major milestone concerning luminescent materials and devices was the realization of gas discharges and electron beams in evacuated glass tubes by Geissler and Braun, as well as the discovery of X-rays by Roentgen. The visualization of gas discharges, cathode rays, or X-rays represents also the first practical uses of luminescent materials. Based on these fundamental results, the first luminescent devices were developed only a few years later, and included cathode ray tubes, fluorescent lamps, and X-ray intensifying screens. Thereafter, research on cathode ray tubes and the relevant phosphors was stimulated during the Second World War for use in radar screens. In addition, the rapid and successful application of fluorescent lamps for illumination purposes, as well as the use of X-rays for medical imaging, intensified research on novel luminescent materials in the first half of the twentieth century. The first luminescent devices contained only a single phosphor material. The emitted light was normally not white, but colored. For instance, blue-emitting $CaWO_4$, which is a highly efficient phosphor, was used in X-ray intensifying screens in Braun's early cathode ray tubes as well as in the first fluorescent lamps by Edison. Lighting became more colorful with so-called zeon or neon tubes. Based on neon, argon, or mercury discharges in combination with different phosphors and colored glass, different colored lamps were realized. These lamps were presented by Claude at the world exhibition in 1937 in Paris and used for advertising. For illumination purposes, however, the emission of white light is required. Some single phosphor lamps were proposed, but their emission was too low. Nowadays, tri-color phosphor mixtures are used in fluorescent lamps worldwide. Fluorescent tubes and compact fluorescent lamps became widely available in the 1950s and early 1990s, respectively. Along with high intensity discharge lamps, they offer a longer life and lower power consumption than incandescent sources and have become the mainstream lighting technology in homes, offices, and public places.

More than a century after the introduction of incandescent lighting and half a century after the introduction of fluorescent lighting, solid state light sources are revolutionizing an increasing number of applications. Whereas the efficiency of conventional incandescent and fluorescent lights is limited by fundamental factors that cannot be overcome, the efficiency of solid state sources is limited only by human creativity and imagination. The high efficiency of solid state sources already provides energy savings and environmental benefits in a number of applications. For example, conventional incandescent or fluorescent lamps rely on either incandescence or discharge in gases. Both phenomena are associated with large energy losses that occur because of the high temperatures and large Stokes shifts involved. Light emitting diodes using semiconductors offer an alternative method of illumination. The operation of LEDs is based on spontaneous light emission in semiconductors, which is due to the radiative recombination of excess electrons and holes that are produced by the injection of current with small energy losses. Subsequently, the radiative recombination of the injected carriers may attain quantum yields close to unity. As a result, compared with conventional lamps, LED-based light sources have superior lifetime, efficiency, and reliability, which promise significant reductions in power consumption and pollution from fossil fuel power plants. Currently, LEDs are widely used as indicators, rear lamps for vehicles, decorated lamps, backlights for cellular phones and liquid crystal displays, and small-area lighting. With advances in the brightness and color rendering properties of LEDs, it is generally accepted that they will replace conventional lamps for general lighting in the near future.

Hundreds of phosphors are available for excitation at 250 nm, the dominant emission band of Hg lamps. In solid state lighting, however, the excitation wavelength is much longer, typically in the range of 380 to 480 nm. New high-efficiency phosphors, which can be efficiently excited at these wavelengths, are now being developed. Whereas high-efficiency yellow phosphors are readily available (e.g., cesium-doped yttrium aluminum garnet-$Y_3Al_5O_{12}$ (YAG) phosphors), the efficiency of red phosphors still lags. As a novel class of inorganic phosphors, oxynitride and nitride luminescent materials have received considerable attention because of their potential applications in solid state lightings and displays. Some of them are discussed in this book.

1.3.2 *Major Applications of Phosphors*

Luminescent materials have been well known for a long time from traditional lighting and display applications, like cathode ray (television) tubes, fluorescent lamps, and also X-ray screens. Exploration of novel materials classes during the last years has enabled development of new applications, such as white LEDs, plasma display panels, bio-markers, and solar cells. Recent progress and future prospects show that luminescent materials remain not only interesting from a scientific point of view but also relevant from an application point of view. For many years to come, scientific challenges as well as societal needs in this research field can be addressed by chemists, physicists, and materials scientists. This book reviews the latest developments on these novel luminescent materials, their preparation, crystal chemistry, characterization, and optical properties. The intention is to focus on the relationship between the luminescence properties on the one hand and the chemical composition and structure on the other.

The research on luminescent materials is stimulated by various applications. Luminescent materials are widely applied today. In the last decades, a large number of luminescent materials based on rare earth ions or rare earth host lattices have been invented. In many cases, rare earth phosphors dramatically improved the performance of the devices in which they are applied (Fig. 1.4).

Major applications of europium and terbium activated phosphors are in emissive displays and fluorescent lamps. Three types of flat luminescent displays are under development: plasma display panels, in which phosphor powders are excited by VUV photons (145–180 nm); field emission displays, in which the luminescent material is bombarded by low-energy electrons emitted by microtips; and thin-film electroluminescence displays, in which electrons injected into the luminescent layer are accelerated under the applied field up to the energies needed to excite luminescent centers [9–11].

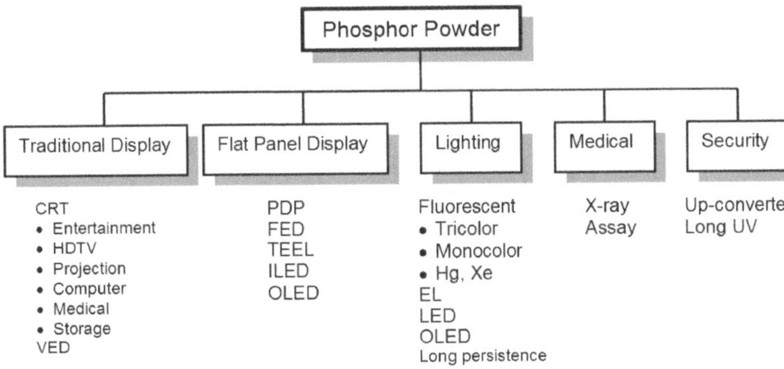

Figure 1.4 Major applications of phosphors.

A more specialized and important application is the use of X-ray intensifying screens for medical purposes [12]. Some X-ray detector systems are based on luminescent materials as well. Research on X-ray phosphors is mainly focused on photostimulable phosphors for digital imaging, the luminescence of which is obtained by optical excitation in an absorption band formed under X-ray irradiation [13]. Scintillator crystals convert the energy of ionizing radiations into UV or visible radiation. They are used for medical diagnostics, scientific experiments, and industrial control (e.g., inspection of baggage and containers, thickness control of metal sheets and paper, geophysical exploration).

During the last 10 years white LEDs have become very important lighting sources, and the importance of LED phosphors for white and colored light generation must be considered an important market driver in the future [14–19].

That's why we choose and take our attention in this book mainly on these important directions in the development of phosphors.

1.4 Rare Earth Activation

1.4.1 *Radiative and Non-radiative Transitions*

Photoluminescent materials, as mentioned above, require a host crystal structure such as $CaWO_4$, $YTaO_4$, Y_2O_3, and $Y_3Al_5O_{12}$ and an activator such as Eu^{3+}, Tb^{3+}, Ce^{3+}, or others. Activators are

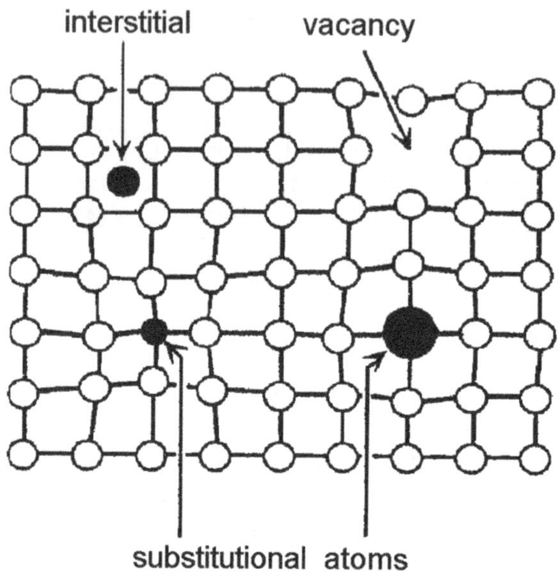

Figure 1.5 A two-dimensional illustration of some point defects that occur in solids.

commonly transition metal, main group, or rare earth (RE) ions from lanthanides. The activator is a dopant that creates substitutional defects in the host structure (Fig. 1.5).

The activator may absorb the excitation energy directly, thereby promoting an electron from the atom's ground state to excited states, or the energy may be transferred from the host to the activator as in X-ray phosphors. In many luminescent materials the situation is more complicated when the exciting radiation is not absorbed by the activator directly, but elsewhere. For example, we can add another ion to the host lattice. This ion may absorb the exciting radiation and subsequently transfer it to the activator. In this case the absorbing ion is called a sensitizer. The direct and indirect excitation via a sensitizer is schematically depicted in Fig. 1.6.

Upon excitation of the sensitizer, the energy is transferred to the lanthanide ion. For efficient energy transfer the sensitizer needs to be in close proximity of the ion and the energy levels of sensitizer and lanthanide ion should match. In principle, the near-infrared emitting lanthanide ions allow the use of sensitizers that absorb

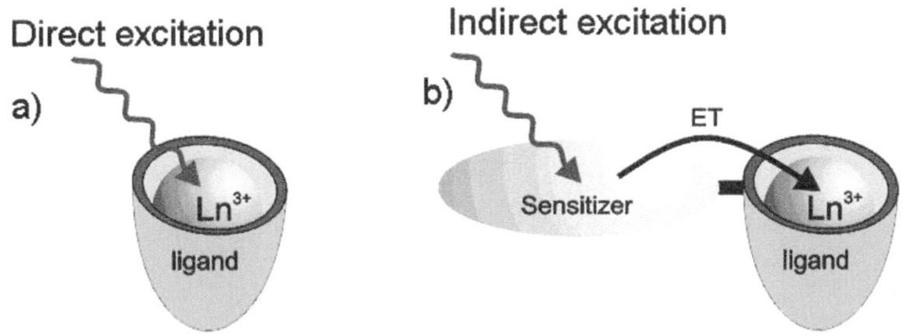

Figure 1.6 (a) Direct excitation of a lanthanide ion that is complexed by a ligand. (b) Indirect excitation of a lanthanide ion via a sensitizer that is bound to the ligand (ET stands for energy transfer).

visible light. This sensitization is attractive because it circumvents the low absorption coefficients and the narrow absorption peaks of the lanthanide ions.

The activator has discrete energy levels associated with it, and these levels are modified to different degrees depending on the type of activator and the local environment of the host structure. For example, activator ions that feel only a small influence from the host lattice will luminesce with narrow line emission characteristic of that particular activator atom. This narrow line emission is common

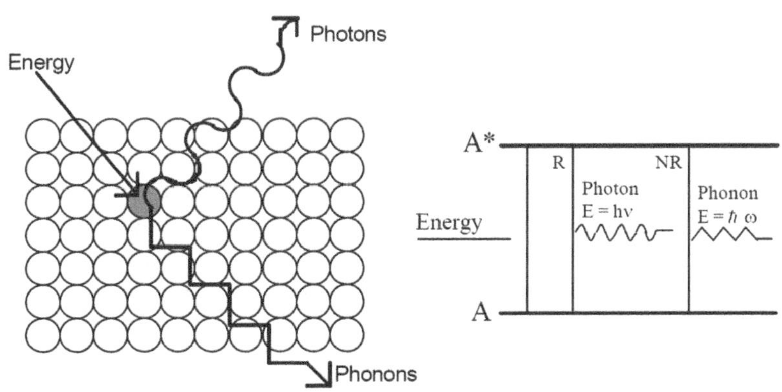

Figure 1.7 (a) An illustration of how an insulating material doped with activators can emit light. The dark circle denotes an activator ion which is surrounded by a host lattice. (b) Simplified energy level diagram of a luminescent ion A, A* being the excited state and the radiative (R) and non-radiative transitions (NR).

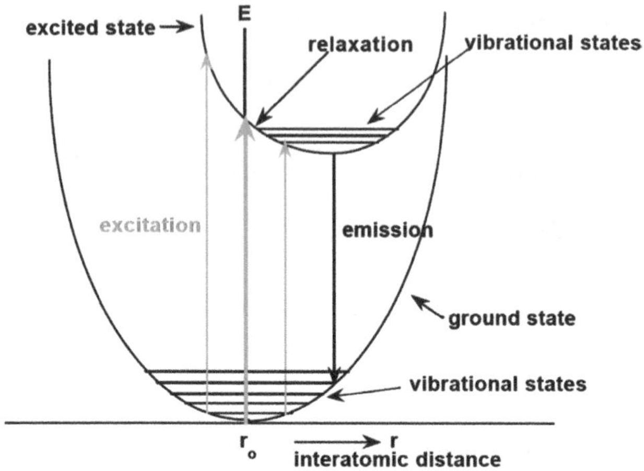

Figure 1.8 Radiative transitions explained by the configurational coordinate model: the excitation of an electron from ground to excited state, relaxation of the excited state electron to a lower energy excited state, and return to the ground state with emission of light.

for rare earth activators such as Tb^{3+}, or Eu^{3+}, whose optical transitions involve only *f*-orbital energy levels. To understand the luminescence processes in ionically bonded phosphors, the configurational coordinate diagram is used [20–23] (Fig. 1.8).

Figure 1.8 shows a commonly used diagram to explain the configurational coordinate model. The horizontal axis represents the average interatomic distance *r* from the activator ion to its neighbors and the vertical axis represents energy *E*. The bottom curve represents the ground state energy of the activator, while the top curve represents the excited state. The activator is excited from its ground state to a higher energy level. Some of this absorbed energy is given up to the host structure as heat, and the activator ion relaxes to a lower energy excited state. The activator then returns to the ground state, releasing its energy by emitting light. It is a radiative transition. Note that the energy of excitation is greater than the energy of emission. The decrease in energy, or the increase in wavelength, is known as the stokes shift.

When the activator ion becomes excited, the curve for the excited state can shift to a new interatomic distance due to changes in the bonding length and the electrostatic charges between the activator ion and the surrounding atoms. Such large shifts are normally seen with transition metal activators due to the outer electrons which are unshielded transitioning to higher available energy states. The horizontal lines in the ground and excited state curves represent the quantized vibrational states, with each separated by an energy ½ $\hbar\omega$. Even at 0 K (n = 0) there is still some phonon energy. The activator ion can vibrate to positions denoted by the lowest vibrational mode. As the temperature increases, more quantized vibrational modes will appear and the larger the potential change in the interatomic distance.

When the luminescent material is excited, absorption occurs when an electron in the ground state jumps to the excited state, denoted by the vertical in line Fig. 1.8. At 0 K, this vertical transition could take place from any position denoted by the lowest vibrational mode, as shown. As the temperature increases, transitions from the ground to the excited state can occur from any existing vibrational mode ($n > 0$) at any position. Once the electron is in the excited state, it can relax along the excited state curve until it reaches the position of the vibrational modes. As it does this, in order to lose energy, phonons are produced until the electron is taken up by one of the vibrational modes. At 0 K, for luminescence to occur, the electron must relax to the minimum of the excited state (n = 0) before it is allowed to transition back to the ground state curve, denoted by the vertical line in Fig. 1.8 labeled emission. The electron will then terminate on the ground state curve somewhere and must relax to the minimum of the ground state curve before being excited again. At higher temperatures, a luminescent transition can occur from any of the available excited state vibrational modes while the vibrating activator is at any position and can terminate anywhere on the ground state curve or at any available ground state vibration mode. It's these vibration modes and the transition from and termination on these available vibration modes that give rise to the width of the emission curves. The less the number of modes, the smaller is the width of the emission line spectra.

Figure 1.9 shows the configurational coordinate model that explains non-radiative transitions. For non-radiative transitions, the ground and excited state curves must come close enough to each other for the electron to be able to transition from the excited state

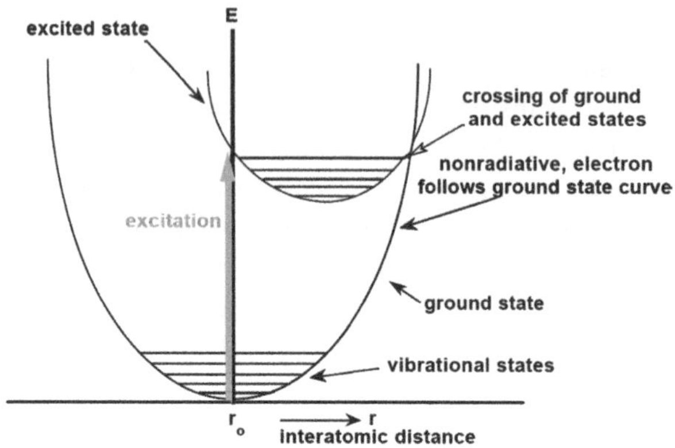

Figure 1.9 Non-radiative transitions explained by the configurational coordinate model.

to the ground state without making a radiative transition. This transition involves the electron occupying a high vibrational state. It can then transition to the ground state curve and relax along the curve back to the lowest energy level of the ground state. This process does not give off emission, but instead creates phonons.

The efficiency of phosphors is much less when they exhibit non-crystalline, or amorphous, phases. Amorphous means that the material does not have any long-range repetitive lattice structure like crystalline solids. The efficiency of an amorphous phosphor can be explained by the configurational coordinate model. When there is not a strong crystal field surrounding the activator ions in the phosphor, Δr becomes large between the ground and excited states, causing them to overlap, which favors the non-radiative emission. In a crystalline solid, the crystal field is stronger due to the atoms sitting in their appropriate sites, which leads to a smaller Δr and thus increases the probability of radiative emission.

The luminance and efficiency of a phosphor are maximized at an optimal concentration of luminescent impurity centers. Initially, as the concentration of the luminescent activator is increased, the luminance also increases since more centers are able to be excited each half-cycle of the driving voltage. Eventually, the luminance usually saturates and then decreases with further concentration increases. This occurs because as the concentration increases the luminescent impurity centers become increasingly closer to other centers. If the distances are small enough, energy may be transferred from one excited center to another center instead of being emitted as light. The energy may still eventually lead to light emission, but the transfer increases the likelihood of a non-radiative transition occurring. Also, when the luminescent impurity is not size- and valence-matched to the substitutional site it occupies, it may be difficult to introduce a large amount of the impurity. Optimal doping levels are thus dependent upon the nature of the phosphor host and luminescent center.

1.4.2 *Peculiarities of Eu and Tb Activation*

We can conclude that rare earth phosphors can be separated into two types: broad band emitting owing to the $5d \rightarrow 4f$ transition (Eu^{2+}, Ce^{3+}) or narrow band emitting owing to the transition between the $4f$ levels (Eu^{3+}, Tb^{3+}, Gd^{3+}, Yb^{3+}, Dy^{3+}, Sm^{3+}, Tm^{3+}, Er^{3+}, Nd^{3+}, etc.).

The luminescence is more intense if the bandgap between the excited state and the highest component of the ground state multiplet becomes larger. For example, Gd→32100 cm^{-1}, Tb→14800 cm^{-1}, Eu→12300 cm^{-1}, Yb→10400 cm^{-1}, Dy→7800 cm^{-1}, Sm→7400 cm^{-1}, Tm→6200 cm^{-1}, Er→5900 cm^{-1}, Nd→4400 cm^{-1}, etc. Gd is a very good emitter, but the $^6P_{7/2} \rightarrow {}^8S_{7/2}$ transition occurs in the ultraviolet region (around 310 nm) that is not convenient for practical application. This explains the prominent role played by Tb and Eu as luminescent probes. Moreover, Eu is a unique activator that gives both the broad band and narrow band emissions depending on its valences. That's why in this book we focus on some different host lattices activated by Eu and Tb. In addition, the co-activation or double activation by these RE elements is also discussed. First, energy can be transferred from one luminescent activator to the other, resulting in more efficient or brighter device operation. If the energy transfer occurs within the Eu-Tb complex so that it works as a donor-acceptor pair due to small energy gap between excited energy levels $^5D_4$ (Tb^{3+}) and $^5D_1$ (Eu^{3+}). Secondly, co-activators can be added to a host material to change the color of the emitted light. Though these Eu and Tb activators have been widely used for a long time, there are many questions that need to be resolved.

This book will discuss the following topics related to Eu and Tb activation.

First, we consider application in calcium tungstate phosphor. The continuous need for high-performance phosphors renewed the interest in this luminescent material with more than a century of history. Self-activated $CaWO_4$:W is an efficient material emitting blue luminescence related to some tetrahedral WO_4 groups of the scheelite host lattice. However, the luminescence properties of this classic phosphor could be essentially varied by doping with RE ions, in particular with Eu^{3+} and Tb^{3+}, and therefore, some new applications including double activation could be proposed. Some of them are considered in section 4.1.

In section 4.2 the discussion is about yttrium niobium-tantalate $Y(Ta,Nb)O_4$ phosphors. The yttrium tantalate ($YTaO_4$) and yttrium niobate ($YNbO_4$) systems are of great interest from a fundamental scientific point of view and also for various spectroscopic applications. The blue light emission from the host lattice could be shifted toward longer wavelengths when the rare earth ions Eu^{3+} or Tb^{3+} replace partially the yttrium ions in these materials. In these phosphors, activated by rare earth elements, the host-lattice emission centers and the rare earth emission centers contribute both to the overall luminescence. The growing interest toward luminescence properties in the vacuum ultraviolet and in the visible spectral range is due to industrial demands for new applications.

Phosphors based on Eu^{3+} doped RE elements such as yttrium, lanthanum, and gadolinium are widely used to provide red light emission for modern display devices. Section 4.3 focuses on the performances of different red-emitting phosphors based on Eu^{3+} doped solid solutions of Y_2O_3–La_2O_3–Gd_2O_3 and Y_2O_2S–La_2O_2S–Gd_2O_2S and YVO_4–$GdVO_4$. These oxides are transparent for the visible light and have no low-energy levels, and the interaction between the RE activator ions and the matrix is minimal. Other promising matrices for phosphor application are RE oxysulphides, such as Y_2O_2S, La_2O_2S, Gd_2O_2S, and Lu_2O_2S. Red-emitting $YVO_4:Eu^{3+}$ is a well-known phosphor, and nowadays this composition is widely used as the red component in cathode ray tubes (CRTs). Due to its good color purity, this phosphor is also considered as a promising candidate for plasma panel displays (PDPs).

Tb^{3+} forms the basis of the very bright phosphor systems that have the highest reported saturated green emission and are being used as the green component in commercial full-color displays and LEDs. The green emission is due to four sharp peaks arising from $^5D_4 \rightarrow {}^7F_J$ transitions, where $J = 3; 4; 5; 6$, with the strongest peak at about 550 nm.

For the purpose of development of highly energy-efficient white light sources, it is necessary to produce highly efficient phosphors, which can absorb excitation energy from blue light emitting diodes and generate emissions. Yttrium aluminum garnet, $Y_3Al_5O_{12}:Ce^{3+}$ (YAG:Ce), and terbium aluminum garnet, $Tb_3Al_5O_{12}:Ce^{3+}$ (TAG:Ce), are used as phosphors for white light emitting diodes using the principle of luminescence conversion. However, the color impression of such conversion on the base of a blue diode and these phosphors is too "cold," and a red component is missing. The common way to improve the emission spectra of lighting devices is with co-activation of the conversion phosphors, for example with Eu^{3+}, which shows an intense line emission in the red. Results of co-doping TAG:Ce with Eu^{3+} are presented in section 4.4. Therefore, it seems attractive to develop new phosphor materials with high efficiency and color saturation or modify well-known compositions by co-doping them with efficient Eu^{3+} and Tb^{3+} activators. These and some new efficient materials for solid state lighting and LED application are also considered in sections 4.5, 4.6, 4.7.

Though many of the trivalent lanthanide ions are important for certain applications, Eu^{3+} and Tb^{3+} ions are the most useful for visible technologies.

There are a number of very useful books and monographs one might consult for a general background on phosphors or for better understanding the field of luminescent materials [2, 8, 20, 24–40]. We also refer our readers interested in comprehensive information specific to the subject presented in this book to some reviews and references on the existing commercial functional devices incorporating luminescent materials [7, 9–19, 41–47]. The special list of phosphor books and reviews is also given in the Appendix: "The Luminescence Literature."

Chapter 2

SYNTHESIS OF PHOSHORS

2.1 General Requirements of Phosphors

The most significant markets for phosphors are in lighting and displays. As technologies within these markets shift – for example from incandescent to fluorescent to solid state in lighting and from CRT to projection to plasma and LCD for displays – so too do the requirements on phosphors for these applications.

The following are general requirements for phosphors:

1. Color (RGB) purity
2. High brightness
3. High efficiency
4. Long-term stability

To achieve these requirements we have to demand much of surface characteristics:

1. High phase purity
2. High crystallinity
3. Homogeneous distribution of the activator ion
4. Spherical shape
5. Fine particle size
6. Narrow size distribution

New Generation of Europium and Terbium Activated Phosphors
Mihail Nazarov and Do Young Noh

www.panstanford.com

7. Non-aggregation characteristics
8. Controlled surface chemistry
9. Good dispersibility
10. Low porosity

For any special application there are predominant requirements and we have to choose one or another chemical route to synthesize an available phosphor.

When fine-grained luminescent material is required the conventional route to fabrication is via a grinding or milling technique. A high-temperature solid state reaction is used to form large grains of the required material, before these grains are reduced in size. This has a serious drawback in that grinding techniques can introduce defects into the material. Defects have serious implications for luminescent materials as they provide non-radiative recombination routes for electrons or holes. In order for the material to be as efficient as possible the number of electron/hole recombinations via the optically active centers must be maximised, i.e. crystal structure is as important as the choice of host lattice and emitting center. As the crystal being ground is reduced in size, the proportion of damaged surface region to undamaged bulk region increases. Typically, as particles are ground below 4–5 μm there is a substantial drop off in luminescence efficiency [48].

For nanoparticle phosphors usually wet chemical synthesis methods such as co-precipitation and sol-gel are applied. Nanometer-sized particles cannot be described simply as an excised fragment of a bulk solid with limited dimensions, mostly because a large fraction of the atoms are localized at the surface of the particles. As a consequence, the physical properties of the nanoparticles are greatly dependent on atoms with a distorted environment, very different from the ones located in the volume of the particles. An investigation of the physical properties of materials in the nanometer size range is very dependent on the synthesis of particles with a very good control of the different parameters which are recognized as having an influence on their properties. The importance of these parameters of course depends on the system under investigation and its physical properties, but in most cases they are related to

- the size of the particles
- their shape
- their surface state
- their dispersion within a host media
- their interactions with other systems (molecules, other particles, surrounding medium).

Specialized synthesis techniques were developed to precisely control particle size (<5 μm), particle size distribution, and particle shape of phosphors. Hydrothermal and solvothermal synthesis, combustion synthesis, and spray processing are a few of these techniques that will be described. The development of colloidochemical synthesis techniques to produce nanocrystals has introduced the opportunity for a new class of optical devices.

Almost all phosphors for X-ray, display, or solid state lighting application are synthesized by traditional solid state reactions also called dry media reactions between raw materials at high temperatures. Solid state and co-precipitation reactions remain the most widely used synthesis techniques for commercial phosphors. The synthesis of solids possessing desired structures, compositions, and properties continues to be a challenge to chemists, material scientists, and engineers.

This chapter will present briefly an overview of the primary methods used to synthesize phosphors, from solid state reaction for micrometer particles (1–10 μm) to colloidochemical synthesis of nanocrystalline (1–10 nm) phosphors.

The readers can find more detailed information in special books [8, 30, 40, 49–53] and in numerous references.

2.2 Solid State Methods

2.2.1 *Solid State Reaction*

A solid state reaction is a chemical reaction in which solvents are not used. In a normal reaction, the reacting agents, also called the reactants, are placed in a solvent before the reaction can take place. These reactants react to form a new substance. After the reaction is completed, scientists are able to remove the new product from the solvent. A solid state reaction, however, allows the reactants to chemically react without the presence of a solvent. Eliminating the solvent from the reaction means that a solid state reaction produces more product than a normal reaction can and the elimination of solvents means that products will cost less. This, in turn, will make those products cheaper to buy. With normal reactions, scientists need to remove the residual solvent from the resulting product after a reaction has finished. Producing materials from a solid state reaction will mean that scientists are able to bypass the purification process.

Though there are many benefits to developing solid state reactions, there are also many drawbacks. The ideal process should result in a homogeneous, or uniform, substance, and some solid state reactions cannot do it. Preparation of a single phase compound is difficult by the conventional solid state method. Hence, doping a low concentration (of the order of 1–3%) of activator has always been delicate. Thus, the limitations of conventional solid state method are as follows:

1. Inhomogeneity of the product
2. Formation of large particles with low surface area, because of which mechanical particle size reduction is required, which introduces impurity and defects
3. Presence of defects, which are harmful to luminescence

Figure 2.1 shows the general concept of a solid state reaction synthesis process.

In this traditional method of solid state reaction the high-purity materials of the host crystal, activators, and fluxes are blended, mixed, and then fired under an appropriate atmosphere. As the product obtained by firing is more or less sintered, it is crushed, milled, and then sorted to remove coarse and excessively crushed particles. In some cases, the product undergoes surface treatments.

2.2.2.1 Preliminary powder preparation

As small amounts of impurities sometimes change phosphor characteristics drastically, raw materials must be purified very carefully

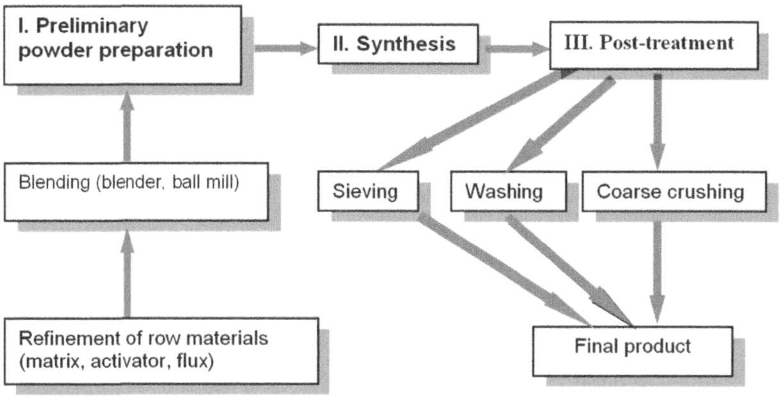

Figure 2.1 Phosphor synthesis process.

and we need select of optimum phosphor material (host, dopant, codopant). The starting materials are weighed out accurately in the required amounts and mixed together using an agate mortar and pestle. Very thorough mixing of the starting materials before firing is extremely important for the successful preparation of a good phosphor. During the grinding, a small amount of acetone is added in order to mix the materials homogeneously. The finely powdered starting materials are mixed into water, alcohol, or any other inert and volatile liquid to uniform slurry. The slurry is dried and the dried material, which sometimes is a more or less hard crust, is powderized in a mortar as well as possible. This method has been used for most of the recipes in this book.

2.2.2.2 Firing and calcinations

The purpose of firing is not only to cause solid state reactions but also to form well-crystallized particles with an appropriate average diameter. The substance added to the raw material mixture to help crystal growth is called a flux. Fluxes are usually compounds of alkali- or alkaline earth metals having low melting points. There are two types of fluxes which are used, volatile and non-volatile. The volatile flux often reacts with the starting material and leaves the reaction mixture spontaneously, through either decomposition or evaporation. Commonly used fluxes include NH_4Cl, NH_4Br, AlF_3, and boric acid. The amounts used are typically small, on the order of few mole percent. Molten salts are used as non-volatile fluxes and generally do not react with the starting materials. As the name denotes, these fluxes are always in the form of a melt. Large amounts of the salts are used (up to 30% by weight of the phosphor material); they generally do not leave the reaction mixture and must be removed by washing with water or other appropriate solvents. Examples of salts used for this purpose are Na_2MoO_4, $Na_2B_4O_7$, Na_2SiO_3, and $Na_4P_2O_7$. If no flux is added, firing has to be performed at very high temperatures (ca. 1400°C). By adding halides, the temperature can be reduced to 1200°C. When the fluxes melt, the surface tension of the liquid helps particles coagulate. The melt also makes it easier for particles to slide and rotate, provides chances of particle-particle contacts, and promotes particle growth. Each flux influences the particle size and the shape in a different way. Therefore, a combination of fluxes is sometimes used to obtain products with desired morphology. Higher firing temperatures and longer firing times result in larger particles. Particle growth is rapid in the initial stage of firing and slows down after a certain period of time.

Phosphor firing requires an electric furnace permitting temperatures between about 500°C and at least 1400°C, better up to about 1600°C. Any desired temperature should be maintained during firing by means of an automatic temperature controller to within ±10°C or better.

The various phosphors are fired either in air or in controlled different atmospheres depending on the materials and the desired reactions. Oxygen-dominated phosphors (oxides, silicates, phosphates, etc.) may be fired in oxidizing (air, O_2), inert (N_2, Ar), or reducing atmospheres (CO, forming gas, H_2, NH_3). CO gas is about the most convenient for firing an oxygen-dominated phosphor in a reducing atmosphere. Some carbon black is placed in a separate open container next to the phosphor container in the firing tube. The air is flushed out by plain O_2 and the gas inlet tube is then closed (but leaves the inlet open). When the charge enters the furnace, some of the carbon black burns to CO *in situ*. Phosphors activated with Tl^+, Pb^{2+}, Sb^{3+}, Mn^{2+}, Mn^{4+}, or Eu^{3+} ions can be fired in air, whereas phosphors activated with Sn^{2+}, Eu^{2+}, Ce^{3+}, or Tb^{3+} ions are fired in a reducing atmosphere. Sulfurization of sulfide phosphors may be achieved by firing either in H_2S or in an inert gas loaded with CS_2. Many of these gases are interchangeable. It makes no difference, for instance, whether a phosphor is fired in N_2 or in Ar.

2.2.2.3 Post-treatment

Fired phosphors often (not always) are slightly sintered cakes that have to be powderized to obtain the desired fine powder. Phosphors coming out the furnace as badly sintered lumps indicate that something is wrong; most likely the firing temperature was too high. The powderizing usually is possible by gentle grinding or milling. Avoid all hard milling because it is likely to damage the phosphor. Weak milling (i.e. separation of coagulated particles into primary ones) changes the efficiency little. However, primary particle destruction lowers efficiency. The possible reasons are that lattice defects created by phosphor crystal destruction act as non-radiative recombination centers, or that a non-luminescent, amorphous layer is formed on the surface of the particles. It is most important to select raw materials, fluxes, and firing conditions so as to avoid strong milling after firing. Even by careful adjustment of materials, fluxes, and firing conditions, the phosphors obtained usually have a broad particle-size distribution. A process is necessary, therefore, to remove both very fine and coarse particles from the phosphor lots; this is done by means of sedimentation, elutriation, or sieving. Sedimentation is usually used to separate very fine particles. In this process, phosphor batches are agitated in water and then left still until larger particles sediment. The remaining suspended, finer particles are removed by decantation. The sedimentation speed can be changed to some extent by adjusting acidity. Elutriation is employed during a wet process, in which both washing and removal of coarse particles are carried out at the same time.

Sieving is used to remove very large phosphor particles after firing. All phosphors should finally be screened to remove undesired big particles. A 200-mesh stainless steel sieve is adequate for most general purposes but some special applications may require finer

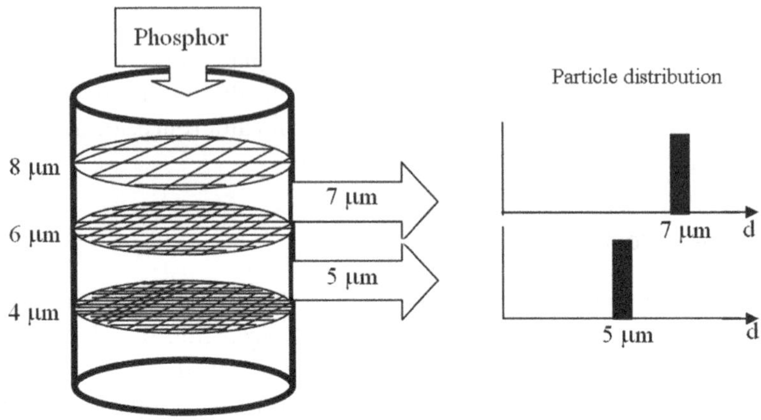

Figure 2.2 Separation of phosphors with multilevel sieves

screens. For obtaining phosphors with homogenous narrow particle distribution the multilevel sieves after chemical route are used [54]. The basic principle of this construction is presented in Fig. 2.2.

The distinctive feature is that upper sieve has a 2 μm mesh more than the lower one. With this process, the homogenous phosphors with optimized particle size can be made.

2.2.2 *Microwave Synthesis Method*

Application of microwave heating technique to inorganic materials technology has been advanced in sintering and joining of ceramics [55]. There are several advantages in microwave heating compared with conventional heating techniques, in view of potentials for (1) energy saving and shorter processing time, (2) improved product uniformity and yields, (3) improved or unique microstructures and properties, and (4) synthesis of new materials.

Recently, there have been growing interests in the use of the microwave heating for the synthesis of inorganic compounds such as metal chalcogenides [56–58], oxide superconductors [59, 60], nitrides [61], and metal halides [62]. Conventional microwave ovens such as domestic microwave ovens and ceramics-sintering furnaces have a multimode microwave, while custom tailored apparatus have a single-mode microwave, where the energy of microwave is focused to a small area [60, 63]. Peelamedu *et al.* have tried to employ a two-phase system, where two kinds of materials which have different microwave absorption efficiencies from each other were used to react with "anisothermaly" [64]. They could obtain to intermediate products which could not be synthesized by conventional heating reactions. In a microwave synthesis, sources are heated over the whole sample quickly and uniformly because microwave energy is immediately absorbed by the sample. Therefore, microwave heating techniques have a significant advantage in the reduction of manufacturing costs, i.e. energy saving and shorter processing time. In addition, they also enable to heat a particular component in the mixtures, and thereby have a potential of new synthetic technique.

Interesting results about the synthesis of $YVO_4:Eu^{3+}$ phosphors using a microwave heating method were reported in the last years [65, 66]. Uematsu *et al.* used oxides Y_2O_3, V_2O_5, and Eu_2O_3 as raw materials for the preparation of phosphor samples of $YVO_4:Eu^{3+}$. The powders were weighted in an appropriate and milled together in a ball mill with an agate ball using an agate container. A fraction of the mixed powder (2.0 g) was pelletized with the dimensions of 15 mm in diameter and about 4 mm in thickness. In order to reduce an escape of heat from the surface of the pellet, the pellet was surrounded by the mixed powder of 1 g. The sample was then put in an insulator box made of silica-wool. A thermocouple sheathed in a stainless steel was contacted to surface of the sample pellet to monitor temperature change of the sample. The sample environment is schematically shown in Fig. 2.3.

The insulator box was placed in a domestic microwave oven (Sharp, RE-T55) operated at 2.45 GHz. Microwave irradiation was carried out at an electric power of 500 W for 15 min. The temperature of the raw materials, a mixture of Y_2O_3, V_2O_5, and Eu_2O_3, irradiated with the microwave goes up about 1113 K immediately in 150 s, followed by a sharp drop in the temperature in spite of irradiating the microwave continuously. The temperature of the sample was recorded by means of a personal computer every 5 s. From the measurement of loss factor at 2.45 GHz, a high-efficiency absorption of microwave energy by V_2O_5 causes a rise in the temperature of the mixture. On the other hand, the low loss factor of the YVO_4 product

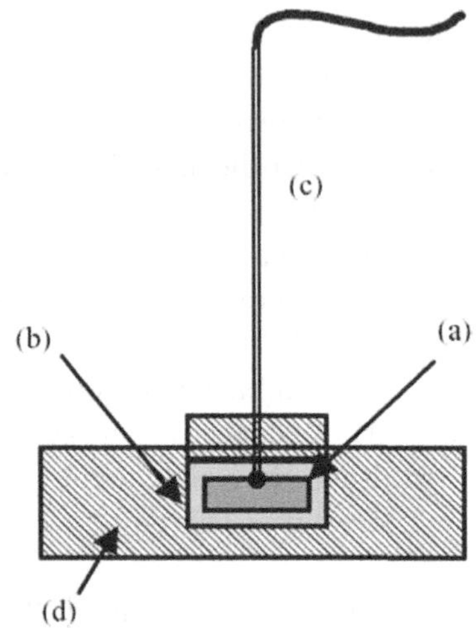

Figure 2.3 Schematic diagrams of the sample environment: (a) pellet of the mixed powder of the starting materials; (b) mixed powder of the starting materials; (c) sheathed thermocouple; and (d) heat insulator made by silica-wool. (Reprinted with permission from [65]. Copyright 2005: Elsevier).

causes a sharp drop in the temperature as soon as the reaction of Y_2O_3 with V_2O_5 is completed. The microwave synthesis can provide a product without sintering as well as good homogeneous distribution of the dopant.

Figure 2.4 shows the scanning electron microscope images of YVO_4 synthesized by the conventional furnace heating method and microwave heating method. In the case where the conventional method was used (Fig. 2.4a), the particles had diameters of around 0.7–2 μm and were highly dispersed. On the other hand, when the microwave heating method was used, the particle size was very small and the particles tended to agglomerate (Fig. 2.4b). However, as the main heating time increased, the particle size increased and the extent of agglomeration decreased (Fig. 2.4c). In fact, the agglomeration of the as-synthesized powder after heat treatment for 10 min at 1200°C was almost eliminated, producing well-formed (bulk-like) particles. When the samples were re-heated for 3 min at 1200°C (Fig. 2.4d), the particles aggregated to form secondary particles due to sintering. In addition, a part of the surfaces of certain particles was smooth, indicating that melting occurred.

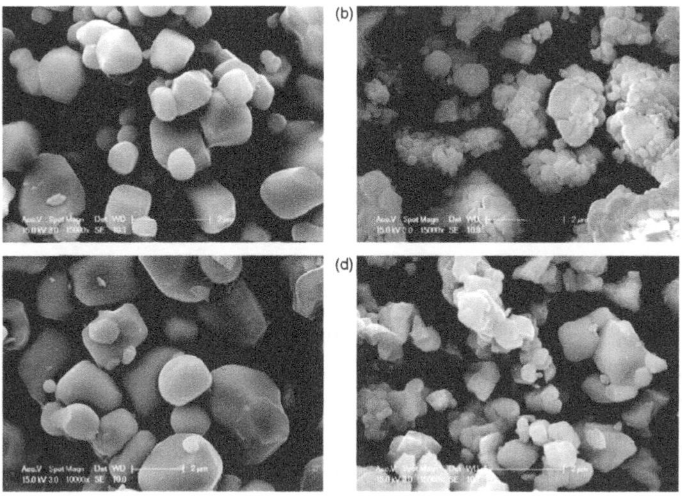

Figure 2.4 Particle sizes of $Y_{0.97}Eu_{0.02}Sm_{0.01}VO_4$ prepared by microwave heating synthesis with different firing temperatures and solid state reaction synthesis: (a) solid state reaction 1200°C for 3 h, (b) microwave heating at 1200°C for 3 min, (c) microwave heating at 1200°C for 10 min, and (d) microwave heating at 1200°C for 3 min, after re-fired at 1400°C for 3 min. (Reprinted with permission from [66]. Copyright 2005: Elsevier.)

The absence of sintering in the case of the samples subjected to microwave heating can be attributed to the shorter heating time. In view of energy saving synthetic method, an ideal system is the one where the heating eases spontaneously as quickly as the reaction is complete.

One of the main problems in solid state reaction is the problem of inhomogeneity. Both the crystallinity and grain uniformity are poor if the treatment parameters are such as not to allow the chemical reaction to be completed. To resolve or mitigate this problem, wet methods of preparation are often used; in the wet method, aqueous solutions of specific constituent metallic salts are employed.

2.3 Wet-Chemical Methods

Wet-chemical techniques are now available for simple oxide nanoparticle phosphor synthesis. These techniques [50, 67] are defined as techniques which do not comprise the normal mixing, calcinations, and grinding operations, because through these methods, the starting materials can be mixed at molecular level. A number of investigations proved that the wet chemical methods could decrease the sintering temperature efficiently [68–73] that is very important to obtain, for example, ultrafine YAG powders at relatively low temperatures. Wet-chemical methods such as co-precipitation [74–79], sol–gel [72, 80–85], hydrothermal [86–88], combustion [73, 89–91] evaporative decomposition [92], reverse micellar [93, 94], surface modifier [95], etc., have been employed to synthesize rare earth activated phosphors. These wet-chemical methods dope rare earth activators uniformly. But calcination is required to get crystalline (required) phosphors. However, to our knowledge, wet-chemical techniques are not available for the synthesis of complex oxide phosphors (blue and green phosphors). Therefore, we would like to emphasize the importance of solution combustion process for the synthesis of both simple and complex oxide phosphors.

2.3.1 *Co-Precipitation*

In these processes the starting materials are dissolved in a solvent, come into contact with each other on an atomic or ionic level, and then co-precipitate forming an intimately commingled starting mixture. As an example, $Y(NO_3)_3$ and $Tb(NO_3)_3$ dissolved in water can be precipitated by the addition of a 2-to-1 solution of oxalic acid dissolved in hot water. The oxalates obtained in the precipitate can be converted to oxides by heating at a relatively low temperature of 800°C. Alternatively, the oxides can also be dissolved directly in hot diluted nitric acid to obtain a treatable precipitate. This method is used in cases in which an insoluble salt can be identified for each of the reaction constituents. Other routes for co-precipitation can involve sulfates or hydroxides.

2.3.2 *Sol–Gel Techniques*

The sol–gel method of phosphor preparation is regarded as a wet method. It is a chemical technique that uses metal alkoxides for the synthesis and production of glasses or ceramics through a series of chemical processes, including hydrolysis, gelation, drying, and thermal treatment. The sol–gel technique was developed from 1864. Graham [96] prepared gels of silica from aqueous salts, while Ebelmen [97] obtained silica gels from metal alkoxides. The potential of the sol–gel process was not appreciated until 1980, when it was "rediscovered" and found to be very useful in synthesizing various materials of practical importance, such as optical glasses, phosphors, and solid state laser materials. Since then the method has received considerable attention and has been investigated extensively [98].

A kind of metalorganic compounds, known as alkoxides of metals, is used as precursors. These metalorganic alkoxides either are in liquid form or are soluble in certain organic solvents. In general, a sol is defined as a colloid of solid particles suspended in a liquid; the particles consist of dense oxide or polymetric clusters formed by the precursors and reagents. A gel, on the other hand, is a composite substance consisting of a continuous solid skeletal structure which results from the gelation of the sol; the gel forms cells which encapsulate colloidal liquids. The addition of a $NH_4(OH)$ or oxalic acid causes the metallic hydroxides or oxalates to precipitate from the mixture; in these cases, the ingredients of the precipitate are in contact with each other at a molecular level and an efficient chemical reaction normally occurs. Through the use of the appropriate reagents, the processes hydrolysis and gelation can be induced to produce homogeneous gels from the mixture of alkoxides. To obtain powder or ceramic samples, the gels can be baked, sintered, and powderized as in other traditional methods.

The sol–gel method is advantageous insomuch as thin films or coatings of the phosphor can be formed on substrates directly and/or the sol–gel can be molded into designated forms. The sol–gel technique presents the following advantages:

(a) High homogeneity of the chemical composition of the materials produced occurs. Molecule-level-homogeneous multicomponent materials can be obtained. Because of the better homogeneity, contributions to the optical spectra of these materials from inhomogeneous sources are generally expected to be smaller than those encountered in unordered systems.

(b) High uniformity of doping ions distribution exists. No "local" concentration quenching will occur because of impurity clustering, and higher doping concentration becomes possible.

(c) Processing temperature can be very low. This allows the doping of fragile organic and biological molecules into porous inorganic materials and the fabrication of organic– inorganic hybrid materials.

(d) The microstructure (porosity and size of pores) of the materials can be controlled. Nano-scale uniform pores can be obtained at intermediate processing temperature, while high-density materials can be produced with higher annealing temperature.

(e) Thin films and multilayers coatings of sol–gel materials can be readily prepared by spinning or dipping methods during the gelation period.

(f) The sol–gel procedures produce little unintentional contamination. No milling and grinding are needed; for example, processes are known to contaminate samples. Fluxes, such as B_2O_3, H_3BO_3, and NH_4Cl, which are commonly used in ceramic technology and contaminate the end products, are no longer needed. In cases where phosphor powders are prepared by the sol–gel method, powderizing may be used and trace of foreign particles can be mix in. This "contamination" does not enter into the lattice and will not affect the intrinsic optical properties of the phosphor.

The technique has the following disadvantages:

(a) The drying and annealing processes have to be slow and deliberate; otherwise cracks and striations will appear in the samples.

(b) It is difficult to completely remove the residual hydroxyls from the sol–gel materials. To get rid of these organic groups, samples have to be annealed above 1000 °C and this may produce undesirable side effects.

The sol–gel method has been widely used to prepare a number of phosphors [99–101] for displays and other materials that are of technical importance. For example, Willi Lehmann reported in 1975 that very efficient luminescence was observed in, what he called, heterogeneous materials, such as SiO_2:CaX_2-Eu^{2+}/Mn^{2+} (where X = Cl, Br, or I), prepared by a solid state reaction [102]. These materials are similar to those found in the nanoclusters in SiO_2 which have been developed recently via the sol–gel method [103–105]. There are two ways to prepare such nanophosphor composites: (a) Alkoxide solutions of the phosphor are prepared from the appropriate precursors and then blended into the sol of SiO_2 (or other matrixes); the geletion process then proceeds as before. (b) Nanophases can be obtained by phase aggregation from doped SiO_2 or other matrices during thermal treatments. Different nanophosphors have been fabricated with this method.

The detailed analysis of sol–gel techniques was kindly provided by Jia [51] and the readers can find it also in special articles.

2.3.3 *Hydrothermal*

Hydrothermal synthesis includes the various techniques of crystallizing substances from high-temperature aqueous solutions at high vapor pressures. The term "hydrothermal" is of geologic origin. Hydrothermal research was initiated in the middle of the 19th century by geologists and was aimed at laboratory simulations of natural hydrothermal phenomena. George W. Morey at the Carnegie Institution and later Percy W. Bridgman at Harvard University did much of the work to lay the foundations necessary to containment of reactive media in the temperature and pressure range where most of the hydrothermal work is conducted. In recent years, commercial interest in hydrothermal synthesis has been revived in part because a steadily increasing large family of materials, primarily ceramic powders, has emerged that can be prepared under mild conditions ($T < 350°C$, $P < 100$ MPa). Hydrothermal synthesis can be defined as a method of synthesis of single crystals which depends on the solubility of minerals in hot water under high pressure. The crystal growth is performed in an apparatus consisting of a steel pressure vessel called autoclave, in which a nutrient is supplied along with water. A gradient of temperature is maintained at the opposite ends of the growth chamber so that the hotter end dissolves the nutrient and the cooler end causes seeds to take additional growth.

Hydrothermal synthesis offers many advantages over conventional and non-conventional ceramic synthetic methods. All forms of ceramics can be prepared with hydrothermal synthesis, namely powders, fibers, and single crystals, monolithic ceramic bodies, and coatings on metals, polymers, and ceramics. One of advantages of the hydrothermal method over other types of crystal growth includes the ability to create crystalline phases which are not stable at the melting point. Also, materials which have a high vapour pressure near their melting points can also be grown by the hydrothermal method. The method is also particularly suitable for the growth of large good-quality crystals while maintaining good control over their composition. Hydrothermal methods have important advantages in the manufacture of powders and films such as nanocrystalline structures with controlled chemical composition, high homogeneity and sinterability. Disadvantages of the method include the need of expensive autoclaves, good quality seeds of a fair size and the impossibility of observing the crystal as it grows.

Some examples of application hydrothermal method for synthesis of YAG phosphor are given by Hakuta *et al.* [106]. Phosphor YAG:Tb ($(Y_{2.7}Tb_{0.3})Al_5O_{12}$) nanoparticles were synthesized by a hydrothermal method at supercritical conditions (400°C and 30 MPa) using a flow reactor. Hydroxide sol solutions formed by stoichiometric aluminum nitrate, yttrium nitrate, terbium nitrate, and potassium hydroxide solutions. Single phase YAG:Tb particles with 20 nm particle size were obtained and the relationship between particle size and experimental variables, including pH, concentration of coexistent ions and hydroxide sol, were investigated.

Rapid and continuous production of phosphor (YAG:Tb) fine particles by hydrothermal synthesis in supercritical water is presented in Fig. 2.5 [86].

Figure 2.5 shows a diagram of the experimental apparatus employed. A premixed aqueous metal salt solution was fed into the apparatus by a high pressure pump at a rate of 2 cm^3 min^{-1}. Aqueous potassium hydroxide solution was introduced by another pump, also at a rate of 2 cm^3 min^{-1}. These two aqueous solutions were combined at a mixing point (MP1) at room temperature. This solution was then added to supercritical water (450–550°C, 8 cm^3 min^{-1}) at mixing point MP2, and thus the solution was rapidly heated to the reaction temperature. The solution was passed through a tube reactor and then cooled by an external water jacket at the end of the reactor.

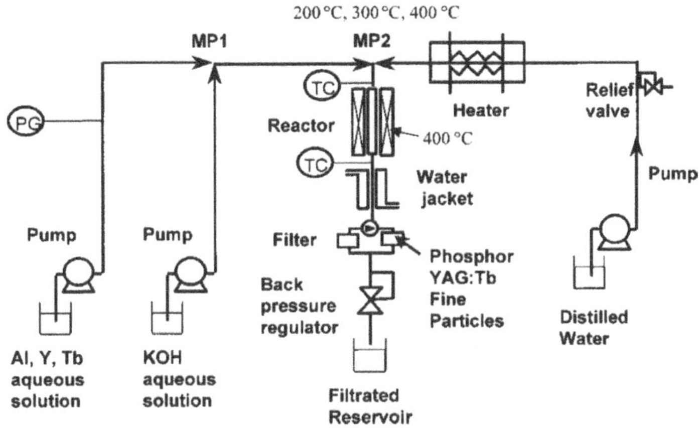

Figure 2.5 Supercritical water flow apparatus. MP1 and MP2 refer to solution mixing points as discussed in the text. PG, pressure gauge; and TC, thermocouple. (Reprinted with permission from [86]. Copyright 1999: Elsevier.)

Particles produced were collected by an inline filter. The pressure of the system was maintained at 30 MPa by a back-pressure regulator (Model 26-1722-24, Tescom Corp., MN, USA). The reaction temperatures employed were 350, 385, 400 and 420°C. Reaction times were evaluated by total flow rate, reactor volume and density of pure water at the reaction temperature and pressure. Reaction times were in the range 40 to 120 s.

At temperatures higher than the critical temperature of water, continuous and rapid (<60 s) production of YAG:Tb fine particles was achieved. No additional thermal treatment was necessary to obtain crystalline products. Particles size and particle size distribution was found to depend on reaction temperature. Higher temperatures gave larger particles, but that work did not clarify the relationship between particles size and process variables. This applied hydrothermal method has the following attractive features: (a) rapid reaction times (<1 min), (b) continuous production of products, (c) highly crystalline products without the need for further thermal treatments, and (d) low solvent (water) contents in the products. Furthermore, this method can be applied to produce complex metal oxides.

The detail information about hydrothermal method and its application can be found in special books [53, 107–109] and a review [110].

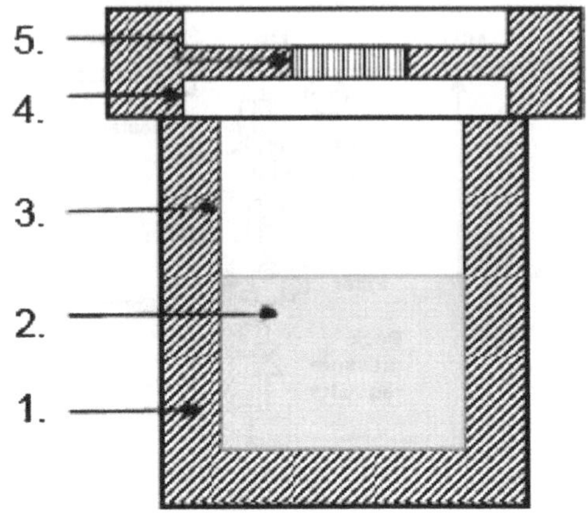

Figure 2.6 (1) Stainless steel autoclave, (2) precursor solution, (3) Teflon liner, (4) stainless steel lid, and (5) spring.

2.3.4 *Solvothermal*

Solvothermal synthesis route is very similar to the hydrothermal route, where the synthesis is conducted in stainless steel autoclave with the only difference is that the precursor solution is non-aqueous (Fig. 2.6). With solvothermal synthesis certain experimental parameters or conditions can easily alter the shape, type, size, size distribution, surface area, and the crystallinity. Some of these experimental parameters may include reaction temperature, reaction time, solvent type, surfactant type, precursor type, etc.

Many research groups have used solvothermal synthesis to make all sort of nanostructure using a range of different experimental parameters and conditions [111–113].

The YAG-based materials are normally synthesized at a relatively high temperature (>1500°C) by a solid state reaction of a stoichiometric mixture of the component oxides with prolonged heating time. Recently, Hukuta *et al.* [86, 106] reported on a hydrothermal method for the rapid production of fine YAG:Tb phosphors in supercritical water (374°C, 22.4 MPa). However, this process requires complicated and expensive equipment due to the high temperatures and pressures required. Inoue *et al.* [83] also synthesized fine YAG powders by the organic solvothermal method. However, the powders prepared by this method consist of aggregations of irregular

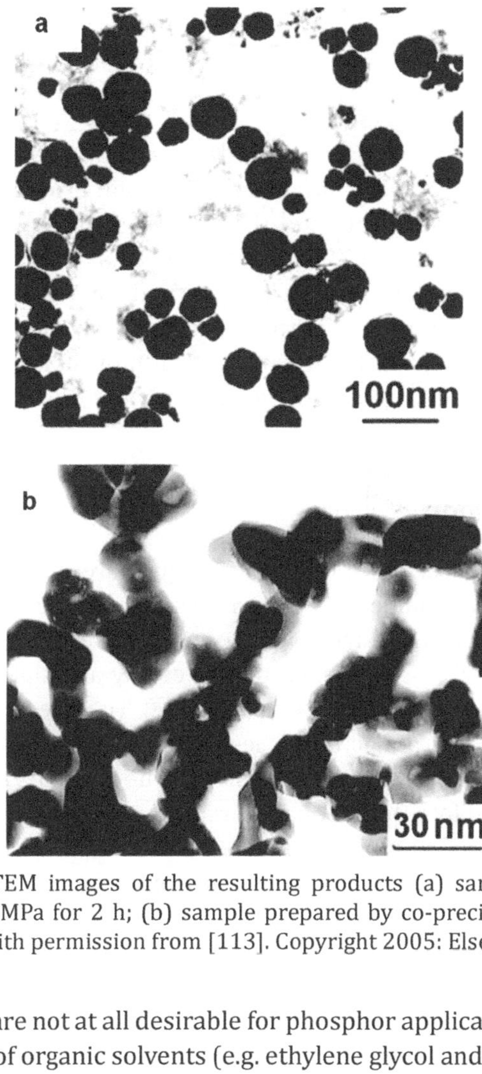

Figure 2.7 TEM images of the resulting products (a) sample prepared at 300°C, 10 MPa for 2 h; (b) sample prepared by co-precipitation route. (Reprinted with permission from [113]. Copyright 2005: Elsevier.)

grains, and are not at all desirable for phosphor applications. Moreover, the use of organic solvents (e.g. ethylene glycol and butancediol) increases the cost of production and may even pollute the environment. Li *et al.* [113] reported for the first time the preparation of YAG:Tb nanosized phosphors via a mixed alcohol–water solvent method using inexpensive inorganic salts as precursors at relatively low temperatures (300°C) and pressures (10 MPa). The advantages of the mixed solvothermal method include the following: low-cost precursors, low-temperature preparation, low operating cost, and an environmentally friendly solvent. The phase development, particle size and morphology, as well as the optical properties of the resulting powders were evaluated; this process requires complicated and expensive equipment.

This study demonstrates that the solvent and reaction temperature strongly affects the kinetics of YAG grain growth and its morphology. The crystallization of the YAG phase increases with rising reaction temperature. TEM analysis shows that the nanosized particles prepared by the solvo-thermal method were spherical and well dispersed in comparison with co-precipitation method (Fig. 2.7). The YAG:0.02Tb particles with spherical morphology and non-aggregation had good photoluminescence characteristics.

2.3.5 *Colloidochemical*

Colloidal chemical precipitation techniques are used to fabricate a wide variety of compounds with a crystallite size in the submicron range. Colloid is a substance composed of extremely small particles of one material (the dispersed phase) evenly and stably distributed in another material (the continuous phase). The size of the dispersed particles (1 nm–1 μm across) is less than that of particles in suspension but greater than that of molecules in true solution. Optically active centers doped into the host nanocrystals are then used to render the crystal luminescent under excitation by UV photons, X-rays, and electrons. In colloid chemistry it is possible to grow perfect nanocrystals. If these can be agglomerated in a controlled fashion to form well-defined spherical precursors, heat treatment can then be used to convert these into highly crystalline particles. The development of new methods for the production of micron or submicron powders may increase the possibilities for better control of the structural parameters.

Many works with different chemical methods have been devoted to the synthesis of nanoparticles. Nevertheless, good control of the structure can be achieved only on dispersion of particles in a solvent, which preserves the accessibility of their surfaces for further chemical treatments of passivation or functionalization. In this field, colloid chemistry has been shown in many different systems to be a very powerful tool for the elaboration of nanoparticles.

Colloidal synthesis of nanoparticles always relies on the precipitation of the material from dissolved precursors in solution [114]. Formation of the solid occurs when the concentration of its constituents goes beyond saturation, and follows the nucleation/growth process well described by the famous diagram of Lamer [115]. The main difficulty is the control of the size and dispersion state of the particles. Rapid aggregation of the particles occurs due to strong van der Waals (vdW) interactions, leading to a precipitate of aggregated nanoparticles. Two main strategies may be used against attractive vdW interactions. The first consists of using the electrostatic repulsion between the particles, resulting from the presence of ionic charges on their surfaces. These charges arise either from the ionization of surface groups (e.g. hydroxyls with acid/base properties) or from the surface complexation of the particles with ionic species (e.g. citrate salts). This strategy can only be achieved in dissociating solvents and with low ionic strengths. It is then used almost exclusively for aqueous syntheses, mostly in the case of oxide materials. The second strategy consists of complexing the surfaces of the particles with an organic agent, which will provide a steric hindrance between the particles keeping them far enough apart that vdW interactions are negligible. The nature of the surface complexing agent may be either a polymer or a surfactant molecule, or more generally any compound which may by grafted onto the surface of the particle and strongly coordinated by the solvent. The resulting thick coordinating sphere of solvent molecules around the particles also ensures steric hindrance and stability of the dispersion. Up to now, only a few studies have concerned the elaboration of luminescent oxide nanoparticles. The first reason is that the excellent emission properties of II–VI nanoparticles were discovered rather recently, so that practical applications of luminescent nanoparticles, justifying more extended research, were clearly identified only a few years ago. The second reason is that most of the well-known oxide phosphors are prepared through conventional ceramic processing techniques and require high-temperature treatments to ensure a good crystallinity of the final compounds. Such treatments are not compatible with the synthesis of well-controlled particles, and usually lead to agglomerated powders with large distributions of grain sizes.

There are nevertheless a limited number of compounds that can be crystallized at low temperatures. Among them are vanadates [116–119] and phosphates [119, 120] that can be prepared by precipitation from soluble precursors and that exhibit high emission efficiencies in the bulk state. Spherical shaped nanoparticles Er^{3+}, Yb^{3+} co-doped Gd_2O_3 phosphor of the average diameter in the range of 8–26 nm have been synthesized by Singh *et al.* [121]. Wakefield *et al.* [122, 123] discuss the fabrication and properties of a variety of nanocrystalline luminescent compounds in particular, yttrium, terbium and europium oxides prepared by colloidal chemical precipitation techniques. Two synthesis routes for the production of sub-micron luminescent materials are used.

- For the production of crystals with a grain size of 20–200 nm a precursor, which is typically "amorphous" or a mix of nanocrystals, is produced by decomposition of urea in a cell containing the correct proportions of the required cations. A typical example is shown in Fig. 2.8a. The precursor can then be fired into crystalline particles as shown in Fig. 2.8b. Note the absence of any defects in the material. The surface region of the crystals is crucial for suitable luminescent efficiency. Examination of the material in a high-resolution transmission electron micrograph shows that the crystals have no surface damage region, as shown in Fig. 2.8c.
- For particles in the size range 2–20 nm, crystalline material is formed directly as nanocrystals. At this size the surface atoms form a significant proportion of the whole and can act as recombination paths. They must therefore be passivated with a surfactant compound in order to retain the desired luminescent properties. Materials such as this, a typical example is shown in Fig. 2.8d, are known as quantum dots since quantum confinement effects come into play within this size range [124].

The advantage of using chemical techniques to fabricate phosphor material is shown in Fig. 2.8. Fig. 2.8b is a bright field transmission electron micrograph of a cluster of 70–100 nm europium doped yttrium crystals. Note that the crystals are structurally perfect, i.e. no defects are found within the crystal structure. Defects tend to migrate to the surface of the crystal during the growth stage. A high-resolution

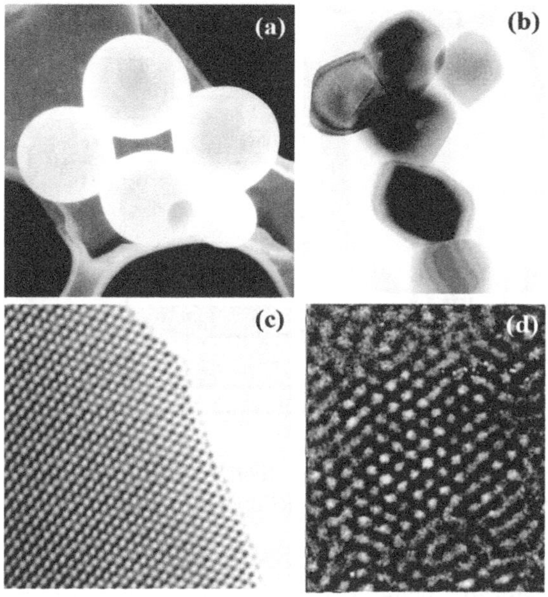

Figure 2.8 (a) Amorphous phosphor precursor particles. Scale: 1 cm = 100 nm; (b) Nanocrystalline Y_2O_3:Eu after firing of precursor material. Scale: 1 cm = 50 nm; (c) Surface of a phosphor nanocrystal. Note the absence of a damage layer. Scale: 1 cm = 2 nm; (d) Tb_2O_3 quantum dot. The particle appears as a regular array of atoms on an amorphous support grid. Scale: 1 cm = 7 Å. (Reprinted with permission from [123]. Copyright: WILEY-VCH Verlag GmbH, D-69469 Weinheim, 2001.)

lattice image of the surface of one of the crystals is shown in Fig. 2.8c demonstrating the absence of any surface defect layer.

Colloidal routes may also be used to produce luminescent particles in the quantum confinement regime, with size and doping control allowing variation of absorption edge energy and luminescence peak position independently.

Additional information about colloidal chemical precipitation techniques can be found in cited references and reviews [119, 125–127].

2.3.6 *Spray–Pyrolysis*

The method of spray drying can also be used to synthesize luminescent materials. In this method, the reactants are first dissolved in a

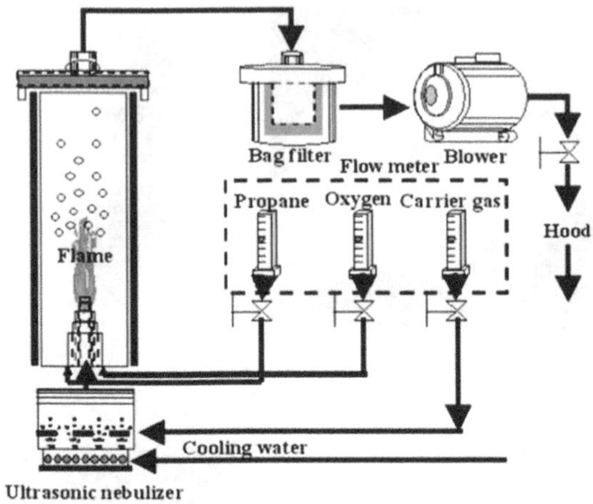

Figure 2.9 A schematic diagram of the flame spray pyrolysis process. (Reprinted with permission from [131]. Copyright 2008: Elsevier.)

solvent, preferably water. Droplets of this solution are then injected into a gas stream and are heated as they are transported downstream. The steam heating results in a very fast evaporation of the solvent or water. As a result, an intimate reaction mixture is formed and deposited downstream from the flow. This method is known as a promising aerosol process which is very simple and suitable for the preparation of multicomponent oxides such as phosphor materials [128–132].

We show two examples. The first one is preparing $BaMg_{0.9}Al_{10}O_{19}:Mn_{0.1}^{2+}$ ($BAM:Mn^{2+}$) phosphor powders by a flame spray pyrolys (FSP) process and compare the SEM images of this phosphor with commercial BAM:Mn prepared by solid state reaction [131]. The second example is a suitable phosphor powder Y_2O_3:Eu with fine grain size and high luminous efficiency for plasma application which has been prepared by spray pyrolysis using an original preparation procedure [132].

The typical flame spray pyrolysis system is shown in Fig. 2.9. It has a droplet generator, a flame nozzle, a quartz reactor, a powder collector, and a blower. A 1.7 MHz ultrasonic spray generator with six resonators is used to generate droplets, which are carried into the high-temperature diffusion flame by oxygen as the carrier gas.

Propane, as the fuel, and oxygen, as the oxidizer, create the diffusion flame. The flame nozzle has five concentric pipes. The droplets generated from the precursor solution are supplied to the diffusion flame through the center pipe by different flow rates of the carrier gas. The temperatures of the diffusion flame and the residence time of the powders inside the diffusion flame are controlled by changing the amounts of the fuel, oxidizer, and carrier gases.

The morphologies of the powders were investigated using scanning electron microscopy (SEM) and shown in Fig. 2.10.

Figure 2.10 shows SEM micrographs of the same phosphor composition prepared by two different methods. Commercial BAM:Mn phosphors have been prepared using the solid state reaction route

Figure 2.10 SEM photos of commercial BAM:Mn prepared by solid state reaction (a) and the BAM:Mn particles prepared by the spray pyrolysis (b). (Reprinted with permission from [130]. Copyright 2005: Elsevier.)

which generally involves a repetitive heat treatment at high temperature and a postmilling process to obtain high crystalline powder. As a result, most commercial phosphor particles have an irregular shape and a large size of about 2–5 μm. The spray pyrolysis is a gas-phase synthesis process and has been used to prepare spherical-shaped phosphor particles because the process has no repetitive heat treatment and easily produces phosphor particles with high phase purity. The BAM:Mn particles prepared by the spray pyrolysis with the polycation solution have a completely spherical shape. Because the morphology of phosphor particles affects the luminous properties, the phosphor particles with spherical shape and dense structure have better luminescence intensity than other ones, especially when they were prepared by the spray pyrolysis.

The powder Y_2O_3:Eu with fine grain size has been prepared by spray pyrolysis using 2.4 MHz ultrasonic pellets which is used to generate droplets of a precursor solution [132]. The experimental setup is shown in Fig. 2.11.

A 0.6 M concentrated solution of yttrium and europium nitrates is used as precursor. Every prepared sample has a 3.5 at% Eu ($Y_{1.93}Eu_{0.07}O_3$) doping concentration. The same doping concentration has been determined by ICP analysis on the commercial phosphor

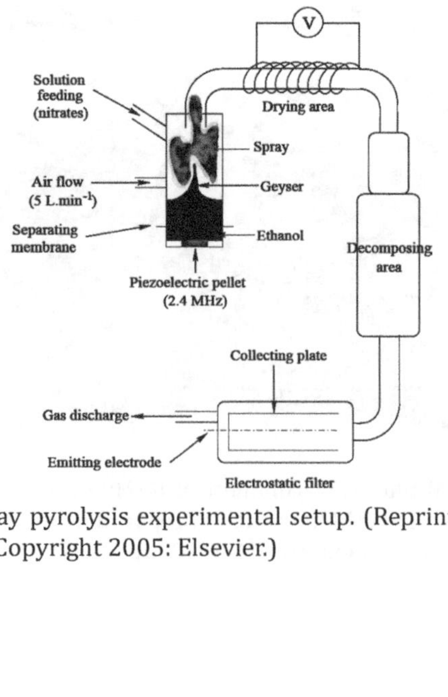

Figure 2.11 Spray pyrolysis experimental setup. (Reprinted with permission from [132]. Copyright 2005: Elsevier.)

Figure 2.12. SEM micrograph of Y_2O_3:Eu as prepared spray pyrolysis particles. (Reprinted with permission from [132]. Copyright 2005: Elsevier.)

(Kyokko) used as a reference in this study. An air flow around 5 L min^{-1} carries away the droplet through a drying area about 100°C; then through a decomposing area about 700°C: Particles are collected with an electrostatic filter. The powder is then annealed in a tubular furnace under air atmosphere from 900 to 1400°C.

Figure 2.12 is an SEM image of spray pyrolysis as prepared powder. The particles are spherical shaped which is typical of spray pyrolysis process and seem to be dense. The D_{v50} – volume median diameter: particles which have a diameter below D_{v50} represent 50% of the total volume – is 2.2 ± 0.2 μm measured by laser size analyzer.

2.3.7 *Combustion Method*

Combustion method, also known as "self-propagating high-temperature synthesis" (SHS), is yet another wet-chemical method which does not require further calcinations and repeated heating. This method was accidentally discovered in 1988 in Prof. Patil's lab in India [133, 134]. It is an exothermic reaction and occurs with the evolution of heat and light. Such a high temperature leads to formation and crystallization of phosphor materials. For any combustion, fuel and oxidizer are required. When the mixture of fuel and oxidizer are ignited, combustion takes place.

Beginning in the late 1980s, combustion synthesis has been investigated as a method to produce homogeneous, crystalline, fine oxide powders, as an alternative to time-consuming solid state reaction and sol–gel processing techniques [133, 135, 136]. Later in the mid-1990s, several research groups began to investigate the use of combustion synthesis for oxide phosphor preparation [91, 137, 138] and found it to be a technique of interest for phosphor synthesis in general. For oxide phosphors, the advantages of combustion synthesis are in its ability to produce well-crystallized, fine particle size powders rapidly without extensive high-temperature (>1500°C) annealing and mechanical separation (e.g. grinding and milling) steps. Grinding has often been implicated in the degradation of luminescent emission intensity through the creation of surface defects that quench the luminescence.

The term "combustion" covers flaming (gas-phase), smouldering (heterogeneous) as well as explosive reactions. The combustion method has been successfully used in the preparation of a large number of technologically useful oxide (refractory oxides, magnetic, dielectric, semiconducting, insulators, catalysts, sensors, phosphors, etc.) and nonoxide (carbides, borides, silicides, nitrides, etc.) materials. In recent years, there has been tremendous interest in the combustion synthesis of materials because it is simple, fast, energetically economic and yields high purity products compared to the conventional routes used to prepare these materials. As it is a high-temperature process, only thermodynamically stable phases can be prepared. At the same time, rapid heating and cooling rates provide the potential for the production of metastable materials with new and unique properties.

A number of review articles [139–143] on the combustion synthesis of oxides and nonoxides have been published during the last decades. Merzhanov, one of the pioneers of SHS has periodically reviewed [141, 142] the developments in SHS and its applications. His recent article [141] discusses the theory and practice of SHS worldwide along with new directions in the field. A two-part review [139, 140] on combustion synthesis of advanced materials by Moore and Feng gives an account of the historical perspectives of SHS, parameters that control the SHS process, and materials that have been prepared and their applications. Various types of SHS reaction and proposed models of reaction are discussed; the thermodynamics and kinetics of SHS reactions are also presented. Important parameters that control combustion synthesis such as the particle size and shape of the reactants, ignition techniques, stoichiometric ratio, processing of reactant particles (green density, i.e. the density of the pellet before sintering), and the adiabatic temperature, which is a measure of the exothermicity of the reaction, have been discussed in detail [139–143]. For this reason no attempt is made here to elaborate on these points.

Chapter 3

PHOSHOR CHARACTERIZATION

This is intended as a brief introduction to some of the common techniques used in materials characterization. It is designed for people who are novices in this field but are interested in using the techniques in their research. Some examples of application techniques to phosphors examination as well as some references are listed below in the end of each paragraph. Extensive and authoritative discussions can be found in the numerous books and journal articles on this subject.

3.1 Morphology Analysis with Scanning Electron Microscopy

The scanning electron microscope (SEM) is a microscope that uses electrons instead of light to form an image. In SEM an electron beam is scanned across a sample's surface. The electrons interact with the atoms that make up the sample producing signals that contain information about the sample's surface topography, composition and other properties. The scanning electron microscope was first constructed by von Ardenne in 1938 when he added scan coils to a transmission electron microscope. The real first scanning electron microscope to employ secondary electron detection was developed by Zworykin and his team in 1942. In our days the scanning electron

New Generation of Europium and Terbium Activated Phosphors
Mihail Nazarov and Do Young Noh

www.panstanford.com

microscope has many advantages over traditional microscopes. The SEM has a large depth of field, which allows more of a specimen to be in focus at one time. The SEM also has much higher resolution, so closely spaced specimens can be magnified at much higher levels. All of these advantages, as well as the actual strikingly clear images, make the scanning electron microscope one of the most useful instruments in research today. With SEM we can investigate:

- **Topography**
  The surface features of an object or "how it looks," its texture; detectable features limited to a few nanometers
- **Morphology**
  The shape, size, and arrangement of the particles making up the object that are lying on the surface of the sample or have been exposed by grinding or chemical etching; detectable features limited to a few nanometers
- **Composition**
  The elements and compounds the sample is composed of and their relative ratios, in areas ~1 μm in diameter
- **Crystallographic Information**
  The arrangement of atoms in the specimen and their degree of order; only useful on single-crystal particles >20 μm

3.1.1 *Electron Beam Interaction with Solid Matter*

When an electron beam is bombarding a solid surface secondary electrons, back-scattered electrons, Auger electrons, cathodoluminescence, and characteristic X-rays are produced (Fig. 3.1).

These main signals provide the greatest amount of information in SEM, but it is rare that a single machine would have detectors for all possible signals. We consider briefly all possible information from different siganals and everybody decides himself what is more important for his investigations.

Secondary electrons (SEs) are emitted from the atoms occupying the top surface and produce a readily interpretable image of the surface. The contrast in the image is determined by the sample morphology. A high-resolution image can be obtained because of the

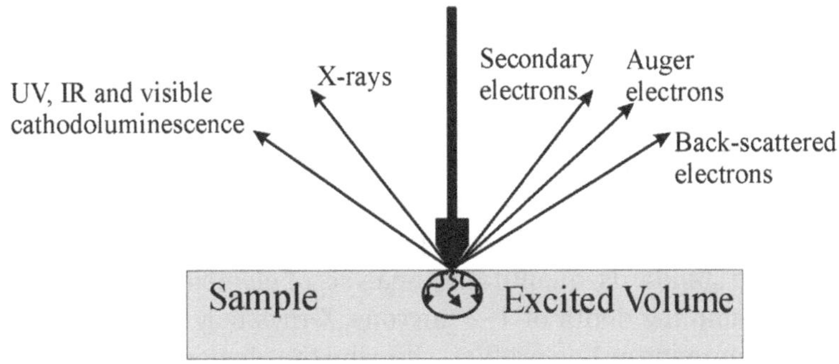

Figure 3.1 Schematic representation of the effects produced from electron beam interaction with solid matter.

small diameter of the primary electron beam. In the most common or standard detection mode the SEM can produce very high-resolution images of a sample surface, revealing details about less than 1 to 5 nm in size. Due to the very narrow electron beam, SEM micrographs have a large depth of field yielding a characteristic three-dimensional appearance useful for understanding the surface structure of a sample. A wide range of magnifications is possible, from about 10 times (about equivalent to that of a powerful hand-lens) to more than 500,000 times, about 250 times the magnification limit of the best light microscopes.

Back-scattered electrons (BSEs) are primary beam electrons which are reflected from atoms in the solid by elastic scattering. The produced contrast in the image is determined by the atomic number of the elements in the sample. BSE arise due to elastic collisions between the incoming electron and the nucleus of the target atom (i.e. Rutherford scattering) and as a consequence, if higher is Z, more BSE are emitted. The image will therefore show the distribution of different chemical phases in the sample. Because these electrons are emitted from a depth in the sample, the resolution in the image is not as good as for secondary electrons. BSE are often used in analytical SEM along with the spectra made from the characteristic X-rays. Because the intensity of the BSE signal is strongly related to the atomic number (Z) of the specimen, BSE images can provide information about the distribution of different elements in the sample.

Characteristic X-rays are emitted when the electron beam removes an inner shell electron from the sample, causing a higher energy electron to fill the shell and release energy. These characteristic X-rays are used to identify the composition and measure the abundance of elements in the sample. Detection and measurement of the energy permits elemental analysis (Energy Dispersive X-ray Spectroscopy or EDS). EDS can provide rapid qualitative, or with adequate standards, quantitative analysis of elemental composition with a sampling depth of 1–2 microns. X-rays may also be used to form color maps or line profiles, showing the elemental distribution in a sample surface.

Auger electron emission is a process similar to the X-ray scenario, but instead of emitting an X-ray, weakly bound outer shell electrons are emitted (typically from the first 2nm of the surface). Outer shell electrons can give bonding information about the samples that is very useful in surface science.

Cathodoluminescence is also similar to X-ray and Auger, but energy is conserved by the emission of a photon in the IR/visible/UV range. Scanning the sample with the incident beam will allow us to collect CL spectra from different regions of the sample, and CL photographs can be obtained.

There are basically two types of detectors in SEM: electron detectors and cathodoluminescent detectors. In general SEM setup (Fig. 3.2) we show all possible detectors compatible with SEM.

SE or BSE detectors are usually enough for topography and morphology investigations of powder samples. Characteristic X-rays and cathodoluminescence we will discuss in more detail in the next paragraphs related with microanalysis and luminescence characterization of materials.

3.1.2 *SEM Images of Phosphors in Back-Scattered Electrons*

The first conclusions about the synthesis can be made from analysis of SEM images in secondary or back-scattered electrons. In Fig. 3.3 $YNbO_4:Eu^{3+}$ phosphors prepared with different fluxes are presented.

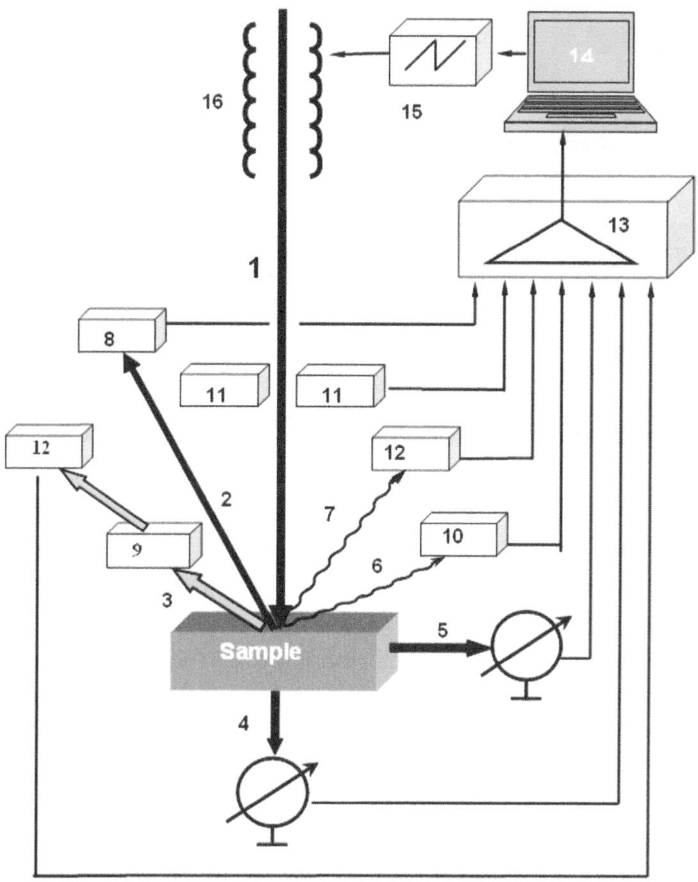

Figure 3.2 SEM setup. Image formation in SEM: 1 – primary electron beam; 2 – back-scattered electrons; 3 – secondary electrons; 4 – transmitted electrons; 5 – absorbed electrons (electron beam induced current, EBIC); 6 – characteristic X-rays; 7 – cathodoluminescence; 8,9,10 – detectors of back-scattered electrons, secondary electrons, and X-rays, respectively; 11 – detector of electron beam induced voltage (EBIV); 12 – photomultiplier; 13 – video amplifier; 14 – cathode ray tube or monitor; 15 – scan generator; 16 – deflecting coils.

The synthesis conditions with LiCl as flux are not optimized, because we can see a lot of small particles covered big ones, and shape of the particles is not perfect. Much better morphology is shown for $YNbO_4:Eu^{3+}$ prepared with Li_2SO_4 flux (right). The particles are clean

Figure 3.3 SEM images in back-scattered electrons of $YNbO_4:Eu^{3+}$ prepared with LiCl as flux (left) and Li_2SO_4 (right).

and shape is rounded. Preliminary results about particle dimensions can be also made from SEM image.

One can find additional information about SEM and its applications in special books and numerous articles and reviews [144–151].

3.2 Cathodoluminescence Characterization

The emission of ultraviolet, visible, or infrared light stimulated by electron bombardment is known as cathodoluminescence (from the cathode rays as J. J. Thompson has described electrons in 1897).

Cathodoluminescence was first studied by William Crookes in 1879. He discovered that $CaWO_4$ scheelite, placed opposite the cathode in a vacuum discharge tube yielded spectacular fluorescence. He concluded that some particle was traveling from cathode to the target causing the fluorescence well before J. J. Thomson discovered the electron. Crookes also studied luminescence from many other minerals (diamond, ruby, sapphire, zircon, etc.). However, systematic observations and discussion of cathodoluminescence did not take place until around 1965, when Smith and Stenstrom studied cathodoluminescence with the microprobe [152]. Early cathodoluminescence studies were carried out with a cathodoluminescence microscope, which is fundamentally a petrographic microscope to which some kind of cathode gun is attached. Subsequently, the electron-probe microanalyzer and, especially, the scanning electron microscope have been utilized to generate high-resolution, high-magnification cathodoluminescence images.

It should be noted that CL-modes being incorporated into scanning electron microscope (CL-SEM) give both the spectral and spatial information with high resolution. In geology, mineralogy and materials science a scanning electron microscope with specialized optical detectors, or an optical cathodoluminescence microscope, is used to examine internal structures of luminescent materials in order to get information on the composition, growth and quality of the material. Cathodoluminescence in semiconductors occurs because the impingement of a high-energy electron beam onto a semiconductor will result in the promotion of electrons from the valence band into the conduction band, leaving behind a hole. When an electron and a hole recombine, it is possible for a photon to be emitted. The energy (color) of the photon, and the probability that a photon and not a phonon will be emitted, depends on the material, its purity, and its defect state. In this case, the semiconductor examined can, in fact, be almost any non-metallic material. In terms of band structure, classical semiconductors, insulators, ceramics, gemstones, minerals, glasses and powders can be treated the same way.

The luminescence of most phosphors comes from a few sites (activator centers) occupied by selected chemical impurities which have been incorporated into the matrix or host solid. Direct excitation of the activator is only possible with ultraviolet and/or visible radiation. High-energy electron beam excitation always excites the host lattice. Because of the complex mode of interaction of cathode rays with phosphors, the energy efficiency of light production by cathodoluminescence is lower than the best efficiencies obtainable with photoluminescence. The main advantage of CL spectroscopy in comparison with PL is the possibility of local measurements with geometrical resolution less than 1 micrometer, while PL is an integrated mode.

The penetration depth of the incident beam and thus the excited volume in the sample depends on the energy of the incident electrons and atomic number of investigated material (Fig. 3.4). If secondary and back-scattered electrons are formed from the

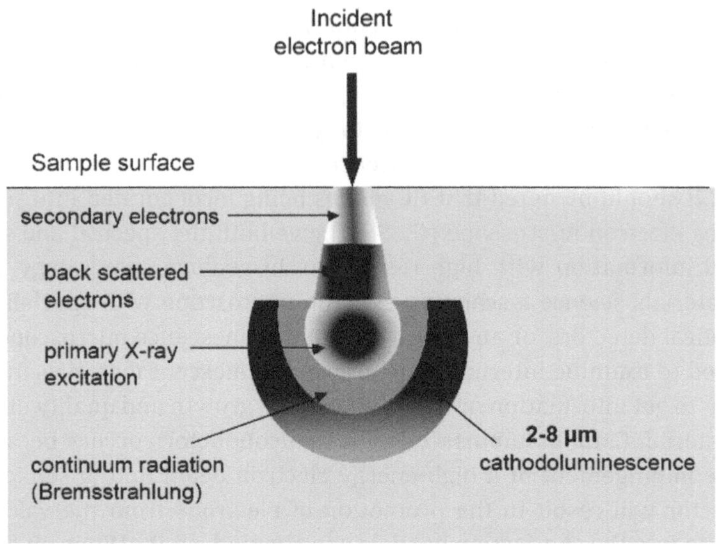

Figure 3.4 Schematic shapes of electron interaction volumes.

subsurface layer, CL is typically generated from depths down to 2–8 microns and it is a volume effect. Thus, the geometrical resolution in cathodoluminescence depends not only on electron beam diameter, but it is determined by CL interaction volume. Practically it can be about 1 μm. A 3D image of the sample is possible to be obtained if the scanning mode and electron energy variation are combined. Usually, a CL mode is applied in phosphor investigation in combination with other modes in SEM, like SE or BSE (see next paragraph).

3.2.1 *Monochromatic Cathodoluminescence*

One of the examples of CL application to the powder sample ZnO:Zn is shown in Fig. 3.5. Dominant CL emission peak for ZnO:Zn is located at 495 nm.

This work has demonstrated the utility of low-voltage scanning cathodoluminescence microscopy for high-resolution spatially resolved characterization of the luminescence properties of cathodoluminescent phosphor.

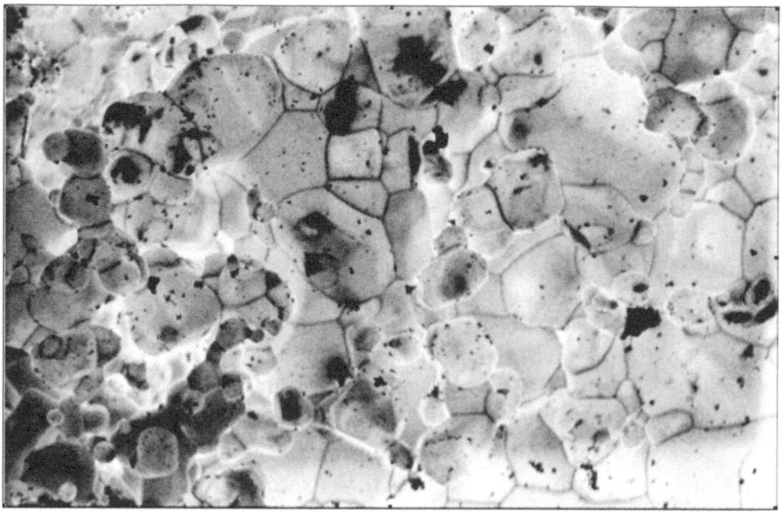

Figure 3.5 CL image ZnO:Zn at 495 nm, 1 kV. Horizontal width of field (HWF) = 58 μm. (Reprinted with permission from [153]. Copyright 2006: Microscopy Society of America).

3.2.2 *Panchromatic or Color Cathodoluminescence*

Among the different CL-SEM modes the most informative is color cathodoluminescence (CCL). The use of color contrast was shown to increase the information content of color image in comparison with its black-and-white counterpart by about two orders of magnitude [154]. Earlier the CCL-SEM mode was used for a complex characterization of different organic, dielectric and semiconductor luminescent materials [154–156]. Hereunder we show one of the examples of CCL application to phosphors in powder form (Fig. 3.6).

Mechanisms leading to the emission of light from phosphors are rather similar at all the forms of excitation. In all the cases a luminescence arises owing to stimulation of substance to an excited state. This excitation involves an input of energy from a source of suitable radiation, e.g. from the electron beam at the CL. In solids centers of luminescence are atoms ions or groups of ions located near lattice imperfections. Such imperfection in the host lattice may be produced by introducing activator or by producing a vacancy. The radiative transition of an excited center to the ground state is accompanied by the emission light.

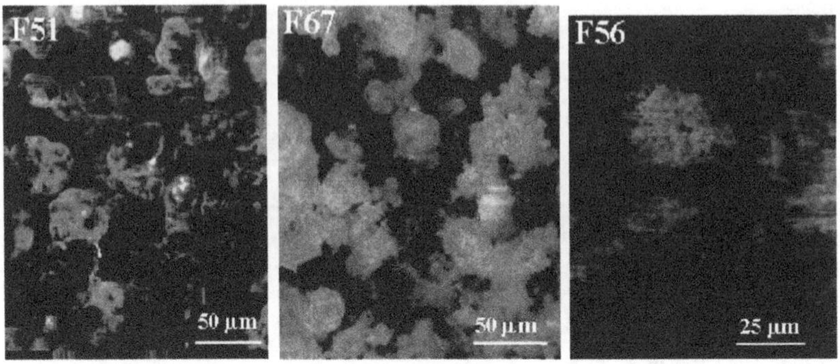

Figure 3.6 Panchromatic images of no activated $CaWO_4$ (F67) and activated with Eu^{3+} (F51) and Tb^{3+} (F56) (original blue color for host lattice emission, red color for Eu incorporations and green for Tb).

A special attachment for CCL-SEM examinations of topography and morphology of powder samples and cathodoluminescent spectrometer for local CL measurements are given in Fig. 3.7.

3.2.3 *Composite Cathodoluminescent and Back-Scattering Electron Contrast in SEM*

A schematic diagram of the (CCL+BSE) detector unit in the SEM is presented in Fig. 3.7. Basic equipment comprises an SEM Stereoscan

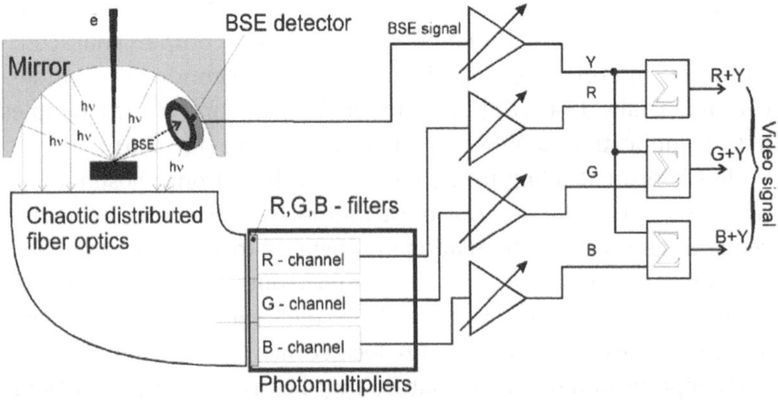

Figure. 3.7 CCL-SEM attachment for color cathodoluminescence. (Reprinted with permission from [157]. Copyright 1997: Wiley Periodicals, Inc.).

MK-IIA integrated with a CCL light collector and a solid state barrier counter for the measurement of BSE current. The BSE detector is located near the specimens under a parabolic mirror. Undistorted CCL and BSE signals are detected simultaneously during electron beam scanning.

Cathodoluminescence is transmitted by optical fibers from the specimen chamber. Later, using three R-, G-, and B-filters, the CL signal is transformed into electric video signals by photomultipliers. These signals are multiplied by preamplifiers which set the amplification ratio and zero value (black level). The BSE detector is integrated with the amplifier and operates at negative bias that increases the electrical and noise stability. The electric potential of the detector body is chosen to be identical with that of the microscope column potential to eliminate its action on the electron beam. The amplifier of the BSE signal also allows varying the amplification factor and the black level of the video signal. After pre-amplification, the R-, G-, B- and BSE (Y)-signals are summed, as noted in Fig. 3.7. As a result we have three video signals: R+Y, G+Y, and B+Y, which are transformed into digital form and are displayed on the computer screen. The technique of signal detection described above measures the final color of the image, as the true color of luminescence on the specimen and the gray level of the image is tuned by the BSE-signal value which reflects the specimen topography. The dynamic interval of the final signal is limited by the capability of the human eye to perceive and balance between CCL and BSE contrast that is fixed by amplification coefficients in the R-, G-, B-, and Y-channels. Using others SEM modes, we can apply the above principle to different signals, for example, the induced current signal (Y-channel) + (CCL mode for RGB-channels), and so forth.

Composite CCL+BSE contrast in scanning microscopy is a new type of SEM mode that is a rather perspective tool for the investigation of luminescence materials. The (CCL + BSE) – SEM images with both fine structure and color information may be able to simplify the observation, interpretation, and analysis of different nature specimens, since the human eye is essentially an organ for observing color and can distinguish thousands of color hues and intensities.

The following literature can be used to further CL reading [158–160].

3.3 Microanalysis

3.3.1 *Electron Probe Micro-Analyzer*

When an electron beam of sufficient energy interacts with a sample target it generates X-rays, as well as derivative electrons (e.g. secondary, back-scattered, auger).

An electron probe micro-analyzer (EPMA) is a microbeam instrument used primarily for the *in situ* non-destructive chemical analysis of solid samples including powder phosphors. It is fundamentally the same as an scanning electron microscope, with the added capability of chemical analysis. The primary importance of an EPMA is the ability to acquire precise, quantitative elemental analyses at very small "spot" sizes (as little as 1–2 microns), primarily by wavelength-dispersive spectroscopy (WDS). The spatial scale of analysis, combined with the ability to create detailed images of the sample, makes it possible to analyze phosphors materials *in situ* and to resolve complex chemical variation within single phases. The electron optics of an SEM or EPMA allow much higher resolution images to be obtained than can be seen using visible-light optics, so features that are irresolvable under a light microscope can be readily imaged to study detailed microtextures or provide the fine-scale context of an individual spot analysis. A variety of detectors can be used for

- imaging modes such as secondary-electron imaging, back-scattered electron imaging, and cathodoluminescence imaging
- acquiring 2D element maps
- acquiring compositional information by energy-dispersive spectroscopy and wavelength-dispersive spectroscopy
- analyzing crystal-lattice preferred orientations

An electron microprobe operates under the principle that if a solid material is bombarded by an accelerated and focused electron beam, the incident electron beam has sufficient energy to liberate both matter and energy from the sample. These electron-sample interactions mainly liberate heat, but they also yield both derivative electrons and X-rays. Of most common interest in the analysis of phosphors materials are secondary and back-scattered electrons, which are useful for imaging a surface or obtaining an average composition of the material. X-ray

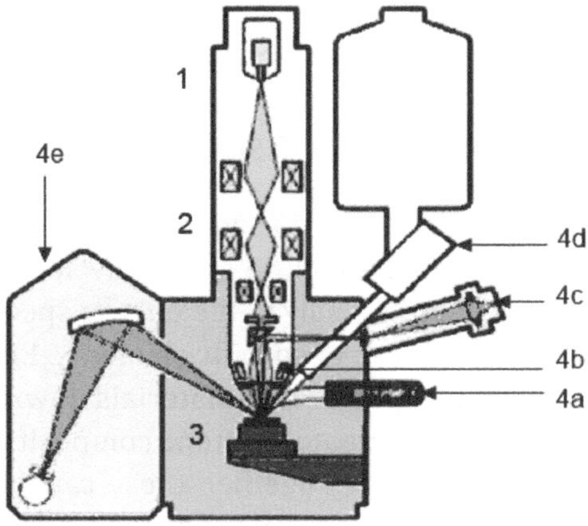

Figure 3.8 Schematic cut-away diagram of a typical microprobe.

generation is produced by inelastic collisions of the incident electrons with electrons in the inner shells of atoms in the sample; when an inner-shell electron is ejected from its orbit, leaving a vacancy, a higher-shell electron falls into this vacancy and must shed some energy (as an X-ray) to do so. These quantized X-rays are characteristic of the element. EPMA analysis is considered to be "non-destructive"; that is, X-rays generated by electron interactions do not lead to volume loss of the sample, so it is possible to re-analyze the same materials more than one time.

EPMA (Fig. 3.8) consists of four major components, from top to bottom:

1. An electron source, commonly a W-filament cathode referred to as a "gun".
2. A series of electromagnetic lenses located in the column of the instrument, used to condense and focus the electron beam emanating from the source; this comprises the electron optics and operates in an analogous way to light optics.
3. A sample chamber, with movable sample stage (*X-Y-Z*), that is under a vacuum to prevent gas and vapor molecules from interfering with the electron beam on its way to the sample; a light microscope allows for direct optical observation of the sample.
4. A variety of detectors arranged around the sample chamber that are used to collect X-rays and electrons emitted from the

sample: 4a-SE detector, 4b-BSE detector, 4c-visible light microscope, 4d-EDS spectrometer, 4e-WDS spectrometer.

3.3.2 *Energy-Dispersive X-Ray Spectroscopy*

An energy-dispersive detector is used to separate the characteristic X-rays of different elements into an energy spectrum, and EDS system software is used to analyze the energy spectrum in order to determine the abundance of specific elements. EDS can be used to find the chemical composition of materials down to a spot size of a few microns, and to create element composition maps over a much broader raster area. Together, these capabilities provide fundamental compositional information for a wide variety of materials.

EDS systems are typically integrated into either an SEM or EPMA instrument. EDS systems include a sensitive X-ray detector, a liquid nitrogen dewar for cooling, and software to collect and analyze energy spectra. The detector is mounted in the sample chamber of the main instrument at the end of a long arm, which is itself cooled by liquid nitrogen. The most common detectors are made of Si(Li) crystals that operate at low voltages to improve sensitivity, but recent advances in detector technology make available so-called silicon drift detectors that operate at higher count rates without liquid nitrogen cooling.

An EDS detector contains a crystal that absorbs the energy of incoming X-rays by ionization, yielding free electrons in the crystal that become conductive and produce an electrical charge bias. The X-ray absorption thus converts the energy of individual X-rays into electrical voltages of proportional size; the electrical pulses correspond to the characteristic X-rays of the element.

EDS cannot detect the lightest elements, typically below the atomic number of Na for detectors equipped with a Be window. Polymer-based thin windows allow for detection of light elements, depending on the instrument and operating conditions. Because the wavelength-dispersive method is more precise and capable of detecting lower elemental abundances, EDS is less commonly used for actual chemical analysis although improvements in detector resolution make EDS a reliable and precise alternative.

3.3.3 *Wavelength-Dispersive X-Ray Spectroscopy*

A wavelength-dispersive spectrometer uses the characteristic X-rays generated by individual elements to enable quantitative analyses (down to trace element levels) to be measured at spot sizes as small as a few micrometers. WDS can also be used to create element X-ray compositional maps over a broader area by means of rastering the beam. Together, these capabilities provide fundamental quantitative compositional information for a wide variety of solid materials. This technique is complementary to energy-dispersive spectroscopy in that WDS spectrometers have significantly higher spectral resolution and enhanced quantitative potential. Many SEM and EPMA instruments have EDS systems mounted to the column, and an EPMA typically has an array of several WDS spectrometers for simultaneous measurement of multiple elements. In typical EPMA applications, EDS is used for quick elemental scans to find out what a material contains, and WDS is then used to acquire precise chemical analyses of selected phases.

A wavelength-dispersive (WD) spectrometer is used to isolate the X-rays of interest for quantitative analysis. There may be a single WD spectrometer horizontally mounted on an electron column (more typical in SEM instruments) or 4–5 spectrometers may be mounted vertically in sequence around the sample chamber (more typical of EPMA).

WDS analysis involves four steps that must work together to achieve optimal results.

1. A beam of electrons is accelerated in an evacuated electron column of an EPMA or SEM to the sample surface with sufficient energy (typically with a potential difference of 15–20 kV) to generate characteristic X-rays for the elements to be analyzed.
2. Once X-rays are generated in the sample, they are selected using an analytical crystal with specific lattice spacing. The geometry of the X-ray generating sample and the analytical crystal is such that they maintain a constant take-off angle. When X-rays encounter the analytical crystal at a specific angle θ, only those X-rays that satisfy Bragg's law are reflected and a single wavelength is passed on to the detector. The wavelength of the

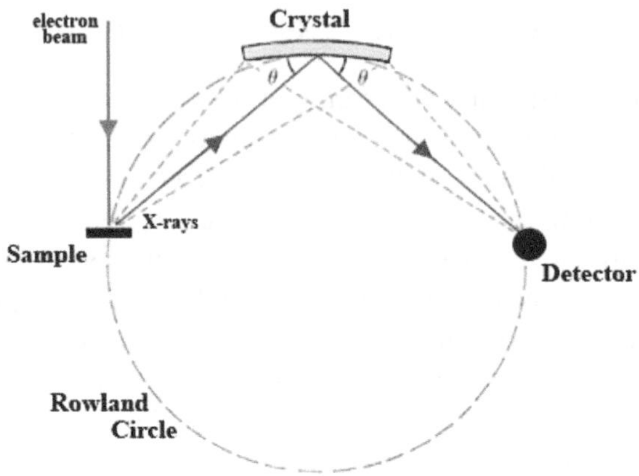

Figure 3.9 Configuration of sample, analytical crystal and detector on the Rowland circle within the WD spectrometer.

X-rays reflected into the detector may be varied by changing the position of the analyzing crystal relative to the sample, i.e. the X-ray source-crystal distance is a linear function of the wavelength. Consequently, X-rays from only one element at a time can be measured on the spectrometer and the position of a given analytical crystal must be changed in order to adjust to a wavelength characteristic of another element. There are two types of analytical geometry currently in use. The first type, the Johann geometry, is constructed by bending the analytical crystal to a radius of $2R$, where R is the radius of the focusing circle, the Rowland circle (Fig. 3.9). The second type, the Johansson geometry, is bent to a radius $2R$ and then ground to radius R, so that all of points of reflection lie on the Rowland circle, which maximizes the collection efficiency of the spectrometer. There is commonly more than a single analytical crystal in a WD spectrometer and, in the case of most EPMA instruments, there are typically multiple spectrometers with a suite of analytical crystals. Several different analytical crystals with dissimilar crystal lattice spacings (*d*-spacings) are normally used for WDS so that the spectrometers can reach all of the elemental wavelengths of interest and it will optimize performance in different wavelength ranges. An EPMA configured with 5 spectrometers, for example, might be set up to run WDS in a sequence to acquire 10 elements from one unknown sample by cycling through each crystal during analysis.

3. X-rays of specific wavelengths from the analytical crystal are passed on to the X-ray detector. The sample, crystal, and detector must lie on the Rowland circle and remain on it for all wavelengths of interest in order to focus X-rays efficiently. Because the sample and take-off angle of the X-rays are fixed, the analytical crystal and detector must both move to remain on the Rowland circle. In addition, the analytical crystal rotates as the X-ray source-crystal distance changes (requiring clever and precise engineering!). Detectors used in WD spectrometers are most commonly gas proportional counter types, in which incoming X-rays enter the detector through a collimator (slit) and thin window, are absorbed by atoms of the counter gas, and then a photoelectron is ejected by each atom absorbing an X-ray. The photoelectrons are accelerated to a central wire such that additional ionization produces an electrical pulse which has amplitude proportional to the energy of the original X-ray photon. The two general types of detector use sealed and gas-flow proportional counters. The sealed proportional counters are typically used for high-energy X-ray lines and have a relatively thick window that prevents leakage of Xe or Xe-CO_2 gas in the detector. The gas-flow proportional counters typically are used for low-energy lines, have ultrathin windows, and use a counter gas (generally, Ar gas with 10% methane) that flows through the detector at a constant rate.
4. Within a given sample, once the X-ray intensities of each element of interest are "counted" in a detector at a specific beam current, the count rates are compared to those of standards containing known values of the elements of interest. In turn, the X-ray intensities must be corrected for matrix effects associated with atomic number (Z), absorption (A) and fluorescence (F). This correction procedure is performed within a computer program that takes the raw counting rates of each element, compares these to standards, computes the ZAF correction (or similar type of correction) and displays the results as a function of the weight % of the oxides or elements.

Thus, measurement of an element's abundance requires exciting an atom to produce X-rays, focusing the X-rays through a crystal

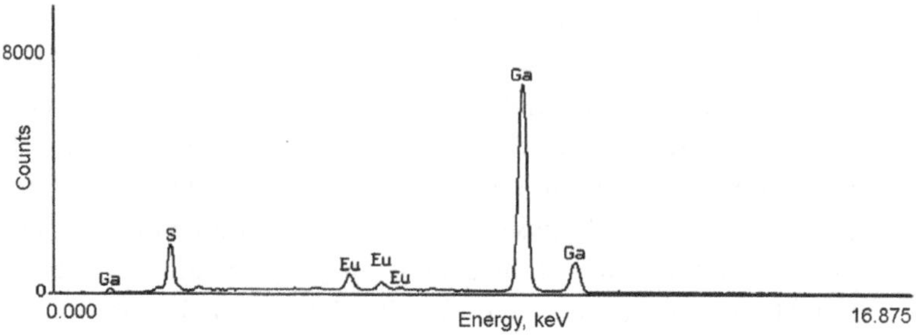

Figure 3.10 EDS patterns corresponding to $MgGa_2S_4:Eu^{2+}$.

spectrometer to a detector, converting the X-rays to photoelectrons, which in turn generate an electrical signal whose magnitude is proportional to the abundance of the element! This multistep process involves many potential breakdowns, but works reliably well to allow for routine analysis.

3.3.4 *Example of Microanalysis Application*

In Figs. 3.10 and 3.11 we present an application of EDS, WDS, and EPMA to $MgGa_2S_4:Eu^{2+}$ phosphor. For quick elemental scans to find out what a material contains, EDS was used. Fig. 3.10 shows that unfortunately, EDS can not detect the light elements, such as magnesium, for example. Thus, the compositions of the sample should also be examined by EPMA.

The EPMA technique allows us to use WDS to acquire precise chemical analyses (Fig. 3.11).

From WDS spectra of $MgGa_2S_4:Eu^{2+}$ it is clearly seen that all elements: Mg, Ga, S, Eu, and O were detected.

The following literature can be used to further explore EPMA, EDS, and WDS [144, 161–165].

3.4 Particle Size Analysis

3.4.1 *Methods of Particle Analysis*

Particle size analysis (PSA) comprises the measurement and the quantitative description of physical properties of particulate matter.

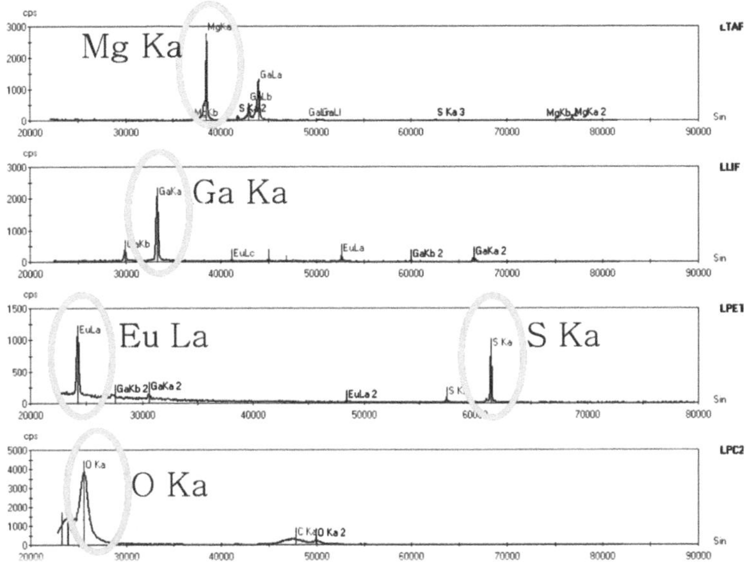

Figure 3.11 WDS spectra of $MgGa_2S_4:Eu^{2+}$.

In PSA every possible physical principle has been used to determine the particle size distribution. Only very few have reached an important position in phosphor application, e.g.:

- sieving
- sedimentation
- laser diffraction

PSA methods related to physical properties and quantity types are presented in Table 3.1:

Table 3.1 Main PSA methods

Method	Particle property	Quantity type
Sieve analysis	linear dimension	mass
Sedimentation	settling velocity	mass
Coulter counter	volume	number
Laser diffraction	distortion of optical field	area

The most important quantitative method for phosphor powders often used in practice is a laser diffraction, which we consider here in detail.

3.4.2 *Laser Diffraction*

Although the effect of particles interacting with light was described by Fraunhofer already early in the last century the idea of measuring particle size distribution with this physical principle could be realized only after reliable lasers were developed providing a light source and microcomputers became powerful enough to calculate the size distribution from the light intensity distribution detected by a semiconductor-detector.

In the early seventies the first instruments for PSA with means of laser diffraction covered a measuring range from coarser than 1 micron up to 200 microns. Until the middle of the eighties only wet dispersion was available. Knowledge of the shape of the sample distribution was required or presumed.

In the mid-eighties a different attitude towards particle size analysis entered the market with capable dry dispersion [166]. Laser diffraction systems no longer required the samples to be prepared according to the measuring system but offered the ability to be adapted to the needs of the products. Furthermore no longer additional information about the distribution shape was required. Latest technology covers a measuring range from below 0.1 μm up to 1 cm.

The classical setup for the particle–light interaction is shown in Fig. 3.12.

The basic physical principal is light scattering at the edges of particles. After focusing the light in the center of a multielement detector the scattering effect leads to signals on the non-central elements of the detector. The optical setup is referred to as Fourier optics [167].

After widening the monochromatic wave of laser light enters the measuring zone. The passing and scattered light is focused by a lens system Fig. 3.13. Generally the smaller the particles are the larger are the resulting diffraction angles. Hence, the focal length determines the measuring range.

The results gained with laser diffraction are the size distribution of equivalent spherical particles, i.e. the result refers to a distribution

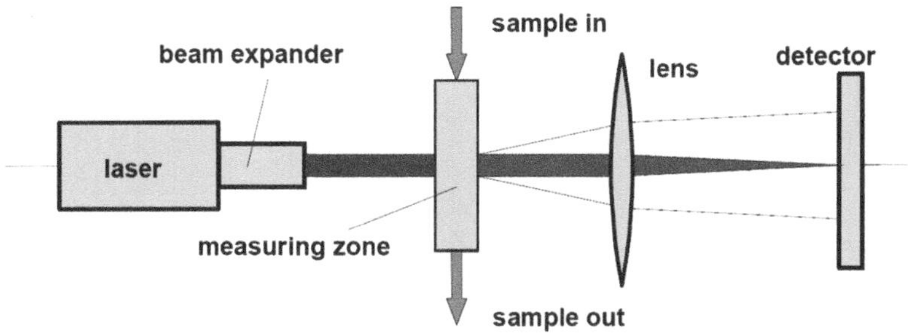

Figure 3.12 Setup of a laser diffraction PSA system.

of spherical particles that interact with the laser light in the same way as the sample. Therefore differences can occur in comparison to sieving, sedimentation or other PSA techniques.

3.4.3 *Cumulative and Density Distribution*

Measures of quantity used in particle size analysis are: cumulative distribution Qr (Fig. 3.14) and density or differential distribution qr (Fig. 3.15) according the graphic presentation of particle size distribution [168].

In a two-dimensional graphic presentation of PSA data, the independent variable, plotted on the abscissa, describes the physical property chosen to characterize the size of the particles. The dependent variable characterizes type of quantity and measure of quantity and is plotted on the ordinate (refer to ISO 9276-1. International Organization for Standardization).

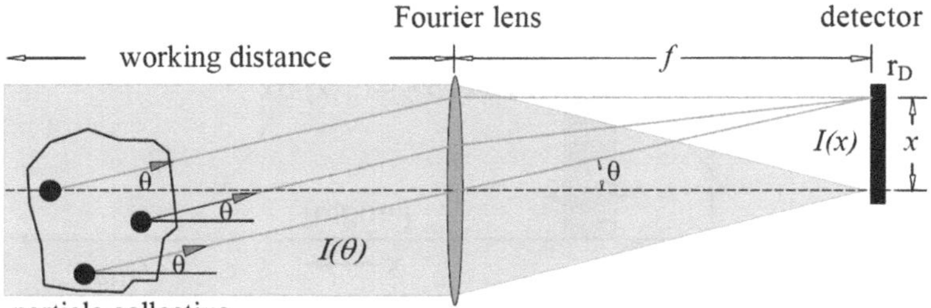

Figure 3.13 Particle size analysis in the parallel beam.

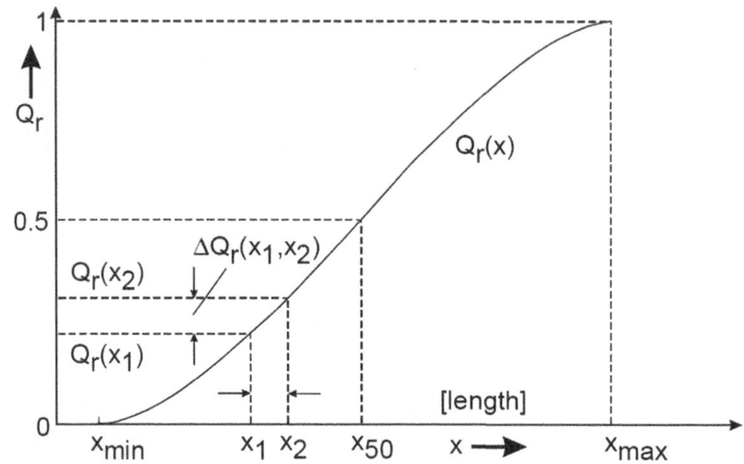

Figure 3.14 Cumulative distribution.

3.5 X-Ray Diffraction Measurements

3.5.1 *X-Ray Generation and Properties*

X-rays are electromagnetic radiation with typical photon energies in the range of 100 eV to 100 keV. For diffraction applications,

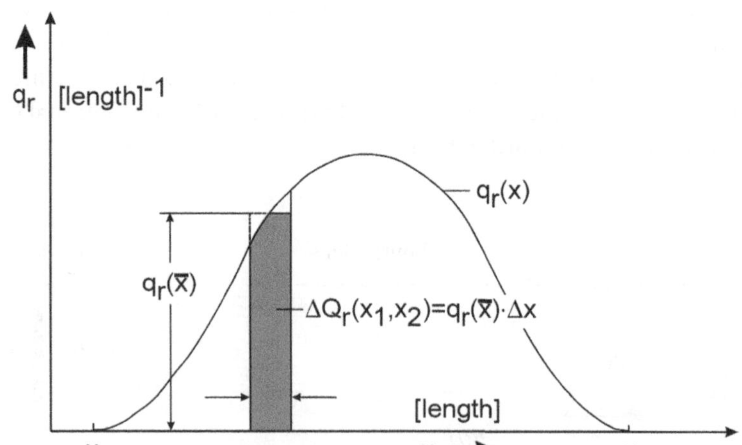

Figure 3.15 Density distribution.

only short wavelength X-rays (hard X-rays) in the range of a few angstroms to 0.1 angstrom (1 keV–120 keV) are used. Because the wavelength of X-rays is comparable to the size of atoms, they are ideally suited for probing the structural arrangement of atoms and molecules in a wide range of materials. The energetic X-rays can penetrate deep into the materials and provide information about the bulk structure.

X-rays are produced generally by either X-ray tubes or synchrotron radiation. In a X-ray tube, which is the primary X-ray source used in laboratory X-ray instruments, X-rays are generated when a focused electron beam accelerated across a high-voltage field bombards a stationary or rotating solid target. As electrons collide with atoms in the target and slow down, a continuous spectrum of X-rays are emitted, which are termed Bremsstrahlung radiation. The high-energy electrons also eject inner shell electrons in atoms through the ionization process. When a free electron fills the shell, a X-ray photon with energy characteristic of the target material is emitted. Common targets used in X-ray tubes include Cu and Mo, which emit 8 keV and 14 keV X-rays with corresponding wavelengths of 1.54 Å and 0.8 Å, respectively. (The energy E of an X-ray photon and its wavelength λ are related by the equation $E = hc/\lambda$, where h is Planck's constant and c the speed of light). Typical scheme of X-ray diffraction setup is shown in Fig. 3.16.

In recent years synchrotron facilities have become widely used as preferred sources for X-ray diffraction measurements. Synchrotron radiation is emitted by electrons or positrons traveling at near light speed in a circular storage ring. These powerful sources, which are thousands to millions of times more intense than laboratory X-ray tubes, have become indispensable tools for a wide range of structural investigations and brought advances in numerous fields of science and technology.

3.5.2 *Lattice Planes and Bragg's Law*

X-rays primarily interact with electrons in atoms. When X-ray photons collide with electrons, some photons from the incident beam will be deflected away from the direction where they originally

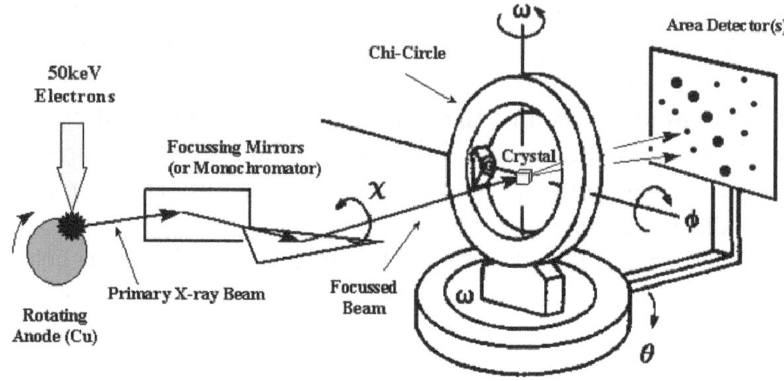

Figure 3.16 Typical scheme of X-ray diffraction setup.

travel, much like billiard balls bouncing off one anther. If the wavelength of these scattered X-rays did not change (meaning that X-ray photons did not lose any energy), the process is called elastic scattering (Thompson scattering) in that only momentum has been transferred in the scattering process. These are the X-rays that we measure in diffraction experiments, as the scattered X-rays carry information about the electron distribution in materials. On the other hand, in the inelastic scattering process (Compton scattering), X-rays transfer some of their energy to the electrons and the scattered X-rays will have different wavelength than the incident X-rays.

Diffracted waves from different atoms can interfere with each other and the resultant intensity distribution is strongly modulated by this interaction. If the atoms are arranged in a periodic fashion, as in crystals, the diffracted waves will consist of sharp interference maxima (peaks) with the same symmetry as in the distribution of atoms. Measuring the diffraction pattern therefore allows us to deduce the distribution of atoms in a material.

The peaks in an X-ray diffraction pattern are directly related to the atomic distances. Let us consider an incident X-ray beam interacting with the atoms arranged in a periodic manner as shown in two dimensions in the following illustrations (Fig. 3.17).

The atoms, represented as gray spheres in the graph, can be viewed as forming different sets of planes in the crystal. For a given

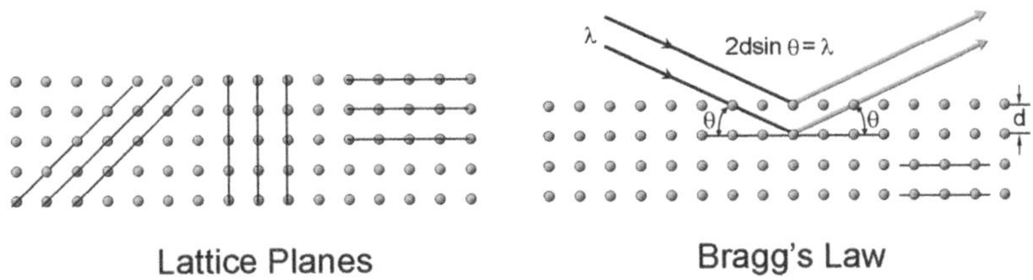

Figure 3.17 Lattice planes (left) and Bragg's law (right).

set of lattice planes with an inter-plane distance of d, the condition for a diffraction (peak) to occur can be simply written as

$$2d\sin\theta = n\lambda \qquad (3.1)$$

which is known as the Bragg's law, after W.L. Bragg, who first proposed it. In Eq. (3.1) λ is the wavelength of the X-ray, θ the scattering angle, and n an integer representing the order of the diffraction peak. Bragg's law is one of most important laws used for interpreting X-ray diffraction data.

It is important to point out that although we have used atoms as scattering points in this example, Bragg's law applies to scattering centers consisting of any periodic distribution of electron density. In other words, the law holds true if the atoms are replaced by molecules or collections of molecules, such as colloids, polymers, proteins, and virus particles.

3.5.3 *Powder Diffraction*

Powder X-ray diffraction (XRD) is perhaps the most widely used X-ray diffraction technique for characterizing materials. As the name suggests, the sample is usually in a powdery form, consisting of fine grains of single crystalline material to be studied.

The term "powder" really means that the crystalline domains are randomly oriented in the sample. Therefore when the 2-D diffraction pattern is recorded, it shows concentric rings of scattering peaks corresponding to the various *d*-spacings in the crystal lattice. The positions and the intensities of the peaks are used for identifying the underlying structure (or phase) of the material. For example, the diffraction lines of graphite would be different from

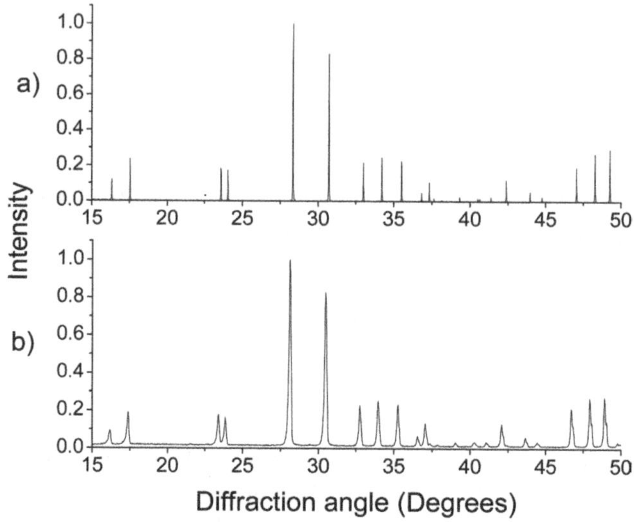

Figure 3.18 A comparison of XRD measurements of $YTaO_4$ phosphor on the basis of synchrotron X-ray diffraction patterns (Pohang Accelerator Laboratory, Korea) (top) and Cu-Kα radiation (λ = 1.54 Å) source generated from a Rigaku rotating anode generator (bottom).

diamond even though they both are made of carbon atoms. This phase identification is important because the material properties are highly dependent on structure (just think of graphite and diamond).

3.5.4 *Example of XRD Application*

A powder XRD scans from the same $YTaO_4$ sample on the basis of synchrotron X-ray diffraction patterns and Cu-Kα radiation (λ = 1.54 Å) source generated from a Rigaku setup are shown below in Fig. 3.18 as a plot of scattering intensity vs. the scattering angle 2θ or the corresponding *d*-spacing. The peak positions, intensities, widths, and shapes all provide important information about the structure of the material.

The detailed information about X-ray diffraction measurements can be found in the numerous books and journal articles on this subject [169–174].

3.6 Spectroscopic Methods in Structural, Morphology and Luminescence Characterization

3.6.1 *Spectroscopy Classification*

Everyone is familiar with various kinds of electromagnetic radiation. The spectrum of visible light has been long observed in nature in the form of the rainbow. In 1666 Sir Isaac Newton demonstrated that radiation from the sun could be split into the component colors of the rainbow using a glass prism. However, the visible region is only a small portion of the total electomagnetic spectrum. Other detection systems reveal radiation beyond the visible regions of the spectrum, which are classified as gamma rays, X-rays, ultraviolet rays, infrared, microwave, and radiowaves (Fig. 3.19).

The radiation from these different regions of the spectrum interact with matter in many different ways:

Gamma rays - Atomic nuclei excited;

X-rays - Inner electron transitions;

Infrared - Outer electron transitions;

Microwave - Molecular rotations;

Radiowaves - Nuclear magnetic resonances.

All these interactions of different kinds of radiation with matter can be used for material characterization and various spectroscopy

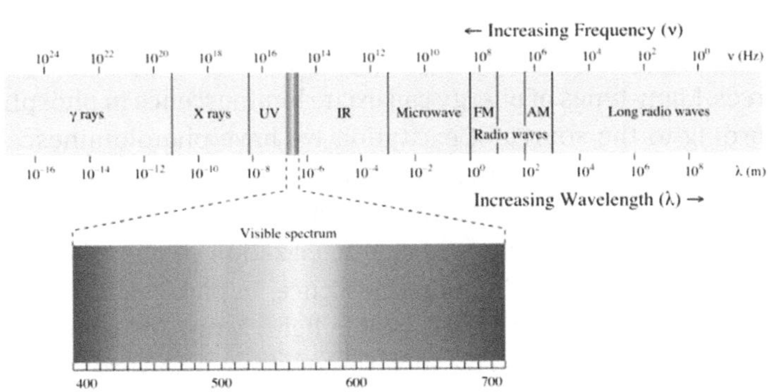

Figure 3.19 Electromagnetic spectrum .

Table 3.2 Electromagnetic spectrum and applied spectroscopy

Frequency range (Hz)	Wavelength range	Type of radiation	Applied spectroscopy
10^{20}–10^{24}	10^{-12}–10^{-16} m	gamma rays	γ-rays excited luminescence
10^{17}–10^{20}	1 nm–1 pm	X-rays	X-rays excited luminescence
10^{15}–10^{17}	400–1 nm	ultraviolet light	UV and VUV excited luminescence
4.3x10^{14}–7.5x10^{14}	700–400 nm	visible light	Photoluminescence
10^{12}–10^{14}	2.5 μm–700 nm	infrared light	Near-infrared spectroscopy (NIRS)
10^{8}–10^{12}	1 mm–2.5 μm	Mid-infrared Far-infrared microwaves	Fourier transform infrared (FTIR) spectroscopy, Raman spectroscopy Terahertz (THz) spectroscopy
10^{0}–10^{8}	10^{8}–1 m	radiowaves	

methods can be proposed. Electromagnetic spectrum, type of radiation and applied spectroscopy can be summarized in the Table 3.2 as follows:

Luminescence spectroscopy describes a group of techniques that measure light emission from samples under a variety of excitation sources. Many types of energy can excite luminescence in phosphors. According to the source of excitation we have photoluminescence, cathodoluminescence, electroluminescence, chemiluminescence, and other types of luminescence. We consider here two the most important types of luminescence characterization of phosphors: photoluminescence and cathodoluminescence. In particular, the photoluminescence is excited by UV, visible light, X-rays, or gamma-rays. NIRS, FTIR, Raman, and THz spectroscopes are also very important for phosphor investigations and we give in this chapter some background, schemes, and examples of application.

Infrared, Raman, and THz spectroscopes are applied for structural analysis and identifying samples and sample components. These spectroscopes are non-destructive, highly sensitive techniques that provide information about impurities, chemical environment, and free-carrier properties. The infrared portion of the electromagnetic spectrum is divided into three regions; the near-, mid- and far-infrared, named for their relation to the visible spectrum. The far-infrared, approximately 400–10 cm^{-1} (1000–30 μm), lying adjacent to the microwave region, has low energy and may be used for rotational spectroscopy. The mid-infrared, approximately 4000–400 cm^{-1} (30–2.5 μm) may be used to study the fundamental vibrations and associated rotational-vibrational structure. The higher energy near-IR, approximately 14000–4000 cm^{-1} (2.5–0.8 μm) can excite overtone or harmonic vibrations. The names and classifications of these subregions are merely conventions. They are neither strict divisions nor based on exact molecular or electromagnetic properties.

Performing both Fourier transform Raman and Fourier transform infrared spectroscopy on a sample gives a complete picture of its bonding structure. Both techniques benefit from high FT sensitivity and can be used with database searches of extensive libraries of FTIR and Raman spectra to identify samples.

We will focus here on very few aspects of the main spectroscopes applied to phosphor investigation, because many textbooks present excellent descriptions of the basis of each spectroscopy. Some references are listed below in the end of each paragraph.

3.6.2 *Photoluminescence Spectroscopy*

Photoluminescence is the spontaneous emission of light from a material under optical excitation. At the PL a substance absorbs radiation of a certain wavelength and then re-emits photons at different wavelengths. It is a contactless, non-destructive method of probing the electronic structure of materials. Light is directed onto a sample, where it is absorbed and imparts excess energy into the material in a process called photo-excitation. Photo-excitation causes electrons within the material to move into permissible excited states. When these

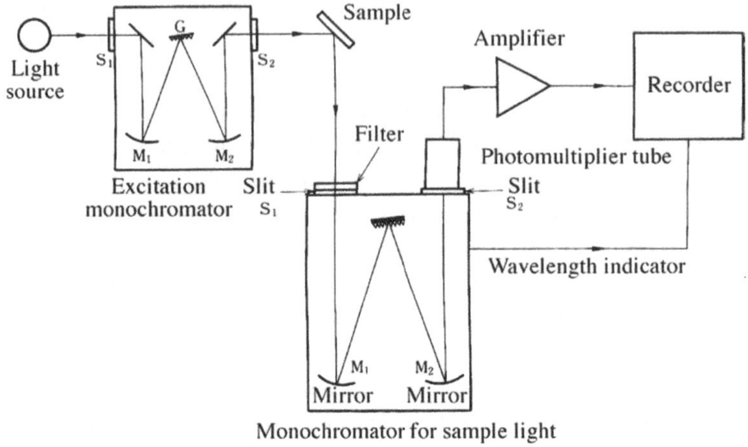

Figure 3.20 Spectroscopic measurement apparatus. (Reprinted with permission from [33]. Copyright 2006: Elsevier).

electrons return to their equilibrium states, the excess energy is released and may include the emission of light (a radiative process) or may not (a non-radiative process). The energy of the emitted light (photoluminescence) relates to the difference in energy levels between the two electron states involved in the transition between the excited state and the equilibrium state. The quantity of the emitted light is related to the relative contribution of the radiative process.

3.6.2.1 Photoluminescence excitation spectroscopy

The apparatus for measuring the spectral characteristics of phosphors is shown in Fig. 3.20. The excitation source consists of the light source and a monochromator, which selects a specific wavelength range from the incoming light. (The monochromator can be replaced by a filter). The light emitted from the sample is analyzed by a monochromator equipped with a light detector. The light detector transforms the photons into electrical signals. After the signals are amplified, they are recorded, typically on a strip chart recorder. It is often convenient to collect all spectral data in the form of digitized electrical signals and to use a computer for further processing the data.

In PL spectroscopy, which is performed at a fixed excitation energy, the luminescence properties are generally investigated, while photoluminescence excitation spectroscopy (PLE), which is carried out at a fixed detection energy, provides mainly information about the absorption properties and probes excited states of optical centers. In this experiment, the sample is excited by photons of the energy ($\hbar\nu$) tuned to be absorbed by specific optical centers. Therefore, this technique requires a tunable excitation source. The high-resolution PLE, i.e. the combination of highly resolved PL spectra as a function of the well-defined excitation energy $\Delta E \sim 1$ meV, provides detailed information about the system under investigation.

3.6.2.2 Measurements in the vacuum-ultraviolet region

The wavelength region between about 0.2 and 220 nm is called the vacuum-ultraviolet (VUV) region; most of the VUV spectrometers need to be evacuated in this region because of the opacity of oxygen in air to this radiation. The VUV part consists usually of a Deuterium lamp as light source, and a vacuum monochromator. The book by Samson [175], despite its age, is still an excellent textbook for beginners of spectroscopy in the VUV region. The book describes details of concave gratings, their mountings, light sources, window materials, detectors, polarizers, and absolute intensity measurements in the VUV region.

The PLE spectrum from $YTaO_4$ as VUV demonstration was measured in the range from 100 to 350 nm using sodium salicylate powder as a reference. The excitation PL spectra were obtained by observing all emission light and calibrated by that of sodium salicylate, which has constant quantum efficiency in 115–300 nm range. A fragment of these spectra is shown in Fig. 3.21.

The band around 220 nm in PLE is associated with the absorption of the host lattices. The TaO_4^{3-} groups can absorb excitation energy through $O^{2-} \rightarrow Ta^{5+}$ charge transfer transition and emit luminescence itself with broad band peaked at 333 nm as shown in Fig. 3.21, or transfer the energy to rare earth luminescent centers, if any.

It is clearly seen from Fig. 3.21 that because the absorption in $YTaO_4$ is shifted to the short wavelength region in comparison with $YNbO_4$, for example, this material can not be excited by 254 nm, and

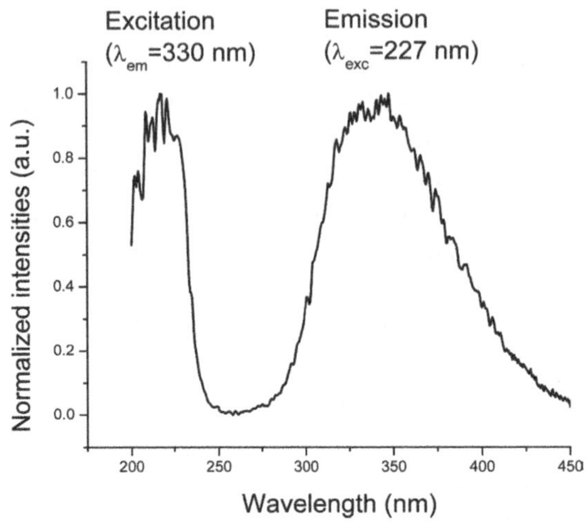

Figure 3.21 The emission and excitation spectra from $YTaO_4$.

more high-energy excitation like X-ray, cathodoluminescence or VUV have to be applied.

More detailed information about photoluminescence under different excitations is given in the next chapter.

3.6.3 *X-Ray Generated Luminescence*

X-ray luminescence can be defined as luminescence from materials which absorb X-rays and convert the absorbed energy efficiently into luminescence, in practice often ultraviolet or visible emission.

Almost immediately after the discovery of X-rays by Wilhelm Conrad Roentgen in 1895, he realized that this particular radiation was not very efficient in exposing photographic film and he initiated a search for phosphors which would absorb the X-rays and more efficiently emit light.

X-ray generated luminescence has some advantages in comparison with luminescence excited by ultraviolet or visible light because of high energy of X-rays and possibility to excite all levels of activator itself and host lattice. Other advantage is in increasing the luminescence because empty levels, which usually exist in any materials, can be readily filled under high-energy excitation and radiative emissions not more trapped by the empty upper levels.

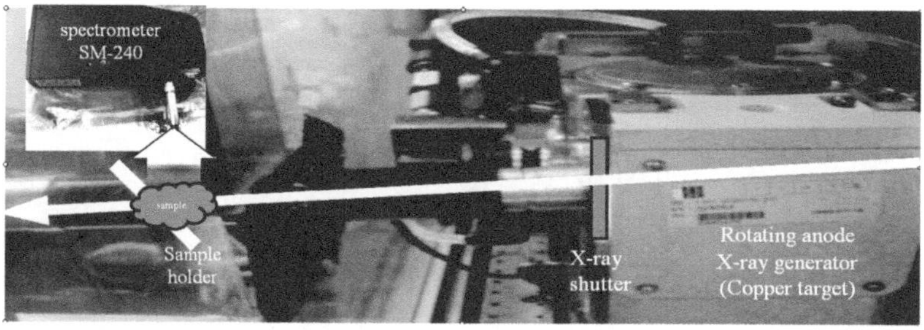

Figure 3.22 Illustration of the real experimental configuration with schematic diagram to study the luminescence of phosphors under X-ray excitation.

The combination of real setup with typical schematic diagram to study phosphor luminescence under X-ray excitation is shown in Fig. 3.22. X-ray beam is formed by the Bremsstrahlung radiation from a Rigaku rotating anode generator for example, operating in the range from 20 to 50 kV. Powder phosphors can be illuminated into the broadband X-ray radiation. The samples are placed at a position close to exit of Be X-ray windows to minimize the loss of X-ray flux by air scattering, and rotated by 45° from the incident X-rays. Since the luminescence signal is irradiated into all directions, collimating lenses are used to collect the light effectively. The emission spectrum is recorded with a spectrometer and processes with computer.

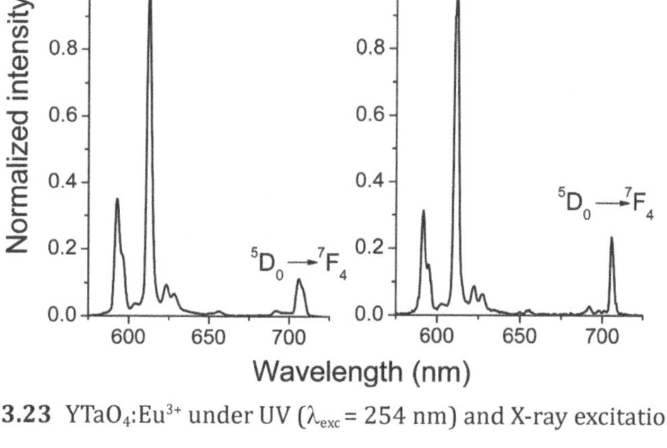

Figure 3.23 $YTaO_4:Eu^{3+}$ under UV (λ_{exc} = 254 nm) and X-ray excitation.

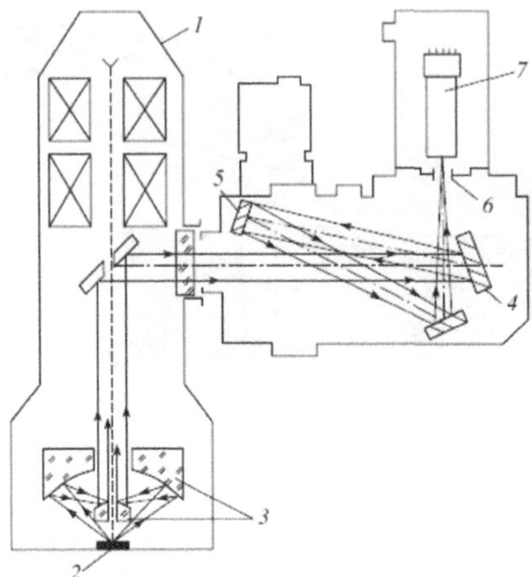

Figure 3.24 Optical scheme of the cathodoluminescent spectrometer: (1) SEM column; (2) sample; (3) reversed Cassegrain objective; (4) diffraction grating; (5) concave mirror; (6) exit slit; and (7) radiation detector. (Reprinted with permission from [177]. Copyright 2004 MAIK "Nauka/Interperiodica").

One of the examples of luminescence intensity of $YTaO_4:Eu^{3+}$ under UV ($\lambda_{exc} = 254$ nm) and X-ray excitation is shown in Fig. 3.23.

The light output of this phosphor increases significantly with increasing energy of excitation. After normalization it is clearly seen that the intensity of weak transitions like $^5D_0 \rightarrow {}^7F_4$ increases also under X-ray excitation because of radiative emissions not more trapped by the empty upper levels [176].

3.6.4 *Cathodoluminescence Spectroscopy*

Cathodoluminescent system presented in Fig. 3.24 incorporates two spectrometers; one of these operates in the near UV and visible ranges, while the other is designed for the near IR. This configuration seems to be optimal in view of the stringent requirements for the spectrometer's size and mass. The optical and mechanical systems of both spectrometers are similar; they differ in diffraction gratings and radiation detectors. The high sensitivity and simultaneously good spectral resolution of the spectrometers (0.1 nm in the visible

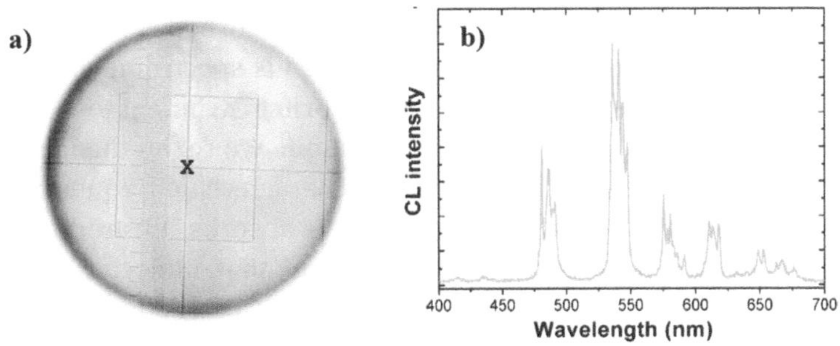

Figure 3.25 Color cathodoluminescence image (a) (original green color) and cathodoluminescence spectra (b) of $YTaO_4:Tb^{3+}$.

and UV ranges and 0.2 nm in the IR range) are achieved thanks to the innovative optical layout of the instrument [178].

In the Fig. 3.25 is given the example of a $YTaO_4:Tb^{3+}$ characterization performed with a CL spectrometer described above and combined with a standard "Camebax". The CL spectra were acquired under accelerating voltage of 15kV at room temperature from the point marked at the left picture. The diameter of beam on the sample was about few micrometers. The beam current was 1.5 nA. The color image (Fig. 3.25a) in real colors was made with defocused electron beam about 200 micrometers. The combination of CL imaging and CL spectroscopy allowed us to get spatially resolved secondary electron images to characterize relatively large parts of the object under study, and to record CL spectra in specific areas.

3.6.5 *Fourier Transform Infrared Spectroscopy*

Fourier transform infrared spectroscopy, or simply FTIR analysis, is a failure analysis technique that provides information about the chemical bonding or molecular structure of materials, whether organic or inorganic. It is used to identify unknown materials present in a specimen, and is usually conducted to complement EDX analysis.

The goal of any absorption spectroscopy (FTIR, ultraviolet-visible [UV-Vis] spectroscopy, etc.) is to measure how well a sample absorbs or transmits light at each different wavelength. The most straightforward way to do this is to shine a monochromatic light beam through a sample, measure how much of the light is absorbed, and repeat for each different wavelength (this is how UV-Vis spectrometers work, for example), or by using a Fourier transform instrument to measure all wavelengths at once. The FTIR technique works on the fact that bonds and groups of bonds vibrate at characteristic frequencies. A molecule that is exposed to infrared rays absorbs infrared energy at frequencies which are characteristic to that molecule. During FTIR analysis, a spot on the specimen is subjected to a modulated IR beam. The specimen's transmittance and reflectance of the infrared rays at different frequencies is translated into an IR absorption plot consisting of reverse peaks. The resulting FTIR spectral pattern is then analyzed and matched with known signatures of identified materials in the FTIR library. When the library of FTIR spectral patterns does not provide an acceptable match, individual peaks in the FTIR plot may be used to yield partial information about the specimen.

Unlike SEM inspection or EDX analysis, FTIR spectroscopy does not require a vacuum, since neither oxygen nor nitrogen absorbs infrared rays. FTIR analysis can be applied to minute quantities of materials, whether solid, liquid, or gaseous.

The schematic below represent the setup of the system.

To use the Fourier transform infrared spectroscopy, a continuum source of light is used to produce light over a broad range of infrared wavelengths. Light coming from this continuum source (IR source, Fig. 3.26) is split in Michelson interferometer into two paths using a half-silvered mirror; this light is then reflected from two mirrors back onto the beam-splitter, where it is recombined. One of these mirrors is fixed, and the second is movable. If the distance from the beam-splitter to the fixed mirror is not exactly the same as the distance from the beam-splitter to the second mirror, then when the two beams are recombined, there will be a small difference in the phase of the light between these two paths. Because of the "superposition principle" constructive and destructive interference exist for different wavelengths depending of the relative distances of the two mirrors from the beam-splitter. It can be shown that if the intensity of light is measured and plotted as a function of the position of the movable mirror, the resultant graph is the Fourier transform of the intensity of light as a function of wavenumber. In FTIR spectroscopy, the light from Michelson interferometer is directed onto the sample. Almost all molecules absorb infrared

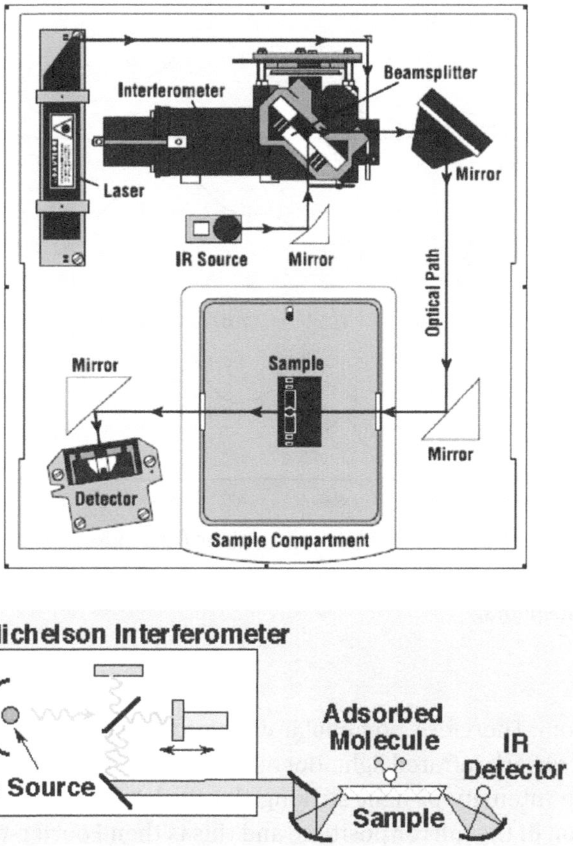

Figure 3.26 Schematic diagrams of a FTIR spectrometer.

light. Only the monatomic (He, Ne, Ar, etc.) and homopolar diatomic (H_2, N_2, O_2, etc.) molecules do not absorb infrared light. Molecules only absorb infrared light at those frequencies where the infrared light affects the dipolar moment of the molecule. In a molecule, the differences of charges in the electronic fields of its atoms produce the dipolar moment of the molecule. Molecules with a dipolar moment allow infrared photons to interact with the molecule causing excitation to higher vibrational states. The homopolar diatomic molecules do not have a dipolar moment since the electronic fields of its atoms are equal. Monatomic molecules do not have a dipolar moment since they only have

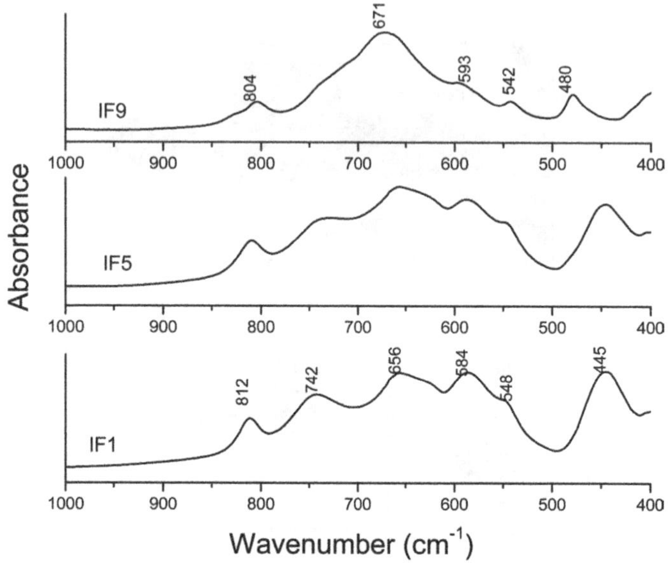

Figure 3.27 FTIR spectra of IF1 ($YTaO_4$), IF5 (Y $Ta_{0.85}Nb_{0.15}O_4$), and IF9 ($YNbO_4$) phosphors.

one atom. Therefore, homopolar diatomic molecules and -monatomic do not absorb infrared light. But all other molecules do!

The intensity of light striking the IR detector is measured as a function of the mirror position, and this is then Fourier-transformed to produce a plot of intensity vs. wavenumber.

Infrared spectroscopy has been highly successful for applications in both organic and inorganic chemistry, including phosphor investigation. We show here as example how FTIR spectroscopy is additionally used in order to put in evidence the crystalline structure of yttrium tantalate-based phosphors (Fig. 3.27).

FTIR spectra registered in the 1000–400 cm^{-1} range show a complex pattern and confirm that the crystalline arrangement of $YTaO_4$, Y $Ta_{0.85}Nb_{0.15}O_4$, and $YNbO_4$ samples is different.

The complexity of the FTIR patterns is associated with the atomic arrangement formed from more or less distorted TaO_6 units, which build the $YTaO_4$ monoclinic modifications. It is known that in M'-type structure, tantalum atoms are in distorted octahedron coordination (six Ta–O bonds), whereas in monoclinic M-type $YNbO_4$, niobium atoms are in tetrahedral coordination (four Nb–O bonds) [179]. The shape of FTIR spectra in the domain of $(Ta_{1-x}Nb_x)O_4$ group vibrations illustrates the change of the number of metal–oxygen bonds with the increase of Nb concentration, in parallel with the structural conversion from M'-$YTaO_4$ to M-$YNbO_4$.

Additional information about FTIR spectroscopy and its applications can be found in special books, manuals, and reviews [180–184].

3.6.6 *Raman Spectroscopy*

C. V. Raman discovered the inelastic scattering phenomenon which bears his name in 1928 and for it he was awarded the Nobel Prize for Physics in 1930. When monochromatic radiation is incident upon a sample then this light will interact with the sample in some fashion. It may be reflected, absorbed or scattered in some manner. Photons interacting with molecules most commonly scatter elastically. This is called Rayleigh scattering. Rayleigh scattered photons have the same wavelength as the incident light. However, approximately 1 out of a million photons are *inelastically* scattered, and it is Raman scattering. Raman scattering produces scattered photons which differ in frequency from the radiation source which causes it, and the difference is related to vibrational and/or rotational properties of the molecules from which the scattering occurs.

With Raman scattering, the incident photon interacts with matter and its wavelength is either shifted lower or higher (red- or blue-shifted, respectively). Red shifted photons are the most common, having been subject to a "Stokes shift." What has happened is that the photon has interacted with the electron cloud of the functional groups bonds, exciting an electron into a virtual state. The virtual state is not a true quantum state of the molecule but it regards as a distortion of the electron cloud arising from the oscillating electric filled of the light in a very short time. The electron then relaxes into an excited vibrational or rotational state (see the diagram in Fig. 3.28). This causes the photon to lose some of its energy and is detected as

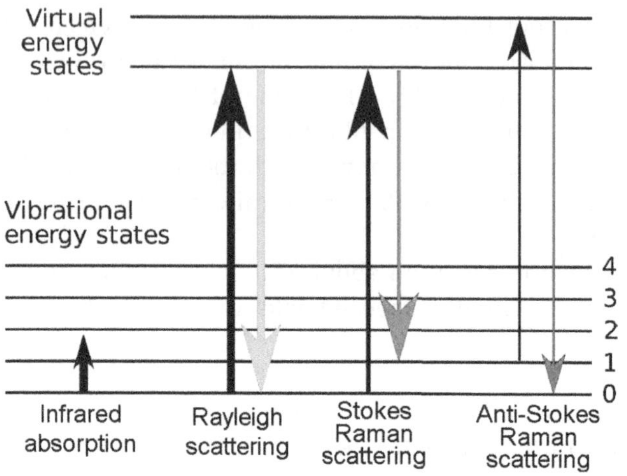

Figure 3.28 Diagram of the Rayleigh and Raman scattering processes. Light scattered from a molecule has several components – the Rayleigh scatter and the Stokes and anti-Stokes Raman scatter. The lowest energy vibrational state $m = 0$ is shown at the foot with states of increasing energy above it. Both the low energy (upward arrows) and the scattered energy (downward arrows) have much larger energies than the energy of a vibration.

Stokes Raman scattering. This loss of energy is directly related to the functional group, the structure of the molecule to which it is attached, the types of atoms in that molecule and its environment.

Figure 3.29 shows the basic processes which occur if a sample is illuminated with a laser beam. At room temperature, most molecules, but not all, are present in the lowest energy vibrational level. Since the virtual states are not real states of the molecule but are created when the laser interacts with the electrons and causes polarization, the energy of these states is determined by the frequency of the light source used. The Rayleigh process will be the most intense process since most photons scatter this way. It does not involve any energy change and consequently the light returns to the same energy state. The holographic filters reject the light scattered without frequency change. The Raman scattering process from the ground vibrational state leads to absorption of energy by the molecule and its promotion to a higher energy excited vibrational state (Stokes scattering). However, due to thermal energy, some molecules may be present in an excited state. Scatter-

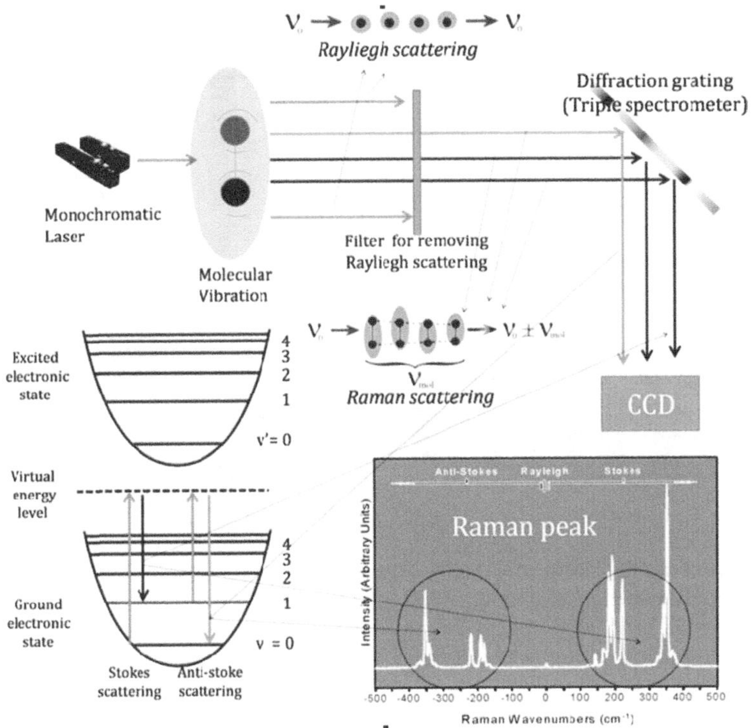

Figure 3.29 Illustration of Raman process.

ing from these states to the ground state (anti-Stokes scattering) involves transfer of energy to the scattered photon. The relative intensities of the two processes depend on the population of the various states of the molecule. At room temperature, the number of molecules expected to be in an excited vibrational state other than any really low-energy ones will be small and as a result anti-Stocks lines in Raman spectra are less intensive in comparison with Stokes peaks.

Raman spectroscopy is a spectroscopic technique used to study vibrational, rotational, and other low-frequency modes in a system. It relies on inelastic scattering, or Raman scattering, of monochromatic light, usually from a laser in the visible, near-infrared, or near-ultraviolet range. The laser light interacts with phonons or other excitations in the system, resulting in the energy of the laser pho-

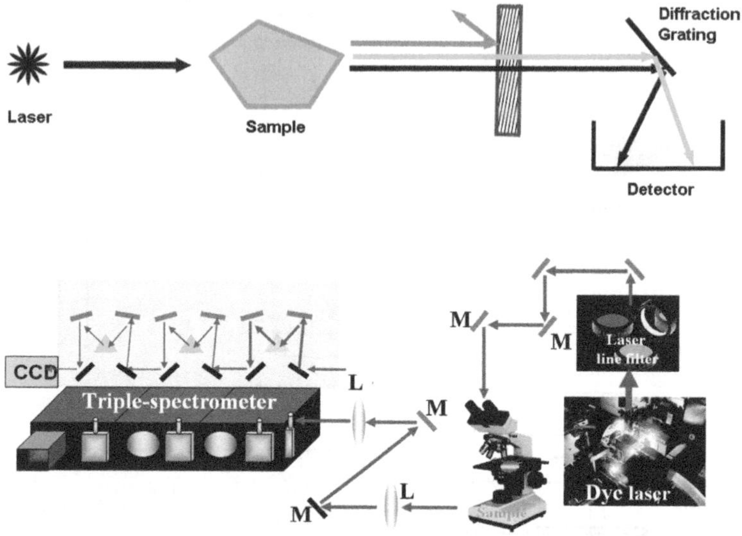

Figure 3.30 Scheme and real experimental setup of triple Raman spectroscope: excitation wavelength 532, 632.8, and 785 nm; resolution 0.2 cm^{-1}; detector TE cooled charge coupled device. Raman scattering signature as close as 20 cm^{-1} to the laser excitation line could be observable.

tons being shifted up or down. The shift in energy gives information about the phonon modes in the system.

The original Raman setup is fairly standard and it is shown in Fig. 3.30.

A Raman system typically consists of four major components:

1. Excitation source – Laser
2. Sample illumination system and light collection optics
3. Wavelength selector – Filter or spectrophotometer
4. Detectors – Photodiode array (PDA), charge-coupled devices (CCDs), or photon-counting photomultiplier tubes (PMTs)

A sample is normally illuminated with a laser beam in the ultraviolet, visible, or near-infrared range. Scattered light is collected with a lens and is sent through interference filter or spectrophotometer to obtain Raman spectrum of a sample. Since spontaneous Raman scattering is very weak the main difficulty of Raman spectroscopy is separating it from the intense Rayleigh scattering. More precisely, the major problem here is not the Rayleigh scattering itself, but the fact that the intensity of stray light from the Rayleigh scattering may greatly exceed the intensity of the useful Raman signal in the close proximity to the laser wavelength. In many cases the problem is resolved by simply cutting off the spectral range close to the laser line where the stray light has the most prominent effect. People use commercially available interference (notch) filters which cut-off spectral range of ±80–120 cm^{-1} from the laser line. This method is efficient in stray light elimination but it does not allow detection of low-frequency Raman modes in the range below 100 cm^{-1}. Stray light is generated in the spectrometer mainly upon light dispersion on gratings and strongly depends on grating quality. Raman spectrometers typically use holographic gratings which normally have much less manufacturing defects in their structure then the ruled once. Stray light produced by holographic gratings is about an order of magnitude less intense then from ruled gratings of the same groove density.

Using multiple dispersion stages is another way of stray light reduction. Double and triple spectrometers allow taking Raman spectra without use of notch filters. In such systems Raman-active modes with frequencies as low as 3–5 cm^{-1} can be efficiently detected.

In earlier times people primarily used single-point detectors such as photon-counting Photomultiplier Tubes. However, a single Raman spectrum obtained with a PMT detector in wavenumber scanning mode was taking substantial period of time, slowing down any research or industrial activity based on Raman analytical technique. Nowadays, more and more often researchers use multichannel detectors like photodiode arrays or, more commonly, charge-coupled devices to detect the Raman scattered light. Sensitivity and performance of modern CCD detectors are rapidly improving. In many cases CCD is becoming the detector of choice for Raman spectroscopy.

The correct selection of the laser wavelength can be an important consideration for Raman spectroscopy. With modern equipment, often several laser wavelengths may be employed so as to achieve the best detection of the Raman signal. For instance, many samples, especially activated phosphors which we study, will be quite fluorescent species. Exciting these samples with a laser in the green (532 nm) may promote this fluorescence, and may swamp any underlying Raman spectrum to such an extent that it is no longer detectable. In this instance, the use of a laser in the red (633 nm) or NIR (785 nm) may provide a solution. With the lower photon energy, a red or NIR laser may not promote the electronic transition (and hence the fluorescence) and so the Raman scatter may be far easier to detect. Conversely, as one increases the wavelength, from green to red to NIR, the scattering efficiency will decrease, so longer integration times or higher power lasers may be required. Thus, it is often most practical to have a number of laser wavelengths available to match the various sample properties one may encounter, be it resonance enhancements, penetration depth of fluorescence. The strong background seen with the green and red lasers swamps the Raman signal, whereas the 785 nm excitation is outside of the fluorescence range, enabling the Raman to be detected.

Raman scattering is a powerful light scattering technique used to diagnose the internal structure of molecules and crystals. Raman scattered light is frequency-shifted with respect to the excitation frequency, but the magnitude of the shift is independent of the excitation frequency. This Raman shift is therefore an intrinsic property of the sample. Because Raman scattered light changes in frequency, the rule of conservation of energy dictates that some energy is deposited in the sample. A definite Raman shift corresponds to excitation energy of the sample (such as the energy of a free vibration of a molecule).

The analysis of the scattered frequencies (Raman spectroscopy) gives information on the material chemical composition, since vibrational information is specific to the chemical bonds and symmetry of molecules. It therefore provides a fingerprint by which the molecule can be identified. In solid state physics, spontaneous Raman spectroscopy is used to, among other things, characterize materials, measure temperature, and find the crystallographic orientation of a sample. As with single molecules, a given solid material has characteristic phonon modes that can help an experimenter identify it. One of examples of Raman spectroscopy is shown in Fig. 3.31.

Other example of Raman application (Fig. 3.32) shows the difference between the M and M' structures of $YTaO_4$. Here the Raman spectra look as a "fingerprint" that is unique for every structure. This is quite important since it is the sole basis for the superior

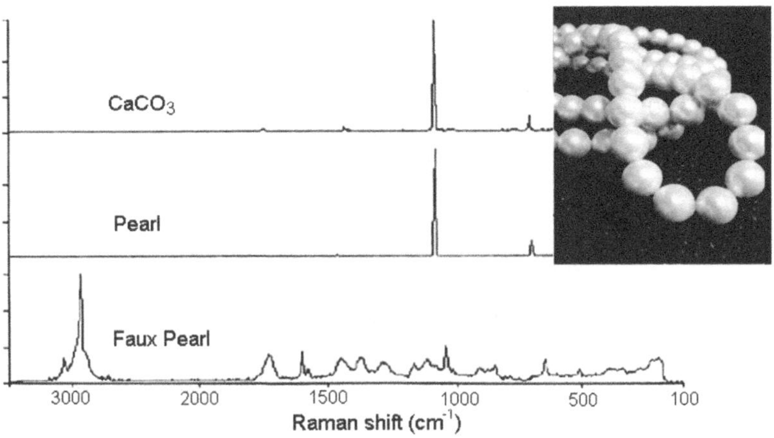

Figure 3.31 The Raman spectra of calcium carbonate (top), natural pearl (middle), and faux pearl (bottom).

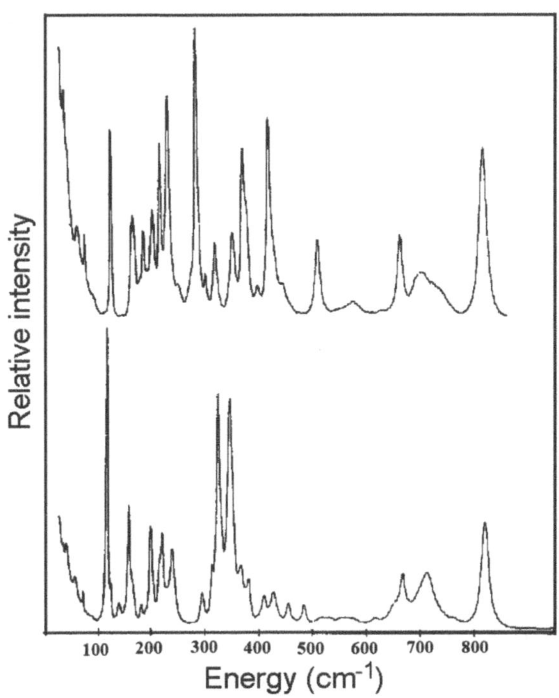

Figure 3.32 Raman spectra of M' (top) and M- $YTaO_4$. (Reprinted with permission from [185]. Copyright 1983: Elsevier).

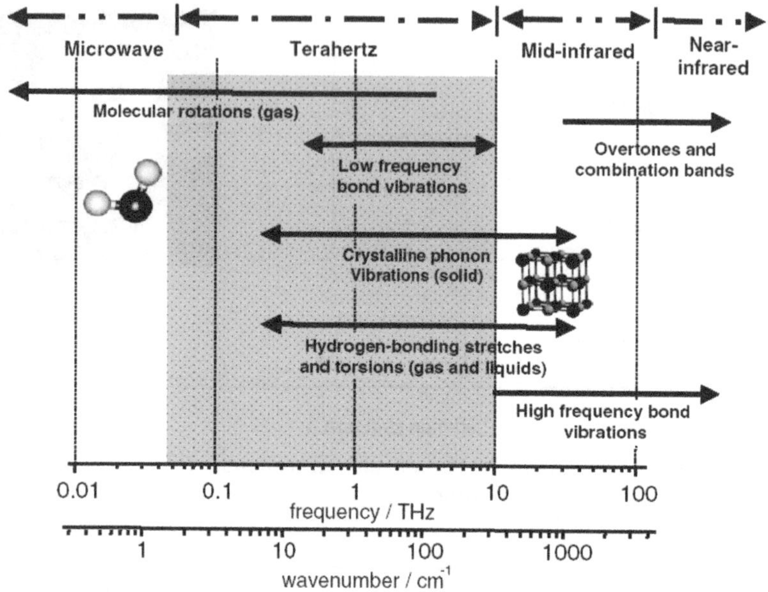

Figure 3.33 The THz region and molecular transitions in the electromagnetic spectrum. (Reprinted with permission from [194]. Copyright 2006: Elsevier).

luminescent efficiency which is nearly two times greater for pure M' $YTaO_4$ than it is for M-$YTaO_4$. This will be discussed in more detail in the next chapter.

One can find more detailed information about the Raman spectroscopy and its applications in many useful books and recently published monographs [186–193].

3.6.7 *Terahertz Spectroscopy*

The terahertz (THz) frequency range lying between the IR and microwave ranges (for reference, 1 THz corresponds to 33.3 cm^{-1}, 4.14 meV, and has a wavelength of 300 μm) is shown in Fig. 3.33 and it is at present the last poorly studied frequency range in the entire scale of electromagnetic waves from radio-frequency to X-ray range.

The THz regime is typically defined as 0.1–10 THz. At the high-frequency end of the electromagnetic spectrum, gamma rays and

X-rays originate from the atomic nucleus and inner electrons, while ultraviolet, visible, and infrared radiation is typically associated with transitions between atomic and molecular states.

The past 20 years have seen a revolution in THz systems, as advanced materials research provided new and higher-power sources, and the potential of THz for advanced physics research and commercial applications was demonstrated. Terahertz technology is an extremely attractive research field, with interest from sectors as diverse as the semiconductor, medical, manufacturing, space and defense industries. Several recent major technical advances have greatly extended the potential and profile of THz systems. Terahertz spectroscopy allows a material's far-infrared optical properties to be determined as a function of frequency. This information can yield insight into material characteristics for a wide range of applications.

Many different methods exist for performing THz spectroscopy. Fourier transform spectroscopy (FTS) is perhaps the most common technique used for studying molecular resonances. It has the advantage of an extremely wide bandwidth, enabling material characterization from THz frequencies to well into the infrared. In FTS the sample is illuminated with a broadband thermal source such as an arc lamp or a SiC globar. The sample is placed in an optical interferometer system and the path length of one of the interferometer arms is scanned. A direct detector such as a helium cooled bolometer is used to detect the interference signal. The Fourier transform of the signal then yields the power spectral density of the sample. One disadvantage of FTS is its limited spectral resolution. Spectral measurements having a much higher resolution may be made using a narrowband system with a tunable THz source or detector. In these systems, the source or detector is tuned across the desired bandwidth and the sample's spectral response is measured directly. Both FTS and narrowband spectroscopy are also widely used in passive systems for monitoring thermal-emission lines of molecules, particularly in astronomy applications. A third, more recent technique is termed THz time-domain spectroscopy (THz-TDS). THz-TDS uses short pulses of broadband THz radiation, which are typically generated using ultrafast laser pulses. This technique grew from work in the 1980s at AT&T Bell Labs and the IBM T.J.Watson Research Center [195, 196]. Although the spectral

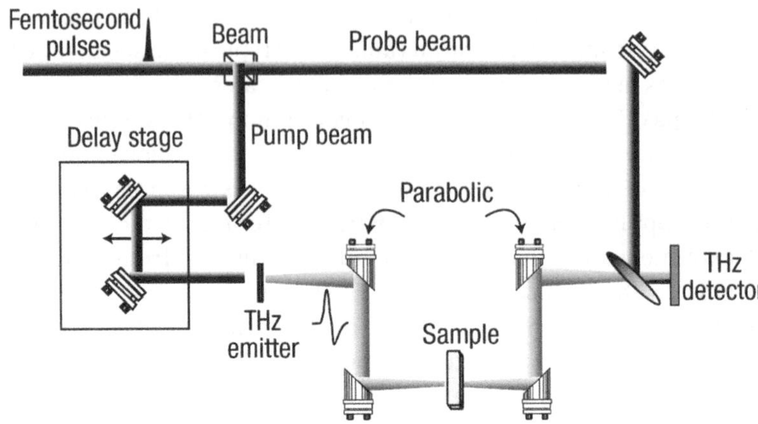

Figure 3.34 Illustration of a THz-TDS pump probe system. The ultrafast laser beam is split into pump and probe beams. The pump beam is incident on the THz emitter to generate THz pulses, and the THz pulses are collimated and focused on the target using parabolic mirrors. After transmission through the target, the THz pulse is collimated and re-focused on the THz detector. The optical probe beam is used to gate the detector and measure the instantaneous THz electric field. A delay stage is used to offset the pump and probe beams and allow the THz temporal profile to be iteratively sampled. (Reprinted with permission from [198]. Copyright 2002: Elsevier).

resolution of THz-TDS is much coarser than narrowband techniques, and its spectral range significantly less than that of FTS, it has a number of advantages that have given rise to some important recent applications. The transmitted THz electric field is measured coherently, which provides both high sensitivity and time-resolved phase information. It is also amenable to implementation within an imaging system to yield rich spectroscopic images. A THz-TDS system is described in Fig. 3.34. Typical THz-TDS systems have a frequency bandwidth between 2 and 5 THz, a spectral resolution of 50 GHz, an acquisition time under one minute and a dynamic range of $\sim 10^5$ in electric field. Signal-processing algorithms may be used to improve the signal-to-noise ratio of the measured signals by almost 30% [197].

THz systems are ideal for imaging dry dielectric substances including paper, plastics and ceramics. These materials are relatively non-absorbing in this frequency range, yet different materials may be easily discriminated on the basis of their refractive index, which is extracted from the THz phase information. Many such materials are opaque at optical frequencies, and provide very low contrast for X-rays. THz imaging systems may therefore find important niche applications in security screening and manufacturing quality control.

An important goal in this context is the application of THz technique to luminescent materials. Terahertz time-domain spectroscopy is a promising tool in the inspection of materials for their optical properties. A THz-TDS waveform transmitted through a material sample is typically rich in information, since its shape is altered by the material's characteristic frequency response. Spectrometers, together with physical models explaining the interaction between a sample and T-ray radiation, are progressively being developed. Measurements of optical constants, in particular, their refractive index and absorption coefficient at terahertz frequencies have been performed extensively using THz-TDS [199]. In this book we will provide such examples in the next chapter and here we give only one example of THz spectroscopy application to phosphor materials.

From Figs. 3.35 and 3.36 we can conclude that THz method is very sensitive to rare earth elements which we use for doping host

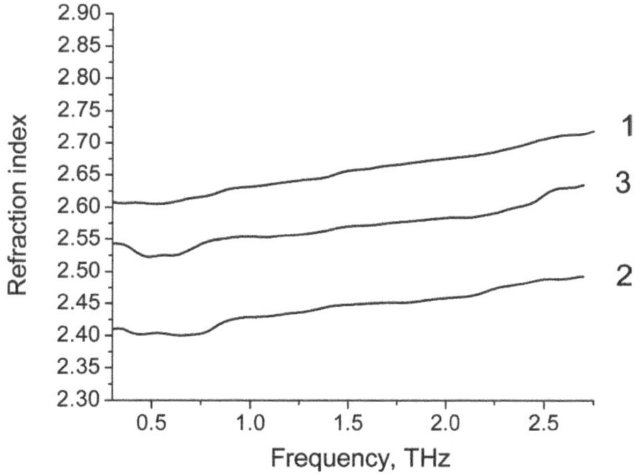

Figure 3.35 Refraction THz spectra of the $YTaO_4$ (1), $YTaO_4:Eu^{3+}$, (2) and $YTaO_4:Tb^{3+}$ (3).

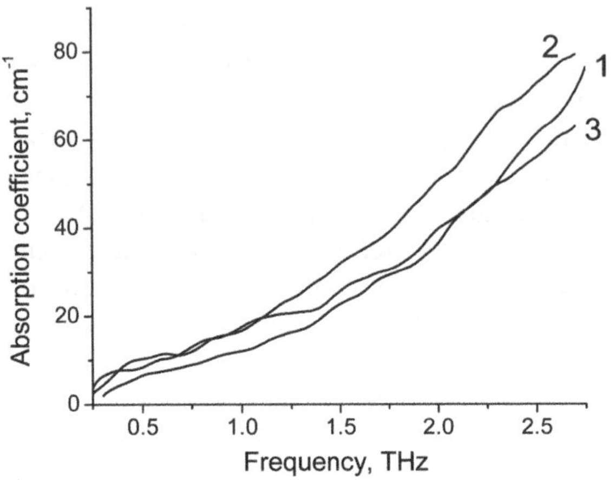

Figure 3.36 Absorption THz spectra of the $YTaO_4$ (1), $YTaO_4:Eu^{3+}$ (2), and $YTaO_4:Tb^{3+}$ (3).

lattice. Absorption coefficients and refraction indexes changed significantly with activator incorporation.

3.7 Computational Materials Science

3.7.1 *Ab initio* Calculations

Computational materials science aims to enhance the communication between experimental materials research, theory and computational work on both existing and new advanced materials and their applications (Fig. 3.37).

Computational materials science presents the most important approaches in this new interdisciplinary field of materials science and engineering. The reader will learn to assess which numerical method is appropriate for performing simulations at the various microstructural levels and how they can be coupled. Modeling and simulation play an ever increasing role in the development and optimization of materials.

Nevertheless, we have to understand that only very few aspects of chemistry can be computed exactly. However, just as not all spectra are perfectly resolved, often a qualitative or approximate

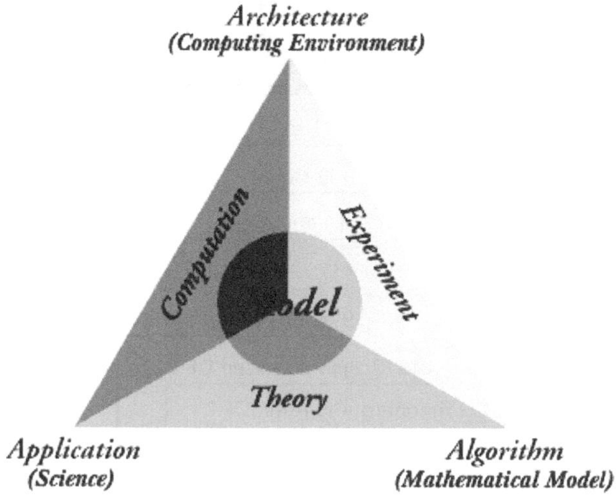

Figure 3.37 Unique model combining experiment, theory, and computation.

computation can give useful insight into chemistry if you understand what it tells you and what it doesn't.

Big success in the development of computer hardware and software enables now to perform the first-principles (*ab initio*) calculations with accuracy comparable with that obtained in experiment. These approaches enable also to predict the behavior of materials in extreme conditions (for example, at very high pressures inaccessible in laboratory conditions), to study dangerous (radioactive, explosive) materials. They are very useful in predicting new materials with interesting properties, which have never been synthesized. Moreover, the first-principles calculations enable to improve our understanding of physical phenomena occurring in known materials. The Nobel Prizes awarded to Robert S. Mulliken (1966) and Walter Kohn and John Pople (1998) can be regarded as acknowledgment of their significant contribution to the development of first-principles methods in physics, chemistry, and materials science (Fig. 3.38).

In physics, a calculation is said to be from first-principles, or *ab initio*, if it starts directly at the level of established laws of physics and does not make assumptions such as empirical model and fitting parameters. For example, calculation of electronic structure using

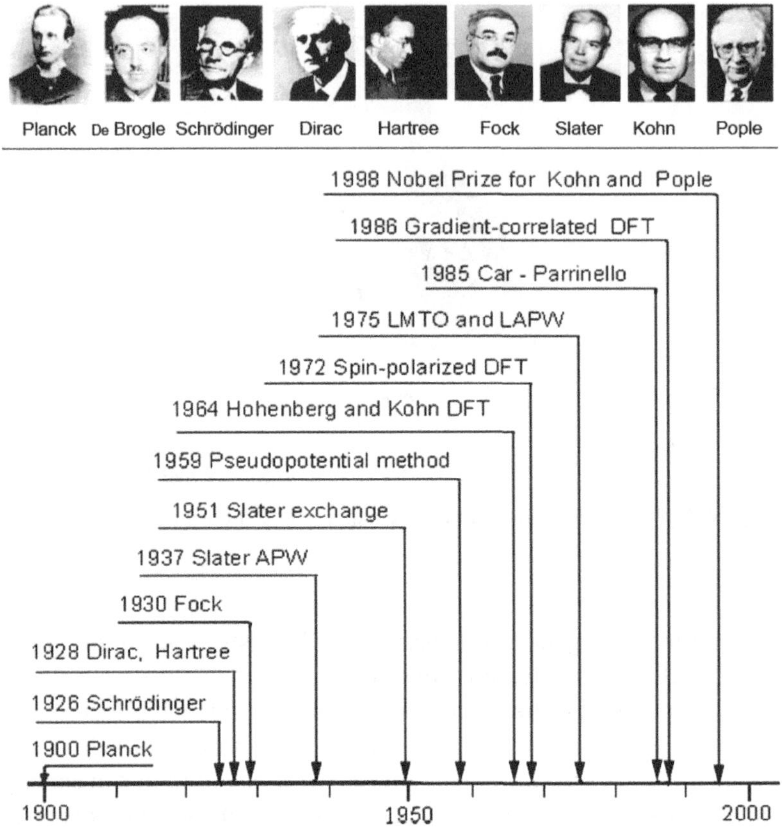

Figure 3.38 Evolution of quantum theory.

Schrödinger's equation within a set of approximations that do not include fitting the model to experimental data is an *ab initio* approach.

The term *ab initio* is a Latin term meaning "from the beginning." This name is given to computations which are derived directly from theoretical principles, with no inclusion of experimental data. Most of the time this refers to an approximate quantum mechanical calculation. The approximations made are usually mathematical approximations, such as using a simpler functional form for a function or getting an approximate solution to a differential equation.

Scientists mainly use three different methods to make calculations:

- Molecular mechanics uses classical physics to explain and interpret the behavior of atoms and molecules. It relies on force-field with embedded empirical parameters and requires experimental data (or data from *ab initio*) for parameters. This method does not calculate electronic properties. Usually it is large system (thousands of atoms).
- Semi-empirical techniques use approximations from empirical (experimental) data to provide the input into the mathematical models. This method uses quantum physics and experimentally derived empirical parameters. It is less demanding computationally than *ab initio* methods and capable of calculating transition states and excited states. At the same time it is less rigorous than *ab initio* methods and requires experimental data (or data from *ab initio*) for parameters. It is medium-sized system (hundreds of atoms).
- *Ab initio*, a group of methods in which molecular structures can be calculated using nothing but the Schroedinger equation, the values of the fundamental constants and the atomic numbers of the atoms present. This method is useful for a broad range of systems because does not depend on experimental data. It is capable of calculating transition states and excited states. A disadvantage is that it is computationally expensive. Usually it is a small system (tens of atoms) requiring rigorous accuracy.

To better understand the computational approaches and simulation packages we propose the next scheme in Fig. 3.39:

Some of the most popular computation programs are summarized below:

VASP: Pseudopotential, ultra-soft projector augmented-wave (PAW), parallel execution in supercomputer.

SIESTA: Localized orbital basis and pseudopotential, parallel execution, very small basis, handles a very large system (nano system).

WIEN97: Linearized-augmented-plane wave (LAPW) method, parallel execution in supercomputer.

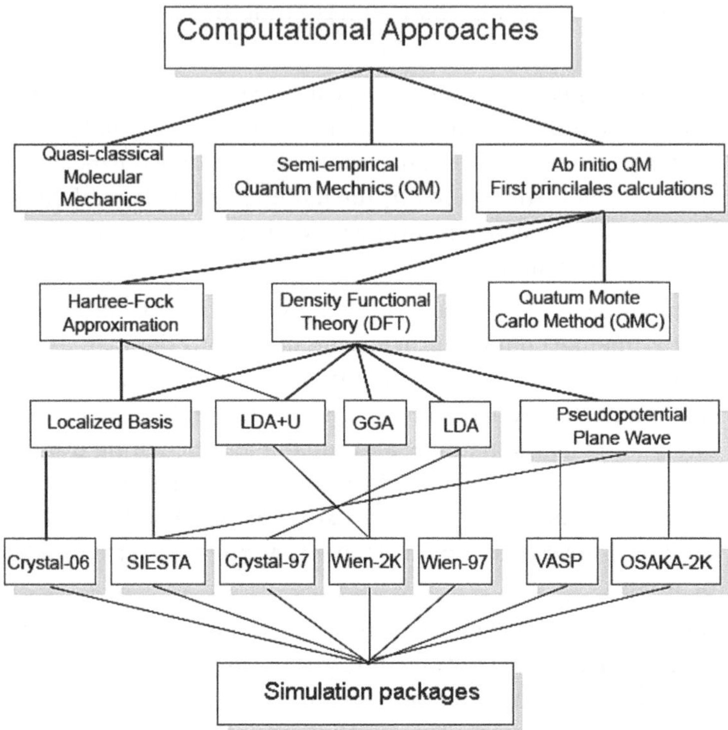

Figure 3.39 Scheme of computational approaches and simulation packages.

WIEN2K: Full-potential linear-augmented plane wave (FPLAPW) approach within the framework of DFT. WIEN2K is based on the method which is among the most precise and reliable ways to calculate the electronic structure of solids.

OSAKA-2K: One of the best pseudopotential methods. This package is a set of program codes which calculate electronic structures of materials. It covers a wide range of calculations from optimization of crystal structure to molecular dynamic simulations, in addition to standard self-consistent calculation and band calculations.

CRYSTAL-98: A package which is based on both density functional and Hartree–Fock theories. The gradient-corrected correlation functional by Lee, Yang, and Parr is used, combined with the Becke 3 exchange functional (B3LYP).

CRYSTAL-06: Linear combination of atomic orbitals (LCAO) formalism. Hybrid Hartree–Fock–DFT Hamiltonians (B3PW and PBE0) is used.

The most common types of *ab initio* calculations are called Hartree–Fock approximation (abbreviated HF), density functional theory (DFT), and the quantum Monte Carlo (QMC) method.

We consider here QMC and HF methods very briefly and pay our attention mostly to DFT methods, because they are among the most popular and versatile methods available in condensed-matter physics, computational physics, and computational chemistry.

3.7.2 *Monte Carlo Calculations*

Quantum Monte Carlo techniques provide a practical method for solving the many-body Schrödinger equation. They are closely related to the well-established classical Monte Carlo methods that have been successfully applied to a wide range of problems involving stochastic behavior ranging from scientific problems, engineering problems and modeling the financial markets. The common link between classical and quantum Monte Carlo techniques is the use of random numbers to evaluate multidimensional integrals. QMC works with an explicitly correlated wave function and evaluates integrals numerically using a Monte Carlo integration. These calculations can be very time consuming, but they are probably the most accurate methods known today.

There are several flavors of QMC such as variational, diffusion and Green's functions. The latter two are projection approaches which dispense with quantum chemical basis sets, but they have to deal with the fermion sign problem, and the related fixed-node approximation. In general, all QMC methods have a good scaling with the number of electrons, enabling relatively large systems to be tackled, but with a computational cost much larger than traditional ab-initio methods based on DFT. The good reviews of Monte Carlo method can be found in Refs. [200–202].

3.7.3 *Hartree–Fock Algorithm*

The Hartree–Fock method is typically used to solve the time-independent Schrödinger equation for a multielectron atom or molecule as described in the Born–Oppenheimer approximation. In the Hartree–Fock method as a starting point, the wave function is used to describe electronic structure. Since there are no known solutions for many-electron systems, the problem is solved numerically. Due to the nonlinearities introduced by the Hartree–Fock approximation, the equations are solved using a nonlinear method such as iteration, which gives rise to the name "self-consistent field method."

In this method the primary approximation is called the central field approximation. This means that the Coulombic electron–electron repulsion is not specifically taken into account. However, its net effect is included in the calculation.

The second approximation in HF calculations is that the wave function must be described by some functional form, which is only known exactly for a few one electron systems. The functions used most often are linear combinations of Slater-type orbitals or Gaussian type orbitals, abbreviated STO and GTO. The wave function is formed from linear combinations of atomic orbitals or more often from linear combinations of basis functions. Because of this approximation, most HF calculations give a computed energy greater than the Hartree–Fock limit.

One of attempts of non-empirical method to obtain the properties of solid is to solve the equation of the many-electron Hamiltonian directly. In practice, this equation is often rewritten through a Slater determinant which is composed of a lot of single-electron wave functions.

In Hartree–Fock approximation only the exchange effect is considered. In many problems, it is known that the exchange term only is not good. Further developments in order to include the correlation effect into account, many methods, such as the configuration interaction by expanding on many Slater determinants and the quantum Monte Carlo method, etc., have been devised [203–206].

Anyway, these approaches are all based upon the wave functions and express the electronic states of solid through the set of wave functions. In the configuration interaction method, the combination of wave functions is very complicated, resulting in severe limitation of the size of problems. Meanwhile, for the many-electron problems, another and very different approach called the density functional theory has been proposed.

3.7.4 *Density Functional Theory*

Density functional theory is a quantum mechanical modeling method used in physics and chemistry to investigate the electronic structure (principally the ground state) of many-body systems, in particular atoms, molecules, and the condensed phases.

In this approach, the electron density is the quantity, from which the theory is developed. To solve one-electron equations which are derived from the density functional theory is much easier than solving general equation. With this theory, the properties of a many-electron system can be determined by using functionals, i.e. functions of another function, which in this case is the spatially dependent electron density. Hence the name density functional theory comes from the use of functionals of the electron density. Electron density is a primary way of describing the system.

DFT has been very popular for calculations in solid state physics since the 1970s. It was impossible to solve the many-body problem quantum mechanically. Only with *adiabatic approximation* (Born–Oppenheimer) and *density functional theory* (Hohenberg and Kohn 1964, Kohn and Sham 1965) it became possible. Since 1980, this method has established a position as one of the main methods of calculating the properties of solid and molecules from the first-principles. However, DFT was not considered accurate enough for calculations in quantum chemistry until the 1990s, when the approximations used in the theory were greatly refined to better model the exchange and correlation interactions. In many cases the results of DFT calculations for solid state systems agree quite satisfactorily with experimental data. Computational costs are relatively low when compared to traditional methods, such as Hartree-Fock theory based on the complicated many-electron wavefunctions. A price of mathematical simplification of the density functional method, which replaces the many-electronic problem by one-electron problem is paid by introducing unknown

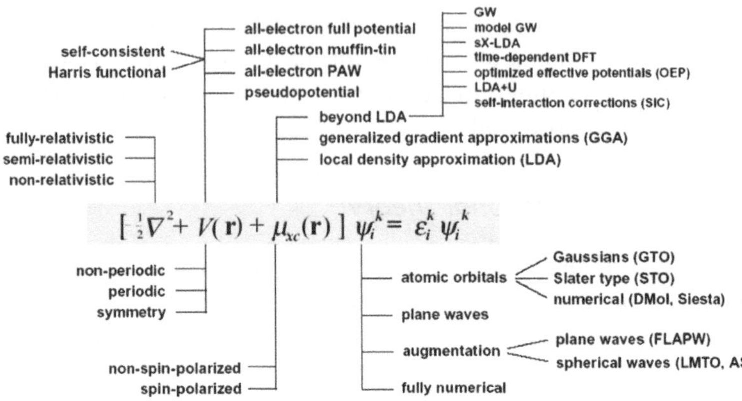

Figure 3.40 DFT Implementations.

functional of exchange and correlation V_{xc} of the charge density. Fortunately, there is an easy approximation for V_{xc}. The most widely used forms of V_{xc} is the so-called local density approximation (LDA) and generalized gradient approximation (GGA) shown in Fig. 3.39.

A generalization of the LDA method for systems with strong Coulomb correlations is proposed as LDA+U [203, 205], which gives a correct description of the Mott insulators. The LDA + U method takes into account orbital dependence of the Coulomb and exchange interactions which is absent in the LDA.

Despite recent improvements, there are still difficulties in using density functional theory to properly describe intermolecular interactions, especially van der Waals forces (dispersion), charge transfer excitations, transition states, global potential energy surfaces, and some other strongly correlated systems, and in calculations of the band gap in semiconductors. The development of new DFT methods designed to overcome this problem, by alterations to the functional or by the inclusion of additive terms, is a current research topic. Some of DFT implementations are summarized and shown in Fig. 3.40.

There are a lot of good reviews about density functional theory [207–210].

To summarize, computational science is a branch of science that generates data which complements experimental data on the structures, properties and reactions of substances. The calculations are based primarily on Schroedinger's equation and include:

1. calculation of electron and charge distributions
2. molecular geometry in ground and excited states
3. potential energy surfaces
4. rate constants for elementary reactions
5. details of the dynamics of molecular collisions

The good side of *ab initio* methods is that they eventually converge to the exact solution once all of the approximations are made sufficiently small in magnitude. However, this convergence is not monotonic. Sometimes, the smallest calculation gives the best result for a given property.

The bad side of *ab initio* methods is that they are expensive. These methods often take enormous amounts of computer time, memory, and disk space.

In general, *ab initio* calculations give very good qualitative results and can give increasingly accurate quantitative results in future.

For details readers are referred to Refs. [211–213].

Chapter 4

PHOSPHORS FOR DIFFERENT APPLICATIONS

4.1 Europium and Terbium Activated $CaWO_4$

4.1.1 *Introductory Remarks*

Calcium tungstate scheelite crystal, $CaWO_4$, is a naturally occurring luminescent mineral that can also be produced synthetically. The optical properties of the scheelite crystal have been studied extensively for decades [20, 21, 214–220]. Its superior luminescence properties led to the practical implementation of $CaWO_4$ crystal as phosphor, scintillation counter, and laser host material. More then 75 years (it is an absolute record for a luminescent material!) calcium tungstate phosphor was used for the manufacture of X-ray intensifying screens. Improvement of the luminescence properties of scheelite-based phosphors has become a primary focus in the luminescent material science.

The incorporation of Eu^{3+} or Tb^{3+} ions into the $CaWO_4$ crystal lattice modifies the luminescence spectrum due to the formation of the emission centers that generate the specific red and green light, respectively [20, 31, 214, 221–224]. Photoluminescence and cathodoluminescence investigations are used to evaluate the performance of Eu^{3+} and Tb^{3+} activated $CaWO_4$ and to put in evidence the interaction between the two activating species. It is to be noted

New Generation of Europium and Terbium Activated Phosphors
Mihail Nazarov and Do Young Noh

www.panstanford.com

additionally that the energy transfer occurs within the Eu-Tb complex so that it works as a donor–acceptor pair due to small energy gap between $^5D_4$ (Tb^{3+}) and $^5D_1$ (Eu^{3+}) energy levels. The two- and three-level models for the energy transfer between the activators that provide the main features of CL and PL experimental data with a quantitative description is discussed in this chapter. The problem of the design of new systems in a controllable way and the improvement of the existing systems require in-depth understanding of the nature of the spectral lines. Due to the formation of the low-symmetry Eu^{3+} or Tb^{3+} luminescent centers with specific coordination surrounding, the optical lines assignation and the elucidation of the influence of the activators nature and host lattice become a complicated task. To the best of our knowledge, no such investigation taking into account an influence of the host lattice has been done for this material hitherto. Bearing in mind the importance of such information for the implementation of phosphors, a theoretical study of luminescence of the double activated phosphor $CaWO_4$: Eu^{3+},Tb^{3+} was initiated [225, 226]. The study of the optical transitions in the double activated calcium tungstate phosphor with the emphasis on the anisotropy of absorption and emission is the main goal of these works. The selection rules and the orientation dependence of the intensities for the optical transitions in the rare earth ions occupying S_4 point symmetry sites in $CaWO_4$ host lattice are discussed. This chapter presents the results of the theoretical analysis as well as some experimental data on the low-temperature PL and CL spectra that are in a qualitative agreement with the theoretical consideration. The angular (polarization) dependence of the two-photon absorption that seems to be an effective tool for the understanding of the complicated optical pattern of the RE ions in low-symmetry sites is analyzed. Emphasis is made on the suggestion of the experimental schemes for the assignation of the complicated pattern of the optical lines in S_4 point symmetry sites. The discussion of some PL experimental data at low-temperature measurements is expected to be helpful for the improvement of the luminescence properties of the double activated $CaWO_4$ phosphors and other RE activated systems.

These $CaWO_4$:Eu^{3+}, Tb^{3+} materials with the controllable properties seem to be promising for the applications in fluorescent lamps, colored lightning for advertisement industries, and other optoelectronic devices.

4.1.2 *Phosphor Preparation*

Described in this chapter, the $CaWO_4$:RE phosphor samples were prepared by solid state reaction from homogeneous mixtures consisting of luminescent-grade $CaWO_4$ as host lattice, WO_3 and Eu_2O_3 (99,99% Jansen Chemical) and/or Tb_4O_7 (Johnson Mattey "Specpure") as activating system, and alkaline salts as flux. $CaWO_4$ used as precursor in phosphor synthesis was prepared from highly purified $CaCl_2$ and Na_2WO_4 solutions [223]. The synthesis mixtures containing flux and equivalent amounts of WO_3 and Eu_2O_3 or Tb_4O_7 were homogenized and fired at 900°C. The prepared powders were carefully washed, dried, and sieved. The incorporation of trivalent RE ions into the $CaWO_4$ crystalline lattice could proceed in two major ways. Either $2RE^{3+}$ replace $3Ca^{2+}$ ions, or a pair of RE^{3+} and M^+ replace $2Ca^{2+}$ ions (where M^+ = alkaline ion). In our sample preparation conditions, alkaline salt acts as a mineralizing agent and also as a source of compensating ion. Based on this principle, a series of synthesis mixtures was prepared according to the general phosphor formula $Ca_{1-2x-2y}\,Eu_xTb_yM_{x+y}WO_4$.

4.1.3 *Luminescence of Eu^{3+} and Tb^{3+} Activated $CaWO_4$ Phosphors*

The low (0.5–1 keV) and high (20–30 keV) voltage cathodoluminescence modes including color CL and PL with different excitation wavelengths (147, 254, 365 nm) were applied to investigate the luminescence properties and the energy transfer processes in $CaWO_4$, activated by trivalent rare earth Eu and Tb ions.

Photoluminescence characteristics were estimated on the basis of emission and excitation spectra registered at room temperature (Xe 500 W lamp) with DARSA PRO 5100 PL System (Diode Array Rapid Scan Analysis – Professional Scientific Instrument Co., Korea). The excitation was performed with a 254 nm UV radiation as well as 365 nm. Low-temperature PL measurements (15 K) were also used to explain the energy transfer mechanism. He-Cd laser with excitation wavelength of 325 nm (55 mW) was applied as a light source.

To verify the luminescence properties of synthesized $CaWO_4:Eu^{3+}$, Tb^{3+} phosphor in a large range of excitations, low-voltage CL was carried out at excitation energy of 0.5–1 keV by using a Kimball Physics FRA2X1-2/EGPS-2X1 electron gun system. High-voltage 5–40 keV local CL measurements were performed with a special CL spectrometer, combined with a standard "Camebax" [178] (Fig. 3.24). Surface investigation and distribution the luminescence centers in powders were carried out with color CL using a special attachment for CCL- SEM [154] (Fig. 3.7).

Both the monochromatic and panchromatic CCL-SEM modes are used. At the monochromatic mode the CL emission was detected at definite wavelength. At the panchromatic CL-mode the CL was recorded at all the visible emission spectrum. By this way, monochromatic CL corresponds to definite transitions between the defect levels attributed to the activator. Spatial distribution of the monochromatic (Fig. 4.1) or panchromatic (Fig. 3.6) CL on the surface under study was recorded in forms of corresponding CCL maps. A variety of colors observed in a panchromatic CCL map shows, therefore, a spatial distribution of all the luminescent centers that occurred by nature or design on the surface of investigated phosphors (Fig. 3.6).

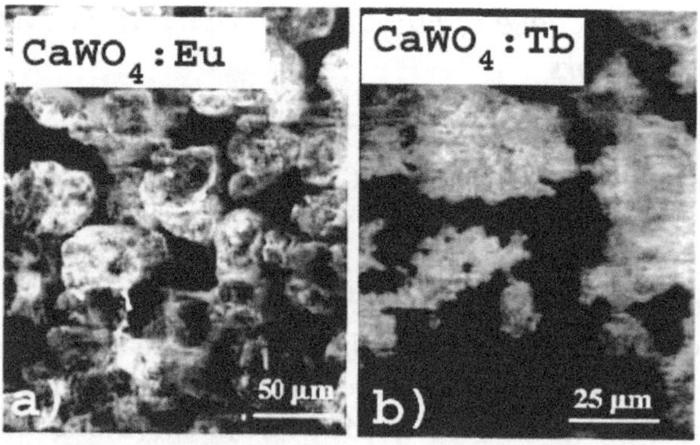

Figure 4.1 Monochromatic images of $CaWO_4$ activated with Eu^{3+} (a) and Tb^{3+} (b) (original, red color for Eu incorporations and green for Tb, see Fig. 3.6).

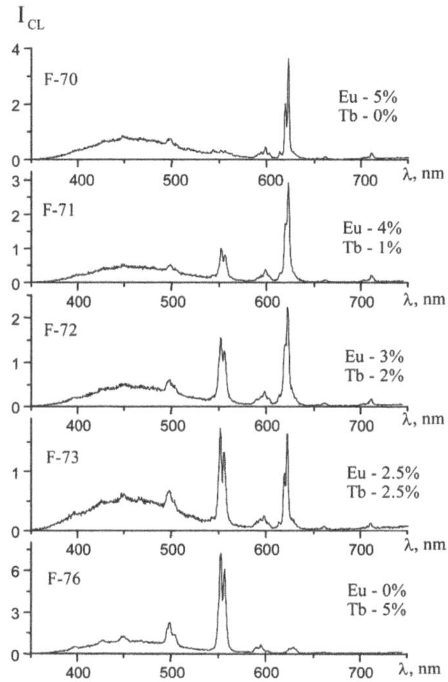

Figure 4.2 CL spectra of $CaWO_4$:Eu^{3+}, Tb^{3+} samples (codes F70-F76) prepared with different activator concentrations.

To explain the nature of CCL and to reveal the spectral composition and its variations with RE activator, the CL and PL spectra were recorded (Figs. 4.2 and 4.3).

The PL and CL spectra of the same material are rather similar; however, there is some difference between the PL and CL related with excitation, generation rate, and volume. In general, the CL spectra are richer because only the electron beam excitation can induce transition between all the defect levels, including those created by surface type of imperfections.

Observed characteristic blue emission at 400–500 nm is related with some tetrahedral WO_4 groups of the scheelite host lattice. The specific green emission bands at 497 and 553 nm can be attributed to electronic transition inside the Tb center. Four red emission bands at 596, 622, 661, 711 nm are caused by transition from emitting level $^5D_0$ to $^7F_1$, $^7F_2$, $^7F_3$, $^7F_4$ for trivalent ions of europium.

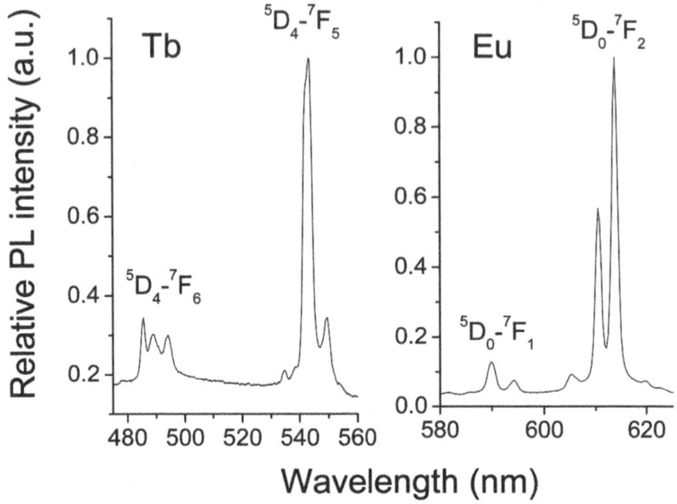

Figure 4.3 Fragments of normalized PL spectra of $CaWO_4$:Eu^{3+}(2.5%), Tb^{3+}(2.5%) phosphor (excitation with 325 nm laser radiation at 15 K).

The dependence of the CL emission intensity on the activator concentration is presented in Fig. 4.2. The variation of the main emission peaks of Eu and Tb with their concentration is clearly seen. As a result, the overall chromaticity also changes, strongly depending on the activator ratio and concentration.

Special attention was given to the CL and PL emissions of samples prepared with equal amounts of Eu and Tb. PL spectra illustrating the main peaks of Eu^{3+} ($\lambda = 612$ nm) and Tb^{3+} ($\lambda = 543$ nm) in $CaWO_4$: Eu^{3+}(2.5%),Tb^{3+}(2.5%) (sample code F 73) are shown in Fig. 4.3.

It is well known that luminescence originating from transitions between 4*f* levels is predominantly due to electric dipole or magnetic dipole interactions [8, 20, 227–230]. When RE ions are incorporated into the host crystalline lattice, electric dipole $f \to f$ transitions becomes partially allowed because of an odd crystal field component. The intensity of electric dipole transitions strongly depends on the site symmetry of the host crystal accommodating the RE ions. On the contrary, magnetic dipole $f \to f$ transitions are only in a little measure affected by the site symmetry, because they are parity-allowed. If RE ions occupy sites with inversion symmetry, optical transitions inside $4f^n$ configuration are strictly forbidden as electric dipole transition. They can only occur as magnetic–dipole transition, which obey the selection rule $\Delta J = 0, \pm 1$. If there is no inversion symmetry at the sites of the RE ions, the electric dipole transitions are no longer strictly forbidden and the transitions with $\Delta J = 0, \pm 2$ are hypersensitive to this effect.

The specific green emission band of Tb centers that is situated at 543 nm is due to electronic transition $^5D_4 \rightarrow {}^7F_5$ inside Tb^{3+} ion with $4f^7$ configuration. In order to obtain an efficient green luminescence, the activator concentration has to be at least 2.5 mol%.

Emission lines of Eu^{3+} ion correspond to transitions from the excited $^5D_0$ level to the ^7F_J ($J = 1, 2, \ldots$) levels of the $4f^6$ configuration (transition from $J = 0$ to $J = 0$ being forbidden, because the total orbital momentum is not changed). The weak emission in the vicinity of 590–600 nm is due to the magnetic dipole transition $^5D_0 \rightarrow {}^7F_1$. The strong emission around 610–630 nm is due to the hypersensitive electric dipole transition of $^5D_0 \rightarrow {}^7F_2$, induced by the lack of inversion symmetry at the Eu^{3+} site. In order to obtain an efficient red luminescence, the activator concentration of Eu^{3+} has to be at least 2.5 mol%.

PL spectra and CL spectra are very similar and thus indicate that the mechanism of energy transfer is the same, whatever the excitation (Fig. 4.4).

Both methods of investigations, namely PL and CL spectroscopy, put in evidence a continuous increase of the red emission band and a considerable decrease of the green emission band. Experimental results showed that no activator quenching effect could be noticed in this concentration range, so that the possible effect of concentration quenching is not to be taken into consideration. To explain the experimental PL and CL results, two- and three-level models taking into account an interaction between Tb^{3+} and Eu^{3+} are proposed (Fig. 4.5). Here we consider a model and theory only for CL excitation, and for PL excitation it will be given in section 4.2.

For *the two-level system* Fig. 4.5(a) (one excited state with population n_1, and the ground state with population n_0) the CL intensity from excited level is determined by the population of this level. The population of the excited state decreases according to Eq. (4.1):

$$\frac{dn_1}{dt} = LJn_0 - \tau^{-1} n_1 \tag{4.1}$$

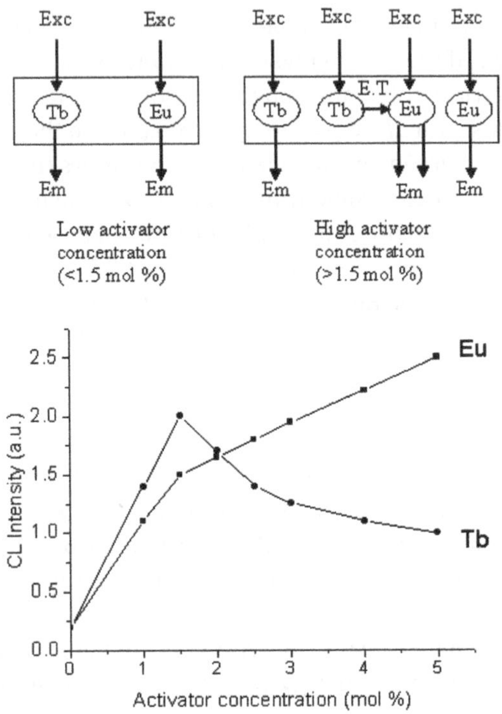

Figure 4.4 Dependence of the CL intensity of the specific Eu^{3+} red band at 612 nm and Tb^{3+} green band at 543 nm on activator concentration.

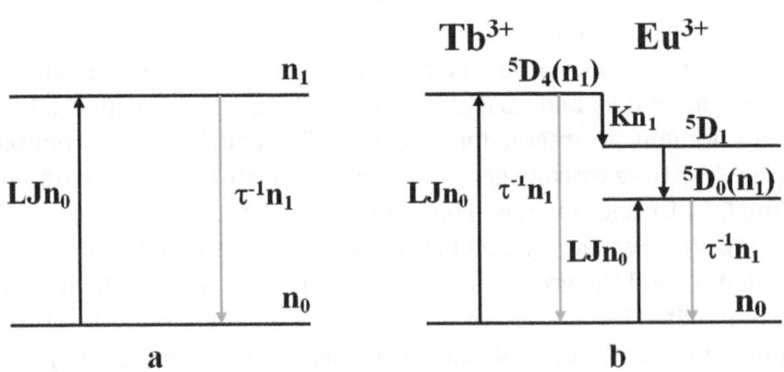

Figure 4.5 Schematic representation of the energy transfer in $CaWO_4$, double activated with Tb^{3+} and Eu^{3+} ions for two-level (a) and three-level (b) models.

where n_0 and n_1 give the number of luminescent ions in the ground (0) and excited (1) state, respectively, τ^{-1} is the probability for spontaneous emission $1 \rightarrow 0$, J is an electron beam density, and L is a quantum yield of CL. Taking into account that for the stationary conditions of excitation $dn/dt = 0$, and that the total number of luminescent ions $N = n_1 + n_0$, Eq. (4.1) becomes

$$n_1 = \frac{LJN}{LJ + \tau^{-1}} \tag{4.2}$$

Expression (4.2) explains the linear dependence of CL intensity on the activator concentration when the concentration is low and the interaction between ions is not to be considered.

At a relative high activator concentration, when different ions are presents, e.g. Tb^{3+} and Eu^{3+}, a more complicated *three-level model* Fig. 4.5(b) (with excited states, namely $^5D_4$ (Tb^{3+}) or $^5D_0$ (Eu^{3+}) and their ground states plus one intermediate state $^5D_1$ for Eu^{3+}) could be proposed. In this situation, the non-radiative transitions $^5D_4 \rightarrow {}^5D_1$ and $^5D_1 \rightarrow {}^5D_0$ are possible. In this case, the Tb^{3+} ion serves as donor and Eu^{3+} ion serves as acceptor, and one additional channel for population of acceptor level appears. Eu^{3+} and Tb^{3+} concentration determine the probability of energy transfer. The most probable energy transfer corresponds to the case when the activators concentrations are equal. One can assume that the probability of this process could be given by the relation

$$K = \frac{N_{\mathrm{Tb}} N_{\mathrm{Eu}}}{N^2} \tag{4.3}$$

where N is the sum of Eu^{3+} and Tb^{3+} concentrations.

The additional energy transfer from excited $^5D_4$ level of Tb reduces the life time of this excited state. For the stationary conditions of excitation, the following expression could be applied:

$$LJn_0 - \left[\tau^{-1} + \left(1/N^2\right) N_{Tb} N_{Eu}\right] n_1 = 0 \tag{4.4}$$

where n_0 and n_1 are populations of ground state and excited state of Tb ions, respectively. Considering that $n_0 + n_1 = N_{\mathrm{Tb}}$, and N_{Tb} +

$N_{Eu} = N$, the expression for Tb^{3+} excited level could be given by Eq. (4.5):

$$I \sim n_1 = \frac{LJN_{Tb}}{LJ + \tau^{-1} + (1/N^2)N_{Tb}(N - N_{Tb})} \tag{4.5}$$

For Eu^{3+} ions the energy transfer from Tb^{3+} means that one additional channel becomes available. When the stationary conditions of excitation are satisfied, expression (4.6) could be written:

$$LJn_0 - \left[\tau^{-1} - (1/N^2)N_{Tb}N_{Eu}\right]n_1 = 0 \tag{4.6}$$

where n_0 and n_1 are the populations of ground state and excited state of Eu ions, respectively. Taking into account that $n_0 + n_1 = N_{Eu}$, and $N_{Tb} + N_{Eu} = N$, the equation for excited Eu^{3+} level could be given by

$$I \sim n_1 = \frac{LJN_{Eu}}{LJ + \tau^{-1} - (1/N^2)N_{Eu}(N - N_{Eu})} \tag{4.7}$$

Comparing Eqs. (4.5) and (4.7), one can conclude that when the interaction between Tb^{3+} and Eu^{3+} is taken into consideration, the luminescence intensity from Eu centers is higher then from Tb ones.

If the condition $LJ >> \tau^{-1} - (1/N^2)N_{Eu}(N - N_{Eu})$ is satisfied, the luminescence intensity could be described as

$$I \sim n = N_E \tag{4.8}$$

Hence, according to the proposed model, the Tb^{3+} emission intensity decreases non-linearly, whereas the Eu^{3+} emission intensity increases linearly with activator concentration. These calculations are in agreement with the experimental data, as shown in Fig. 4.4. When Eu^{3+} and Tb^{3+} were simultaneously introduced in equal amounts at relatively low concentration (up to 1.5%), the linear dependence can be noticed and the mutual activators interaction needs not to be accounted. The simple two-level model could be used in this case (Eqs. (4.1–4.2)). At higher concentration the activators interaction has to be taken into consideration to explain the results, and Eqs. (4.5–4.8) are to be applied. The proposed model can be used also with photoexcitation, including UV and VUV (see section 4.2).

In addition, other interesting properties of $CaWO_4$:Eu^{3+}, Tb^{3+} phosphors were revealed by CL and PL measurements. The chromaticity

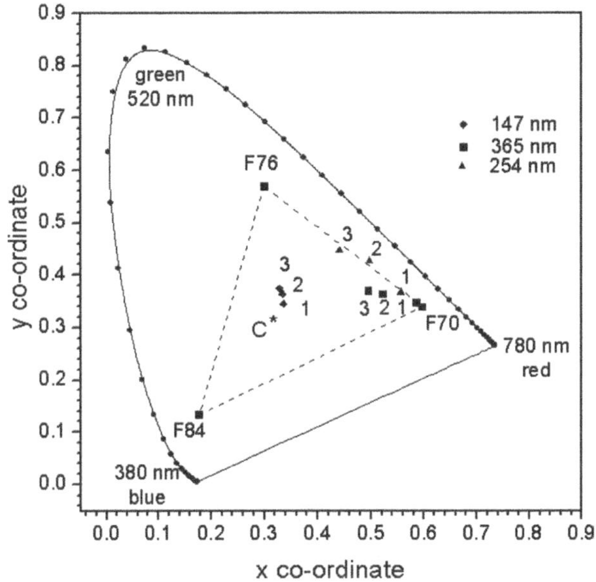

Figure 4.6 Chromatic map of some double activated samples (1–3) under different PL excitations, in comparison with self-and single activated calcium tungstate under 254 nm excitation. 1 – F83 ($CaWO_4$: Eu (5%), Tb(5%)); 2 – F73 ($CaWO_4$: Eu (2.5%), Tb(2.5%)); 3 – F85 ($CaWO_4$: Eu (1.5%), Tb(1.5%)); F70 – $CaWO_4$:Eu (5%); F76 – $CaWO_4$:Tb(5%); F84 – $CaWO_4$:W.

coordinates of luminescence shown by investigated samples under different excitations were determined. The chromaticity CIE diagram presented in Fig. 4.6 illustrates the various luminescence colors of some samples.

Every natural color could be identified by (x, y) coordinates that are disposed inside the "chromatic shoe" representing the saturated colors. Luminescence colors of self-activated ($CaWO_4$:W), terbium activated ($CaWO_4$:Tb) and europium activated ($CaWO_4$:Eu) phosphors are placed in the blue, green or red corner, respectively. The chromatic co-ordinates of the luminescence of these "simple" phosphors are in a little measure affected by the energy of exciting radiation.

Different situation occurs when calcium tungstate is double activated with terbium and europium ions. In this case, the luminescence color position depends not only on the activator concentration and ratio, but also, on the excitation energy. When excitation is performed with radiation whose energy is not high enough to excite the host lattice (λ_{exc} = 365 nm, for example), the interaction between Eu^{3+} and Tb^{3+} follows the proposed model and PL colors are located between the two extreme points corresponding to F70 and F76 samples. For high excitation energy such as VUV (λ_{exc} = 147 nm) radiation and cathode ray, PL and CL color co-ordinates are disposed well inside the F70-F76-F84 triangle, and a more complicated process of energy transfer with the host lattice participation, could be envisaged. In certain conditions, white light can be observed.

4.1.4 *Analysis of Optical Lines*

X-ray diffraction investigations show that all prepared powders are single phased materials and possess uniform and well-formed crystalline lattice with tetragonal $CaWO_4$ arrangement (scheelite-type). The refinement of $CaWO_4$ crystalline structure performed by Zalkin and Templeton [219] gave precise data. The space group for the tetragonal crystals is 88 ($I4_1\,a$, C_{4h}^{6}) with tungsten in 4a position, having a site symmetry S_4 (Fig. 4.7).

It is surrounded by four oxygen ions in a nearly regular tetrahedron at distances of 0.178 nm. Calcium ions have eight oxygen neighbors, four at 0.244 nm and four at 0.248 nm forming a

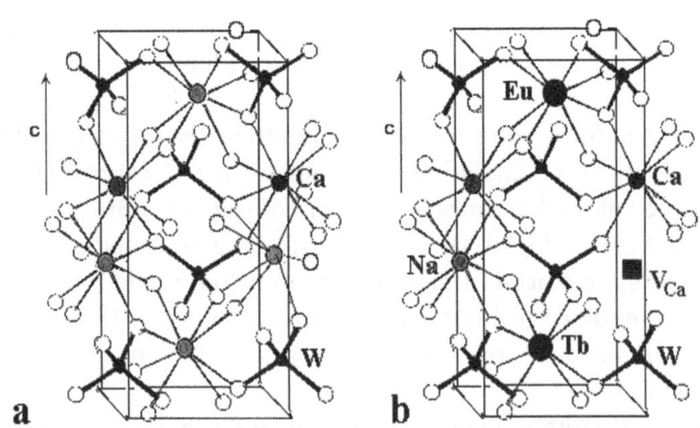

Figure 4.7 Crystal structure of undoped (a) and Tb^{3+}-Eu^{3+} double activated (b) calcium tungstate (charge compensation in the positions of Na^{+} and Ca vacancies is also suggested).

distorted cube. Each tungsten ion is surrounded by four calcium ions, the Ca–W distance being 0.387 nm. The replacement of one or more calcium ions by Eu^{3+} or Tb^{3+} or charge compensating ions give rise to a distortion of the WO_4 tetrahedron due to changing of the position of the bridging oxygen ions.

The analysis of the optical lines is based on the calculation of the selection rules for the electronic transitions and on the study the angular dependencies for the two-photon transitions. First, we focus on the former problem and consider allowed dipole type transitions. The terms *LSJ* of the $4f^{6}$-shell of Eu^{3+} and f^{8}-shell of Tb^{3+} free ions correspondingly belong to the even representations Dj^{+}. Our aim is to find the selection rules and polarization of the lines in the crystal field of the host lattice for the following transitions:

$$^5D_0 \rightarrow {}^7F_6, {}^7F_5, {}^7F_4, {}^7F_3, {}^7F_2, {}^7F_1, {}^7F_0\,(Eu^{3+}),$$

$$^5D_4, {}^5D_3 \rightarrow {}^7F_6, {}^7F_5, {}^7F_4, {}^7F_3, {}^7F_2, {}^7F_1, {}^7F_0\,(Tb^{3+}).$$

According to the set of the irreducible representations (irreps) of S_4 resulting from Dj^{+} (Table 4.1), each *LSJ*-multiplet is split into the Stark components designated as follows: A and B, one-dimensional irreps of S_4; E, two-dimensional representation (in Malliken's notations) [231].

Table 4.1 Crystal field multiplets arising from *LSJ* terms of Eu^{3+} and Tb^{3+} ions in S_4 symmetry

Terms *LSJ*	Scheme of the reduction of Dj^{+} in the point symmetry group S_4
$^7F_6$	$3A + 4B + 3E$
$^7F_5$	$3A + 2B + 3E$
$^7F_4, {}^5D_4$	$3A + 2B + 2E$
$^7F_3, {}^5D_3$	$A + 2B + 2E$
$^7F_2$	$A + 2B + E$
$^7F_1, {}^5D_1$	$A + E$
$^7F_0, {}^5D_0$	A

The transitions within the *f*-shells are parity forbidden, odd components of the crystal field make them partially allowed due to the admixture of sites with the opposite parity (Judd-Ofelt mechanism) [228, 232, 233].

Due to the absence of the inversion symmetry in the system under consideration, the electric dipole mechanism of the optical transitions proves to be the most efficient. The components of the electric dipole moment operator $\boldsymbol{d}$ correspond to the following irreps of S_4: B (d_z), E (d_x, d_y) with z axis being parallel to S_4 axis. Crystal field splitting leads to the anisotropy of the intensity of the light emission and absorption: the component d_z gives rise to the light emission (or absorption) in z polarization (π polarization), the components d_x, d_y are related to x and y polarizations (σ polarization). This assignment will be used in the analysis of the selection rules and polarization for the allowed transitions between Stark components.

For the electric dipole transitions, the following rules can be derived: (1) $A \leftrightarrow A$ and $B \leftrightarrow B$ transitions are strictly forbidden (since M_z component of the magnetic dipole moment corresponds to A, these transitions are allowed only as magnetic dipole ones); (2) $A \leftrightarrow B$ transitions are allowed and the optical lines are π-polarized; (3) $A \leftrightarrow E$ and $B \leftrightarrow E$ transitions are also allowed and the corresponding lines are σ-polarized; 4) $E \leftrightarrow E$ transitions are allowed and π-polarized. It is to be noted that a definite polarization is related to each allowed transition in S_4 symmetry so that mixed polarization (σ and π) in one optical line is forbidden, this feature is peculiar to the S_4 symmetry optical centers. Figure 4.8 illustrates the results for Eu^{3+}, showing schematically the fine structure and the polarization rules caused by the crystal field splitting for the case of $^5D_0 \rightarrow {}^7F_J$ optical lines. For example, the spectral pattern for $^5D_0 \rightarrow {}^7F_6$ transition is expected to consist of seven lines as follows: four lines are π-polarized ($A \rightarrow 4B$ transitions), three lines ($A \rightarrow 3E$) are σ-polarized and 3 transitions of $A \rightarrow A$ type are forbidden.

Table 4.2 summarizes the data regarding the polarization rules for all crystal field lines arising from the above-mentioned transitions in Eu^{3+} and Tb^{3+} ions.

The above consideration is based on the symmetry arguments so that the mutual disposition of the Stark levels is conventional

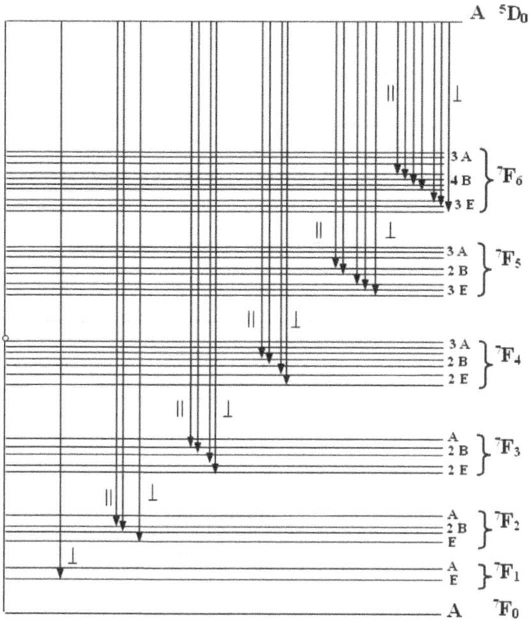

Figure 4.8 Splitting of $^{2S+1}D_J$ multiplets for Eu^{3+} in the crystal field of the S_4 symmetry site.

and aims to illustrate the expected number of optical lines and their polarization. Some of the experimental CL (Fig. 4.9) and PL (Figs. 4.10 and 4.11) data are presented in order to illustrate the optical lines of europium and terbium ions.

Crystal field splitting for the RE ions is usually small as compared to the gaps between *J*-multiplets belonging to a definite *LS*-term [234, 235] so that one can correlate the crystal field states with the definite *LSJ* terms. In this way the series of the lines observed in the ranges 380–580nm and 590–770nm can be attributed to the crystal field spectra of Tb^{3+} and Eu^{3+}, correspondingly. The transitions to the levels $^7F_5$ and $^7F_6$ in Eu^{3+} spectrum are weak and not shown in Fig. 4.11. Due to the relatively high concentration of the emitting ions and to the electron–phonon coupling, the optical lines are broadened so that the observed spectrum can not be reliably resolved into the crystal field components according to the Tables 4.1 and 4.2. In particular, the $^5D_0 \rightarrow {}^7F_1$ transition in Eu^{3+} gives rise to three lines

Table 4.2 Polarization rules for the allowed electric dipole intracenter transitions in Eu^{3+} and Tb^{3+} ions in S_4 symmetry

Terms *LSJ*	Eu^{3+}	Tb^{3+}	
	$^5D_0$	$^5D_4$	$^5D_3$
$^7F_6$	$3\sigma + 7\pi$	$32\sigma + 21\pi$	$26\sigma + 16\pi$
$^7F_5$	$3\sigma + 2\pi$	$25\sigma + 18\pi$	$19\sigma + 14\pi$
$^7F_4$	$2\sigma + 2\pi$	$20\sigma + 16\pi$	$16\sigma + 12\pi$
$^7F_3$	$2\sigma + 2\pi$	$14\sigma + 12\pi$	$12\sigma + 8\pi$
$^7F_2$	$\sigma + 2\pi$	$11\sigma + 10\pi$	$9\sigma + 6\pi$
$^7F_1$	σ	$7\sigma + 4\pi$	$6\sigma + 4\pi$
$^7F_0$	–	$2\sigma + 2\pi$	$2\sigma + 2\pi$

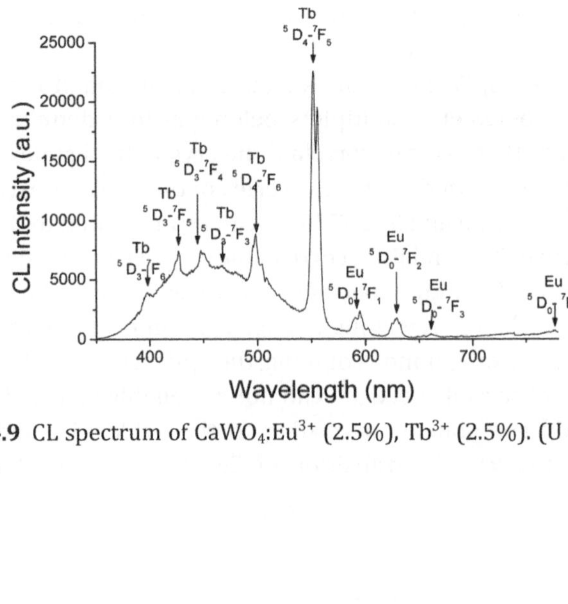

Figure 4.9 CL spectrum of $CaWO_4:Eu^{3+}$ (2.5%), Tb^{3+} (2.5%). (U = 20 kV, I = 100 nA).

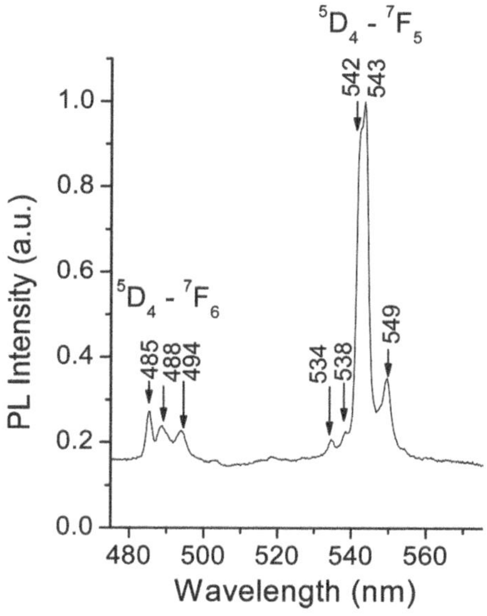

Figure 4.10 Fragment of the emission PL spectrum (15 K) of $CaWO_4$:Eu^{3+} (2.5%), Tb^{3+} (2.5%) sample, illustrating some specific lines of terbium (excitation with He-Cd laser with $\lambda = 325$ nm).

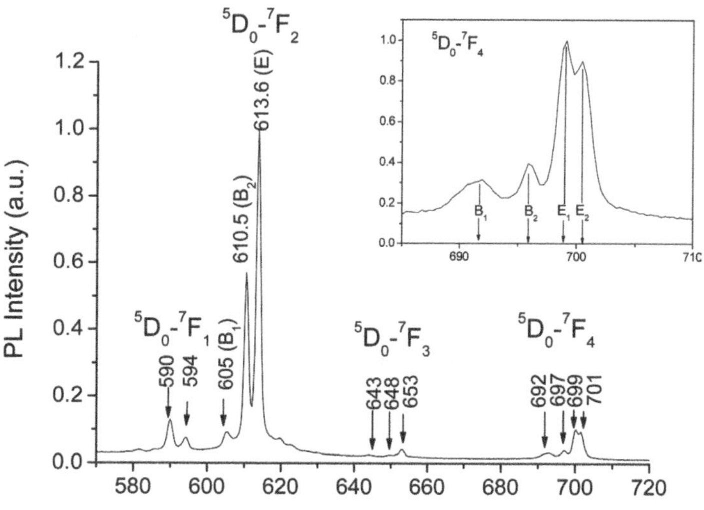

Figure 4.11 Fragment of the emission PL spectrum (15 K) of $CaWO_4$:Eu^{3+} (2.5%), Tb^{3+} (2.5%) sample, illustrating the specific lines of Eu^{3+}.

meanwhile only two lines are expected from the crystal field analysis. The remaining line can be associated with the vibrational satellite or can be the result of the decreasing of the crystal field symmetry, due to the charge compensation in a neighboring site, at the same time the random crystal fields of the charge compensating Ca vacancies results in the broadening of the optical lines. Term $^7F_2$ is split into 4 components ($A + 2B + E$) but only 3 transitions from $^5D_0$ are allowed ($A \rightarrow 2B$ and $A \rightarrow E$), Table 4.2. Experimental observation shows the transition $^5D_0 \rightarrow {}^7F_2$ in Eu^{3+} exhibits three poor resolved lines (Fig. 4.9) in agreement with the crystal field consideration. The band centered at 653 nm can be associated with the four allowed lines belonging to the transition $^5D_0 \rightarrow {}^7F_3$ (Tables 4.1 and 4.2), but poor resolution into the components and low intensity of the band does not allow to make a reliable assignation. Figure 4.11 (inset) shows that $^5D_0 \rightarrow {}^7F_4$ line is split into 4 components in agreement with the theoretical result.

4.1.5 *Polarization Selection Rules*

Regarding the complicated crystal field spectra of rare earth ions, the two-photon laser spectroscopy should be mentioned as an additional power tool in studying these spectra (see basic papers [236, 237] and Refs. [238, 239]. This approach seems to be efficient if the experimental data on single crystal samples are available. The advantage of the two-photon spectroscopy in the study of the crystals doped with the RE ions was clearly demonstrated in study [240] by the direct calculation of the intensities of two-photon transition $(^7F_6)^1A_{1g} \leftrightarrow (^5D_4)^1A_{1g}$ in Tb^{3+} in the elpasolite lattice. Two-photon transitions in Gd^{3+} were observed [241] in cubic elpasolite $Cs_2NaGdCl_3$. The theoretical analysis of the two-photon absorption processes in RE ions [242] with the application to Tb^{3+} in $Cs_2NaTbCl_3$ should be mentioned.

First of all, the restriction implied by the selection rules for one-photon transitions can are removed. The two-photon transitions do not change the parity of the combined electronic terms so that $4f \leftrightarrow 4f$ transitions become allowed without inclusion of the odd component of the crystal field in the perturbation theory. The remaining rules are implied by the point symmetry. For instance, the forbidden one-photon $A \leftrightarrow A$ transition becomes allowed in the two-photon process, if both photons are polarized along S_4 axis $B(d_z) \times B(d_z) = A$. This transition is also allowed if both photons are polarized in the plane. This can be seen from the decomposition of the direct product $E(d_x,d_y) \times E(d_x,d_y) = 2A + 2B$ that contains A term. In the same time, this transition can be referred to as the forbidden if one photon is polarized along C_4 axis and the second one - in plane ($E(d_x,d_y) \times B(d_z) = E$ does not contain A). This property can serve as a tool for the assignment of this particular transition. More general and detailed information can be obtained from the analysis of the polarization dependence of two-photon transitions, i.e. dependencies of the intensities from the directions of polarization vectors in two laser beams. Taking advantages from the point symmetry arguments, the dependence $I_{\Gamma_1\Gamma_2}(\boldsymbol{u}_1\boldsymbol{u}_2)$ for the dipole type two-photon transitions in S_4 point group can be deduced and presented in a simple form [239]:

$$I_{\Gamma_1\Gamma_2}(\boldsymbol{u_1 u_2}) = \sum_{\Gamma} \left| V_{\Gamma}^{\Gamma_1\Gamma_2} \right|^2 \sum_{\gamma} \left| V_{\Gamma_\gamma}(\boldsymbol{u}) \right|^2 \tag{4.9}$$

where: Γ_1 and Γ_2 are the electronic terms in crystal field (irreps of point symmetry group), $\boldsymbol{u}_1$ and $\boldsymbol{u}_2$ are the polarization vectors ($l_1 = u_{1x}$, $m_1 = u_{1y}$, $n_1 = u_{1z}$, etc.) of two photons whose components are transformed as dipole moment components, $V_{\Gamma\gamma}(\boldsymbol{u}) \equiv V_{\Gamma\gamma}(\boldsymbol{u}_1,\boldsymbol{u}_2)$ are the symmetry adapted components of the second rank tensor constructed from the components of the vectors $\boldsymbol{u}_1$ and $\boldsymbol{u}_2$. The $V_{\Gamma}^{\Gamma_1\Gamma_2}$ parameters are independent (from the symmetry point of view) and they are related to the irreps Γ defined as a direct product $\Gamma \subset \Gamma_1 \times \Gamma_2$, symbol γ enumerates the basis set of the irrep Γ. Within the phenomenological approach so far used these parameters are to be considered as the semiempirical constants that can be determined by the comparison of the calculated angular dependencies with the measured ones. The tensors $V_{\Gamma\gamma}(\boldsymbol{u})$ can be built with the aid of the Clebsch-Gordan decomposition:

$$V_{\Gamma\gamma}(\boldsymbol{u}) = \sum_{\gamma_1\gamma_2} u_1(\Gamma_d\gamma_1)u_2(\Gamma_d\gamma_2)\langle \Gamma_d\gamma_1\Gamma_d\gamma_2 \,|\, \Gamma\gamma \rangle \tag{4.10}$$

where $\langle\Gamma_d\gamma_1\Gamma_d\gamma_2|\Gamma\gamma\rangle$ are the Clebsch–Gordan coefficients for the point group under consideration [231], Γ_d the irreps which the dipole moment components belong to, and $u_1(\Gamma_d\gamma_1)$ and $u_2(\Gamma_d\gamma_2)$ the components of the polarization vectors (γ_1 and γ_2 enumerate the basis of Γ_d). In Eq. (4.10), only those representations Γ should be considered that belong to the intersection $\Gamma \subset \Gamma_1 \times \Gamma_2 \cap \Gamma_d \times \Gamma_d$, this is just the condition for the two-photon transitions to be allowed. The number of the irreducible representations Γ belonging to the intersection $\Gamma \subset \Gamma_1 \times \Gamma_2 \cap \Gamma_d \times \Gamma_d$ determines the number of the independent semiempiric parameters $V_\Gamma^{\Gamma_1\Gamma_2}$ (reduced matrix elements for the two photon transition) in the equation for the angular dependence. The application of the mentioned procedure [236, 237] leads to the following results (they are given only for the transitions involved in the luminescence spectrum of Eu^{3+}, Fig. 4.8.):

Transitions A↔A:

$$I_{AA}(\boldsymbol{u}_1, \boldsymbol{u}_2) = \left[n_1n_2 + c_1(l_1l_2 + m_1m_2) + c_2(l_1m_2 - m_1l_2)\right]^2 \quad (4.11)$$

Transitions *A*↔*B*:

$$I_{AB}(\boldsymbol{u}_1, \boldsymbol{u}_2) = \left[(l_1l_2 - m_1m_2) + c_3(l_1m_2 + m_1l_2)\right]^2 \quad (4.12)$$

Transitions *A*↔*E*:

$$\begin{aligned} I_{AE}(\boldsymbol{u}_1, \boldsymbol{u}_2) = & \left[(n_1l_2 + l_1n_2) + c_4(m_1n_2 - n_1m_2) + c_5(n_1l_2 - l_1n_2)\right]^2 \\ & + \left[(m_1n_2 + n_1m_2) + c_4(n_1l_2 - l_1n_2) + c_5(n_1m_2 + m_1n_2)\right]^2 \end{aligned} \quad (4.13)$$

In Eqs. (4.11)–(4.13), c_i are the semiempirical constants that are proportional to the $V_\Gamma^{\Gamma_1\Gamma_2}$ parameters. If the experimental data reflecting the dependence of the luminescence intensity on the angle of polarization vectors are available, these relations enable us to make reliable spectral lines assignments with the transitions between the certain Stark components. The results becomes more descriptive when one polarization vector (let say, $\boldsymbol{u}_1$) is directed along a specific axis so that the intensity depends on the angles $\vartheta_2 \equiv \vartheta$ and $\varphi_2 \equiv \varphi$ defining polarization vector $\boldsymbol{u}_2$ of the second photon. Let us consider two configurations of the optical absorption experiments: scheme I – one of the photons, let us say, 1, is polarized along $S_4(z)$ axis, and scheme II – photon 1 is polarized along x and the photon 2 is polarized in the plane xy.

The transitions $A \leftrightarrow A$ and $B \leftrightarrow B$ that are forbidden in one-photon dipole type absorption, becomes allowed in two-photon spectroscopy. Within the scheme I the intensities of the $A \leftrightarrow A$ and $B \leftrightarrow B$ transitions prove to be proportional $cos^2\vartheta$, so they are axially symmetric and vanish when one photon is polarized along C_4 axis and the second one is in the plane xy. If $\boldsymbol{u}_1 \parallel x$(scheme II), one can find the following angular dependence of the intensity I_{AA} (the same expression is valid for I_{BB} with different constants, c'_1 and c'_2):

$$I_{AA} = \left(c_1\, Cos\varphi + c_2\, Sin\varphi\right)^2 \qquad (4.14)$$

In the scheme 2 the intensities of $A \leftrightarrow A$ ($B \leftrightarrow B$) transitions are anisotropic, they are periodic functions of the angle between vectors $\boldsymbol{u}_1$ and $\boldsymbol{u}_2$; the angles φ_{max} and φ_{min} corresponding to the maximum and minimum intensity depend on the interrelation between the semiempirical parameters c_1 and c_1.

As distinguished from the case of $A \leftrightarrow A$ ($B \leftrightarrow B$) transitions, the two-photon $A \leftrightarrow B$ transitions are strictly forbidden in the configuration I of the two-photon experiment. Within the scheme II one obtains again a periodic function of the angle φ:

$$I_{AB} = \left(Cos\varphi + c_3 Sin\varphi\right)^2 \qquad (4.15)$$

The intensities of singlet-doublet transitions $A \leftrightarrow E$ (and $B \leftrightarrow E$ with different parameters) in the configuration I are found to possess the following dependence from the angles ϑ and φ:

$$I_{AE} = Sin^2\vartheta\left(1 + c_4^2 + c_5^2 + 2C_5\, Cos2\varphi - 2c_4 c_5\, Sin2\varphi\right) \qquad (4.16)$$

The intensity of $A \leftrightarrow E$ ($B \leftrightarrow E$) transitions, eq.(4.16), is anisotropic with respect to the rotation of $\boldsymbol{u}_2$ in the xy plane and vanishes when both photons are polarized along $S_4(z)$ axis. The last observation has a clear explanation: in fact, the doublet E can be associated with the two states with $M = \pm 1$ projections of the pseudoangular momentum

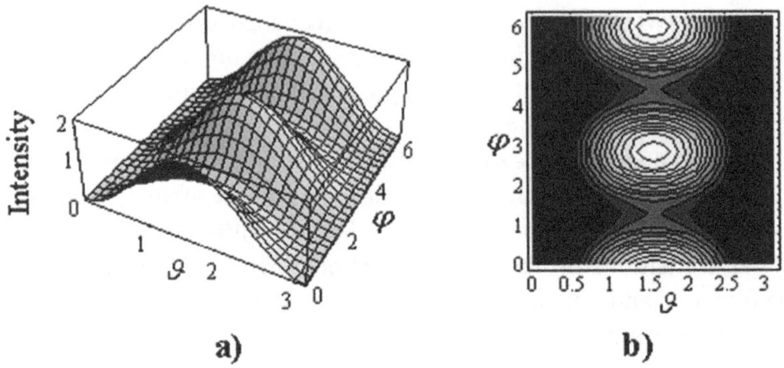

Figure 4.12 The dependence $I_{AE}(\vartheta,\varphi)$, $c_4 = 0.6$, $c_5 = 0.3$: (a) 3D plot, (b) contour plot.

$L = 1$ (S_4 being the axis of quantization), whereas $M = 0$ is for the singlet A. The transition $A \leftrightarrow E$ is allowed if the two-photon states possess $M_{phot} = \pm 1$ that is an impossible value when both photons are polarized along $S_4(z)$ axis, as in the last case $M_{phot} = 0$. The $M_{phot} = \pm 1$ state can be attained within scheme I for which the intensity increases with the increase of the angle ϑ. The same consideration shows that $A \leftrightarrow A$ ($\Delta M = 0$) transitions ($\Delta M = 0$) are forbidden when $\boldsymbol{u}_1 || S_4$; in fact, in this case $M_{phot} = 0$, $M_{2phot} = \pm 1$, so that $M_{phot} = \pm 1$.

As distinguished from the previous cases, the intensity I_{AE} is not axially symmetric and has two distinct maxima in the ($\vartheta\varphi$) plane (Fig. 4.12), the intensity vanishes when both photons are polarized along $S_4(z)$ axis or in the xy plane (scheme II) .

Finally, one can find that within the scheme I, the intensity of the $E \leftrightarrow E$ doublet-doublet transitions is isotropic in the plane xy, $I_{EE} \approx Cos^2\vartheta$, whereas providing the condition of the scheme II, one can find the intensity that is anisotropic in the plane:

$$I_{EE} = a + bCos2\varphi + cSin2\varphi \qquad (4.17)$$

where a, b, and c are constants.

The symmetry properties governing the intensity of the two-photon transitions are useful in the identification of the optical lines of the low-symmetry luminescent centers, especially when they are combined with the results of one-photon experiments data on single crystals. In the case under consideration, selecting both polarization vectors along S_4 axis one can observe only $A\leftrightarrow A$ and $B\leftrightarrow B$ lines that are forbidden in one-photon dipole absorption, the rest of the spectrum is expected to disappear. Selecting $\boldsymbol{u}_1 \parallel S_4$ and $\boldsymbol{u}_2 \perp \boldsymbol{u}_1$ one can separate $A\leftrightarrow E$ and $B\leftrightarrow E$ lines, these lines are expected to be σ-polarized in one-photon absorption. The transitions A↔B are π-polarized in one photon spectra but the corresponding lines are expected to disappear in two-photon spectra if one of the photons is polarized along S_4 axis. The angular dependence is peculiar to each kind of transition.

The rules describing polarization in two-photon transitions are related to the purely electronic lines, whereas the optical pattern involves also vibronicaly assisted transitions. Usually for the 4*f*-shells of the rare earth ions that are well screened from the surrounding, the vibronic interaction is weak so that the condition for the observation of the well-developed vibronic structure is favorable. That is why in many cases it is difficult to distinguish between the electronic lines and their vibronic satellites in the overall spectrum. In these cases, the angular dependence of the light absorption (in one- and two-photon processes) in the region of the vibronic satellites can serve as an additional tool for identification of the lines. This problem will be discussed elsewhere.

4.1.6 *Ab initio* Calculations

For *ab initio* calculations of electronic structure of any material it is necessary to determine only the lattice parameters and atomic positions. These data for $CaWO_4$ were taken from XRD measurements (Fig. 4.13) and presented in the Table 4.3.

Single crystals of $CaWO_4$ with scheelite structure crystallize in tetragonal space group C_{4h}^6 ($I4_1/a$), and it has two molecular units in the primitive cell. The Ca^{2+} and W^{6+} sites have S_4 point symmetry. The Ca^{2+} is eightfold coordinated by O^{2-} in the form of octahedra made up of elongated and compressed tetrahedral. The WO_4 groups have the shape of nearly regular tetrahedra, slightly compressed along the S_4 axis. The oxygen sites have only trivial point symmetry and the crystal has an inversion center [243].

Electronic band structure, and density of states calculations are performed via the full potential linear-augmented plane wave

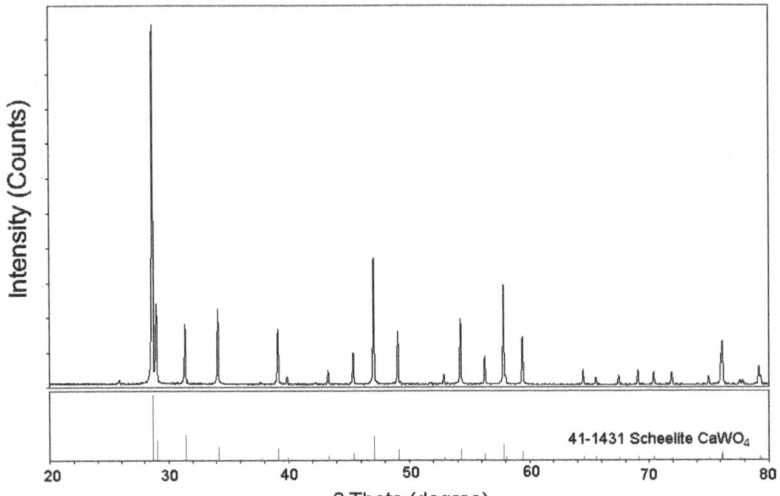

Figure 4.13 The XRD patterns of powder $CaWO_4$ in comparison with the reference ICSD data.

approach, as implemented in the WIEN2k code [244], within the framework of DFT (the details are described also in Chapter 3).

Band dispersions for $CaWO_4$ are plotted along three symmetry directions within the Brillouin zone and presented in Fig. 4.15. The valence band maxima and conduction band minima are located at the Γ point, so that the gap is direct at Γ and its value is 4.2 eV, which is very close to the experimental value deduced from the optical absorption edge. Zhang et al. calculated a 4.09 eV direct band gap using a linearized augmented plane wave DFT method, well known to underestimate the energy gap value [245].

Table 4.3 Atomic positions and lattice parameters in $CaWO_4$

$CaWO_4$	*Atom*	*x*	*y*	*z*	*F*
$a = 0.524$ nm	O	0.25	0.10	0.05	1
$b = 0.524$ nm $c = 1.135$ nm	Ca	0	1/4	5/8	1
$\alpha = 90°$ $\beta = 90°$ $\gamma = 90°$	W	0	1/4	1/8	1

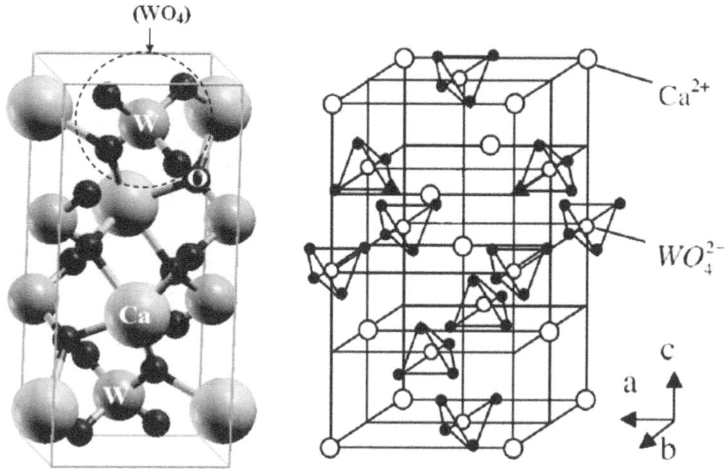

Figure 4.14. Crystalline structure of $CaWO_4$ with undisturbed WO_4^{2-} complexes.

In Fig. 4.16 the total DOS, as well as partial DOS, are combined together with PLE measured at room temperature. This comparison helps us to identify and explain the nature of excitation bands and to determine correctly the band gap in $CaWO_4$.

Only main *p*-like state for oxygen and *d*-like states for calcium and tungsten are presented in Fig. 4.16. The valence band (VB) of $CaWO_4$ mostly consists from O 2*p* states. Conduction band (CB) is divided into two parts. The lower conduction band (from 4 to 7 eV) is mainly composed of W 5*d* states and the upper conduction band (7–10 eV) involves contribution mainly from Ca 3*d* states. The charge transfer gap between W and O is clearly shown and optical band gap for powder $CaWO_4$ can be determined from excitation spectra and DOS calculation about (4.3 ± 0.2) eV.

Calcium and tungsten in partial DOS explain completely the nature of all the bands in excitation spectra. The band C in the PLE spectra is associated with the absorption of the host lattice, which involves mostly the transitions between 3*d*-like states of Ca and 2*p*-like states of O or Ca^{2+}–O^{2-} charge transfer transition. Band B is a hybrid band and it is composed from both [W–O] and [Ca–O] charge transfer transitions. Band A is associated only with

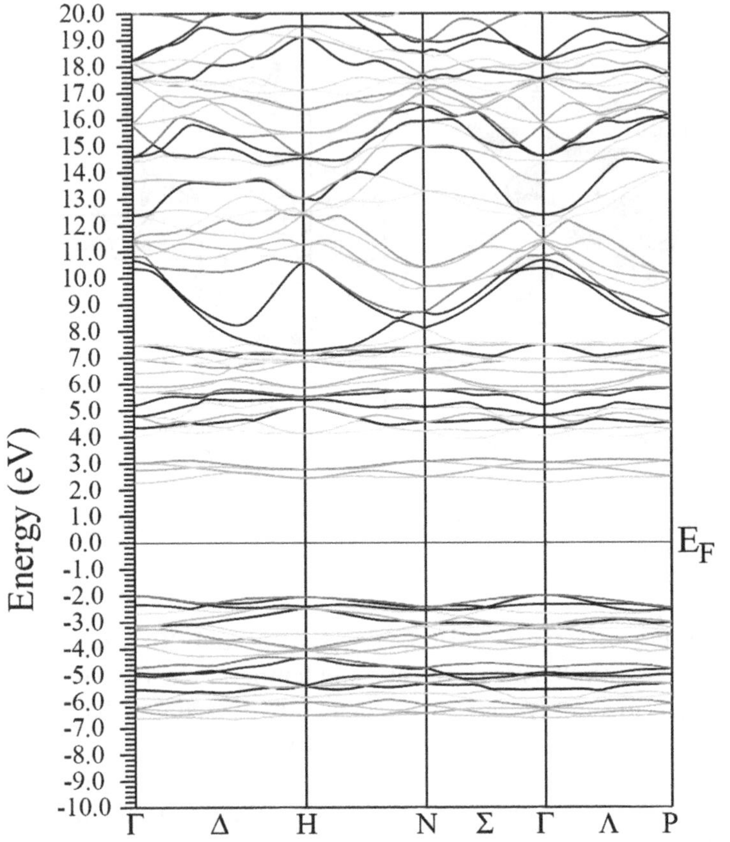

Figure 4.15 Electronic band structure for $CaWO_4$. The zero energy (E = 0 eV) corresponds to the Fermi level.

W 5*d* states and 2*p*-like states of O. The WO_4^{2-} groups can absorb excitation energy through $O^{2-} \rightarrow W^{5+}$ host charge transfer transition. Electrons excited to the lower conduction band and holes in the valence band quickly relax to self-trapped excitons and undergo geminate recombination, yielding intrinsic blue luminescence in pure crystals.

If the high-energy X-ray, electron beam, or VUV excitations are applied, it would be quite reasonable to assume that the excitation energy is absorbed first by the host lattice, which involves the

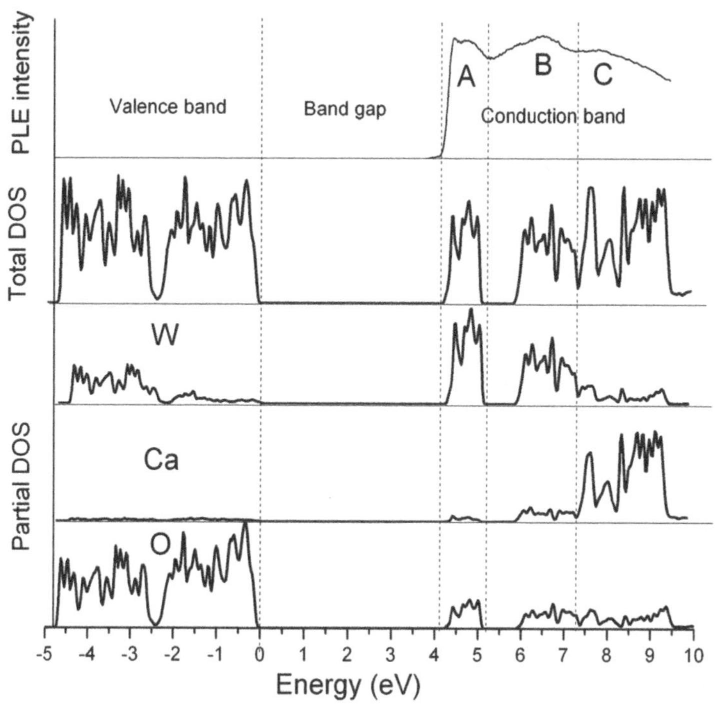

Figure 4.16 $CaWO_4$; PLE spectra under excitation with λ_{em} = 440 nm at 285 K. Below is the Total density of states of $CaWO_4$ and Partial density of states of W, Ca, and O.

transition between 3*d*-like states of Ca and 2*p*-like states of O. The absorbed energy may then be transferred to WO_4^{2-} groups and last transferred to the activator center if any, or emit a blue light. This process is shown in Fig. 4.17.

Figure 4.17 depicts a simple model illustrating the excitation and emission processes in $CaWO_4$. For $CaWO_4$ sample, under 250 nm excitation, a charge-transfer transition from the O^{2-} ion to the W^{5+} metal ion occurs. To excite Ca–O bonds a 150 nm excitation is used. After excitation, the non-radiative transitions from host lattice to WO_4 groups occur and then give the broad blue emission band peaked at about 430 nm.

These data are in good agreement with experimental results.

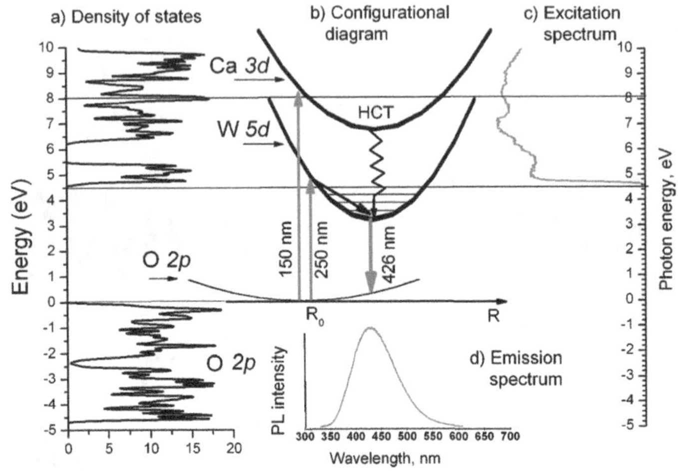

Figure 4.17 Density of states of $CaWO_4$ (a), configurational diagram (b), excitation (c), and emission spectrum (d) of $CaWO_4$ at 285 K.

4.1.7 *Concluding Remarks*

To summarize the results of this paragraph the following is to be mentioned.

The double activation of $CaWO_4$ with Tb^{3+} and Eu^{3+}, in the presence of alkaline salts as flux, gives the possibility to synthesize phosphors with variable chromatic emissions. Variable luminescence colors could be obtained when calcium tungstate phosphors containing 0–5 mol% activator ions were exposed to relatively low excitation energies, such as 365 or 254 nm ultraviolet radiations. Under high energy excitation such as VUV (147 nm) radiation or electron beam, white light could be observed.

The combination of PL and CCL-SEM methods for the complex examination of non-activated and activated by Eu^{3+} and Tb^{3+} $CaWO_4$ is very useful. For example, CL-images obtained in SEM in real colors can be used for a rapid examination of spatial and spectral characteristics of powder materials, produced by different technologies. CL and PL spectroscopy put in evidence the mutual interaction between the two activating ions. The proposed two- and three-level models taking into account an interaction between Tb^{3+} and Eu^{3+} are in good agreement with the experimental results. When Eu^{3+} and Tb^{3+} are simultaneously introduced in equal amounts at relatively low concentration (up to 1.5%), the linear dependence can be noticed and the mutual activators interaction needs not to be accounted. At higher concentration the activators interaction has to be taken into consideration and the results show that the Tb^{3+} emission intensity decreases non-linearly, whereas the Eu^{3+} emission intensity increases linearly with activator concentration.

Using the group-theoretical approach, the selection rules for the transition between Stark components of Tb^{3+} and Eu^{3+} in the actual symmetry of the crystal field as well as the polarization for the allowed transitions have been calculated and analyzed. Splitting of $^{2S+1}D_J$ multiplets for Eu^{3+} and Tb^{3+} in the crystal field of the S_4 symmetry site was demonstrated. It was also shown that the analysis of the angular dependence for the two-photon transitions in the case under consideration can serve as an efficient tool for the assignment of the optical lines in single crystal samples. This theory can be applied for a better understanding of the transfer mechanism in rare earth activated phosphors as well as for the improvement of their properties.

Due to very good chromatic variability, $CaWO_4$:Eu^{3+}, Tb^{3+} powders are potentially utilizable in the manufacture of fluorescent lamps for advertising signs, or other color rendering devices.

4.2 RE^{3+} Activated Y(Ta,Nb)O_4-Based Phosphors

4.2.1 *Introductory Remarks*

It is well known that non-activated $YTaO_4$ and $YNbO_4$ are efficient X-ray phosphors. X-ray phosphors are solid state inorganic materials used mainly in medical X-ray imaging applications [12, 246]. The role of these phosphors is to reduce the exposure of the patient to X-rays while preserving the structural features of the X-ray image. This is accomplished by converting every X-ray photon absorbed by a phosphor screen into hundreds of visible- or UV-light photons which are then recorded by a detector, such as a piece of photographic film. A good X-ray phosphor must meet challenging prerequisites [12, 52, 246]: good X-ray absorption in the diagnostic medical energy range (15–100 keV), high luminescence efficiency, emission in the green to near-UV region, proper crystallite size and shape, air and water stability, and easy large-scale production. Such phosphors with improved properties are used in films/screen cassettes, and also in electronic detector systems such as computed radiography, computed tomography, and fluoroscopy [12, 20, 247, 248].

While the luminescence of $CaWO_4$ was studied as early as 1897, fluorescence properties of the rare earth tantalate based phosphors were not mentioned in the literature until 1964 [249]. Later on, extensive luminescence studies with the M (fergusonite) type rare earth tantalates (where Ln = Gd, Y, and Lu) were carried out by Blasse and Bril [250]. These authors were primarily concerned with the energy and charge transfer processes of excitation via UV as well as electrons, and they recognized the NbO_4 group to be an efficient luminescent center in the fergusonite $YTaO_4$ host, where tetrahedral local site symmetry was assumed.

These tantalate and niobate phosphors exhibit satisfying luminescence whenever excited by UV light, cathode radiation, or X-ray.

The blue light emission from yttrium niobium–tantalate Y(Ta,Nb) O_4 phosphors under X-ray excitation, is associated with TaO_4^{3-} and NbO_4^{3-} groups from the host crystalline lattice [251]. Such luminescent emission could be shifted toward longer wavelengths when rare earth ions such as Eu^{3+} or Tb^{3+} replace partially the yttrium ions in the host crystalline lattice. In this case, Eu^{3+} and Tb^{3+} emission centers are created and can generate the corresponding red and green luminescence, respectively. The general luminescence of Eu^{3+} or Tb^{3+} in tantalates and niobates has been reported previously [12, 20, 250–253]. The luminescent properties of rare earth ions activated by the mixed niobate, tantalate, and vanadate compounds have also been reported. Brixner et al. have investigated the luminescence of NbO_4^{3-} in $YTaO_4$, as well as the luminescence of Tb^{3+}, Eu^{3+} activated $YTaO_4$, $GdTaO_4$, under X-ray excitation [185]. Li *et al.* have successfully synthesized $YTaO_4$:Nb, Eu and studied the influence of Nb content on the luminescent properties of Eu^{3+} [254]. Additionally, the luminescent properties of Eu^{3+} and Dy^{3+} in $RVNb_2O_9$ (R = Y, Gd, La) have been systematically investigated by Chi *et al.*[255]. Some recent publications with separated activation by Eu^{3+} or Tb^{3+} yttrium niobate systems [256–258] and yttrium tantalate systems [179, 259–261] have been appeared in the last years. However, to the best of our knowledge, no work has been reported using doubly activators (Eu^{3+} and Tb^{3+}) for $Y(Ta,Nb)O_4$ phosphors.

Performances of these phosphors are related to composition, crystalline structure, particle dimensions and luminescence properties of powders. The preparation of $YTaO_4$-based phosphors is achieved by solid state reaction route, from synthesis mixtures containing appropriate metallic oxides. The thermal synthesis is flux-assisted by alkaline salts in order to improve the crystalline structure of the $YTaO_4$ host lattice as well as the formation of the emission centers. It is well known that, the synthesis conditions strongly influence the degree of the crystalline order and particle size and morphology that determine the luminescent performances of phosphors e.g. yttrium tantalate-based powders [24, 30]. Moreover, due to the synthesis mechanism and the complexity of Y_2O_3-Ta_2O_5 system, the crystalline homogeneity of $YTaO_4$ powders is difficult to achieve [262, 263] and the incorporation of rare earth ions into the well-organized $YTaO_4$ crystalline matrix is difficult to keep under control.

Some peculiar physical and chemical aspects related to the formation of self activated, terbium and europium activated yttrium niobium–tantalate $Y(Ta,Nb)O_4$ phosphors, during the flux-assisted solid state reaction are analyzed in this chapter. The influence of the flux nature and concentration on the host lattice formation and activators incorporation is investigated. The correlation between composition, microcrystalline parameters and photoluminescence of phosphors is established and shown in numerous figures.

The first reason to continue the investigation of these phosphors is to study the luminescence mechanism of Eu^{3+} and Tb^{3+} ions in the tantalate and niobate host lattices. The other reason is to develop efficient luminescent materials that could be used not only in X-ray intensifying screens, but also in mercury-free fluorescent lamps and plasma display panels.

The present book continues our previous works related to the synthesis and properties of $YTaO_4$-based phosphors with or without rare earth doping ions [176, 262, 264–266].

In this chapter, we show our calculations of the crystallographic data for Y(Ta,Nb)O_4 phosphors doubly activated by Eu^{3+} and Tb^{3+} using X-ray diffraction measurements, and also investigate the luminescence properties of Y(TaNb)O_4:Tb^{3+},Eu^{3+} phosphors under UV, VUV, CL, and X-ray excitation.

Even though $YTaO_4$ and $YNbO_4$ itself are well-known self-activated phosphors, their electronic structure, optical absorption mechanism, and optical transitions have not been clearly understood. We present our calculations of the selection rules in Y(TaNb)O_4:Tb^{3+}, Eu^{3+} for the optical transitions in different polarizations because polarization selection rules and optical transitions in this type of materials have not been studied yet in detail.

To better understand the luminescence mechanism, crystallographic structure, and absorption mechanism of these phosphors, first-principles calculations, THz time-domain, infrared, and Raman spectroscopy have been carried out. The Raman spectroscopy for investigation non-activated $YNbO_4$ were first applied by Blasse [267]. Up till now the vibrational spectra of $YNbO_4$ double activated by Eu^{3+} and Tb^{3+} have not been reported. These compounds have a slight distortion of the structure and it seems interesting to investigate the influence of activators on the Raman spectra. Spectroscopy in THz frequency range is sensitive to crystalline lattice state and to intramolecular transition of the sample. No result has been reported yet on application the THz frequency range for investigation $YTaO_4$ and $YNbO_4$-based phosphors. From THz measurements it is clear seen that the optical properties doped phosphors changed significantly with activator incorporation. We also discuss the interpretation of Raman, THz, and infrared results with luminescence and crystallographic data.

4.2.2 *Self-Activated Y(Ta,Nb)O_4 Phosphors*

Undoped $YTaO_4$ and $YNbO_4$ themselves have been well known, since a long time ago, as self-activated phosphors. Historically, Ferguson [268] in 1957 was first to describe both natural fergusonite (an yttrium niobium–tantalate) as well as synthetic $YTaO_4$ correctly as crystallizing in monoclinic symmetry with the probable space group *I*2/*a*. $YTaO_4$ exhibits three crystal structures. A high-temperature tetragonal form (T) with the scheelite structure (as discussed for

$CaWO_4$, Fig. 4.7) distorts via a second-order phase transition to a monoclinic (M) structure having the fergusonite structure. Another monoclinic structure, called M′, can be synthesized directly at lower temperatures (below −1400°C). M′ transforms to T at approximately 1450°C and then to M upon cooling. It is the M′ modification that is used in phosphor screens. As a new structure type M′ YTa04 was first described by Wolten [269], although with an "R" value of 0.15, the structure was somewhat tentative. Later on Brixner and Chen [185] were able to reduce the "R" value to 0.034 and confirm the structure. It consists essentially of cube-like 8 coordinated Y polyhedra which can best be described as distorted square antiprisms. The average Y–O distance is 2.355 Å. The tantalum atoms are in a distorted octahedral coordination with four shorter Ta–O bonds at 1.86 and 1.95 Å, and two longer ones at 2.23 Å. The unit cell parameters are a = 5.30, b = 5.45, c = 5.11 Å and β = 96.5°. Tantalum is in a 4 + 2 coordination environment with oxygen (Fig. 4.18). These distorted TaO_6 units share edges with one another to form strings. Yttrium, in eight-fold coordination with oxygen, interconnects the Ta–O strings.

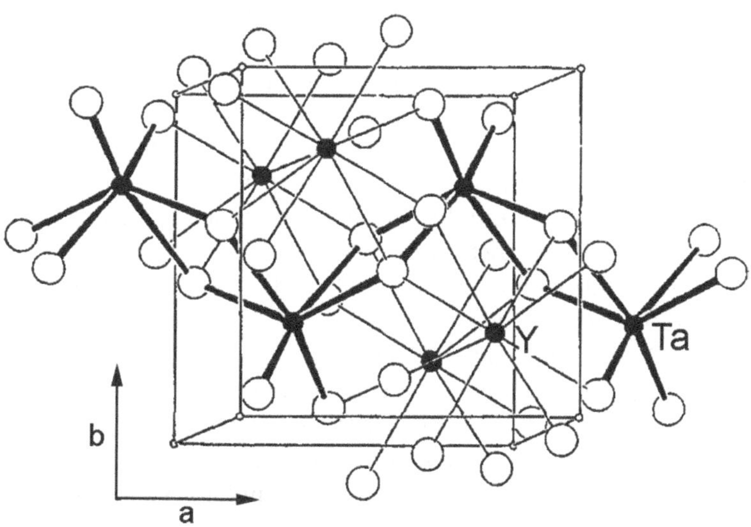

Figure 4.18 Monoclinic structure of M′-TaO_4 emphasizing a section of an edge-sharing chain of distorted TaO_6 units. Chains are interconnected by yttrium in eightfold coordination with oxygen. Metal atoms are shaded. (Reprinted with permission from [246]. Copyright 1995: Elsevier.)

The octahedral Ta coordination is the major difference to the M modification, also known as fergusonite structure. This modification is more closely related to the $CaWO_4$ (scheelite) structure. The main difference for the M and M′ $YTaO_4$ structures is ascribed to a more efficient charge-transfer process for the M′ structure, where the average Ta–O distances are closer than in the M structure. The chemistry as well as the structures of these tantalates are therefore quite complex and for the purpose of the present book we shall restrict our discussion to the M′ form of $YTaO_4$, as it is the only structure exhibiting superior luminescent properties.

The preparation of $YTaO_4$ phosphor is usually achieved by solid state reaction route, from synthesis mixtures containing different metallic oxide sources. The monoclinic crystalline structure of the $YTaO_4$ host lattice as well as the emission center formation is substantially improved when the thermal synthesis is flux-assisted by some inorganic salts like Li_2SO_4, LiCl, or Na_2SO_4. Using a flux facilitates the reaction. In the flux absence, the formation of single phase M′-$YTaO_4$ requires a more severe thermal treatment temperature higher than 1300°C and multiple thermal stages [12, 246]. A typical chemical reaction can be written as

$$Y_2O_3 + Ta_2O_5 \rightarrow 2\ M'\text{-}YTaO_4$$

The chemistry involved in this reaction is discussed in more detail below. As in the case of $CaWO_4$, M′-$YTaO_4$ is self-activated, and this phosphor emits broad band ultraviolet light peaking at 330 nm (Fig. 4.19). Activating with Nb, i.e. M′-$YTa_{1-x}Nb_xO_4$ with x = 0.005–0.020, shifts the emission from UV (330 nm) to blue (410 nm) [185] as seen in Fig. 4.19.

The UV and blue light emission from yttrium niobium–tantalate $Y(Ta,Nb)O_4$ phosphors under X-ray excitation is associated with TaO_4^{3-} and NbO_4^{3-} groups from the host crystalline lattice [251] (Fig. 4.19). Our measurements for non-activated host lattices $YTaO_4$ and $YNbO_4$ under VUV, CL and X-ray excitation show similar results and will be given in this chapter later.

The NbO_4 group in the fergusonite tantalates is an efficient luminescent center. The pioneer works in investigation this luminescent center were carried out by Blasse and Bril [250, 270]

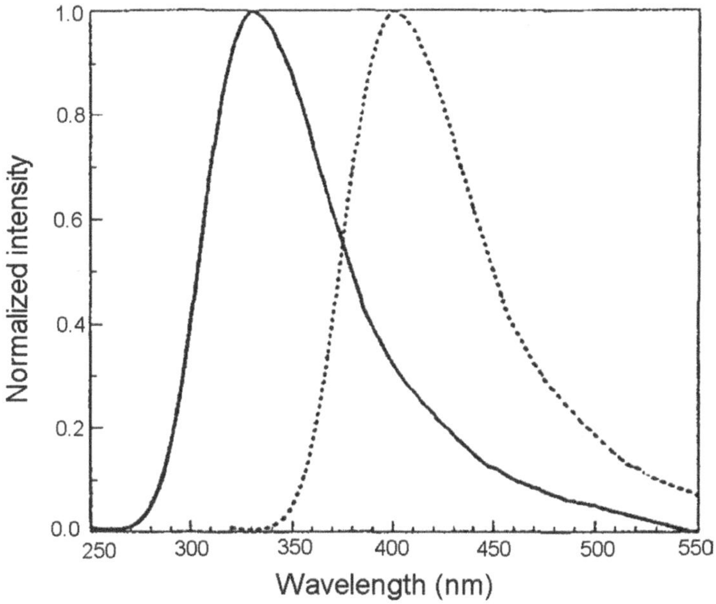

Figure 4.19 X-ray excited (Mo 30 kVp) emission spectra of M′-$YTaO_4$ and M′-$YTa_{0.98}Nb_{0.02}O_4$ showing broad band UV emission centered at 330 nm (solid line) and blue emission centered at 410 nm (dashed line), respectively. (Reprinted with permission from [246]. Copyright 1995: Elsevier.)

and energy transfer from TaO_4 to NbO_4 has been observed. Nevertheless, the development of synthesis methods to improve the luminescence properties of $YTaO_4$:Nb and to study the synthesis conditions that strongly influence the crystalline order degree, particle size, and morphology remains actual.

Method of making $YTaO_4$:Nb phosphor with improved brightness using lithium chloride–lithium sulfate eutectic flux was proposed by Reddy [271].

Hristea *et al.* [262] investigated the influence of firing regime and flux nature on the crystalline structure and particle morphology. The authors used solid state reaction, from homogeneous mixtures consisting of raw oxide precursors Y_2O_3 (99.9%), Ta_2O_5 (Optipur), Nb_2O_5 (99%), and Li_2SO_4 (99%) and/or Na_2SO_4 (99%) as a flux. The stoichiometric amounts of Y_2O_3, Ta_2O_5, and Nb_2O_5 and 15–50 wt% alkaline sulfates were ball-milled with acetone and dried at 70°C. The reactive oxide-flux mixture was prepared in a planetary mill

(Pulverisette type; agate balls and mill-beaker set). The powders mixture was calcined in air at 1200–1400°C, for 4 h and slowly cooled to the room temperature. Finally, phosphors samples were water washed, dried, and sieved.

During the thermal synthesis stage, $YTaO_4$:Nb phosphors are formed in the presence of flux, which allows the growth of particle size and assists greatly in the formation of the compound. The lithium/sulfate flux reacts with the oxides mixture and produce intermediate compounds that are more reactive than the starting oxides [12, 246]. Due to its peculiar behavior, the flux could be considered as catalyst or reactive flux [272]. The formation of $YTaO_4$:Nb phosphors could be described by the following equations:

$$Y_2O_3 + (1-x)Ta_2O_3 + xNb_2O_5 + xM_2SO_4 \rightarrow 2(1-x)MTaO_3 + 2xMNbO_3 + Y_2O_2SO_4,$$
$$2(1-x)MTaO_3 + 2xMNbO_3 + Y_2O_2SO_4 \rightarrow 2YTa_{1-x}Nb_xO_4 + M_2SO_4$$

where M = Li, Na and $x = 0$–1

Phosphor composition and thermal synthesis regime as well as the photoluminescence intensity and crystalline phases of niobium activated yttrium tantalate phosphors are presented in Table 4.4. General formula of phosphors is $YTa_{1-x}Nb_xO_4$, where $x = 0$–1. The photoluminescence intensity at peak position was calculated from the PL spectra, in comparison with an internal standard (T4 sample). For the structural composition, the crystalline phases that are in small amounts are written between brackets.

Luminescence color and intensity of $YTaO_4$-based phosphors strongly depend on the niobium concentration, flux nature and calcination temperature (Table 4.4).

All powder samples are white in color and almost all present blue luminescence under 254 nm excitation. The observed luminescence has a short decay time and it does not last after the excitation stops.

Yttrium tantalate phosphor is an ultraviolet emitter (λ_{em} = 335 nm) whose luminescence corresponds to a charge transfer transition involving TaO_4 groups. Replacing part of the tantalum ions with niobium ions, NbO_4 centers are formed that generate a blue-shifted luminescence emission (385–395 nm). Under 254 nm excitation, in $YTa_{1-x}Nb_xO_4$ only the niobate group and not the host lattice is excited.

Table 4.4 General characteristics of some niobium activated yttrium tantalate samples [262]

Phosphor sample		Synthesis conditions		General properties	
Code	Formula	*T* (°C)	Flux nature(w/w)	PL	Phase
T1	$YTaO_4$	1200	50%Na_2SO_4	2	M′
T2	$YTa_{0.95}Nb_{0.05}O_4$	1200	50%Na_2SO_4	80	M′
T3	$YTa_{0.90}Nb_{0.10}O_4$	1200	50%Na_2SO_4	82	M′
T4	$YTa_{0.85}Nb_{0.15}O_4$	1200	50%Na_2SO_4	100	M′
T5	$YTa_{0.80}Nb_{0.20}O_4$	1200	50%Na_2SO_4	90	M′+M
T6	$YNbO_4$	1200	50%Na_2SO_4	110	M
T7	$YTa_{0.85}Nb_{0.15}O_4$	1200	30%Li_2SO_4	125	M′
T8	$YTa_{0.85}Nb_{0.15}O_4$	1200	15%Li_2SO_4:15%Na_2SO_4	83	M′
T9	$YTa_{0.85}Nb_{0.15}O_4$	1200	30%Na_2SO_4	61	M′+0
T11	$YTa_{0.85}Nb_{0.15}O_4$	1300	30%Li_2SO_4	116	M′+m
T12	$YTa_{0.85}Nb_{0.15}O_4$	1300	15%Li_2SO_4:15%Na_2SO_4	102	M′+m
T13	$YTa_{0.85}Nb_{0.15}O_4$	1300	30%Na_2SO_4	70	M′+m
T15	$YTa_{0.85}Nb_{0.15}O_4$	1400	30%Li_2SO_4	77	m+t
T16	$YTa_{0.85}Nb_{0.15}O_4$	1400	15%Li_2SO_4:15%Na_2SO_4	75	m
T17	$YTa_{0.85}Nb_{0.15}O_4$	1400	30%Na_2SO_4	67	m

Note: M′ = M′-$YTaO_4$; M = M-$YNbO_4$; m = M-$YTaO_4$; t = T-$YTaO_4$; o = orthorhombic Y_3TaO_7.

Non-activated yttrium tantalate phosphor cannot be practically excited by 254 nm, as illustrated by the performances of sample T1. The partial or total replacement of tantalum with niobium atoms strongly improves the sensitivity to UV radiation resulting in a substantial intensification of the PL emission (samples T2–T6). Yttrium niobate (sample T6) shows the strongest emission (λ_{max} = 395 nm) behaving as a good self-activated phosphor. From

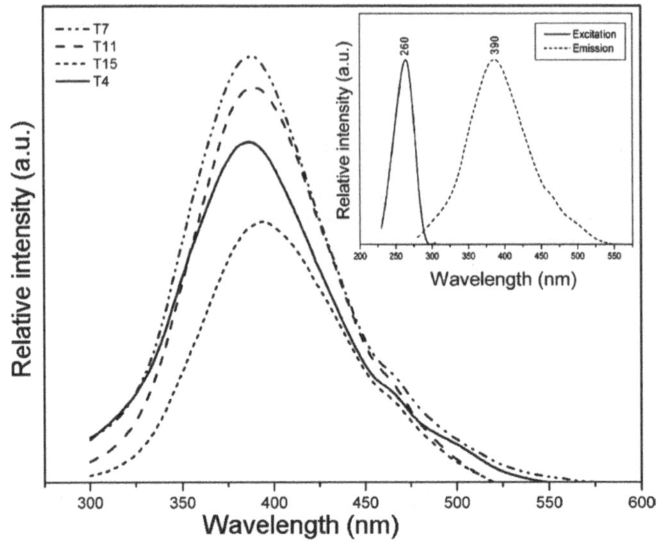

Figure 4.20 Emission spectra of $YTaO_4$:Nb samples ($\lambda_{exc} = 254$ nm) prepared with Li_2SO_4 as a flux.

niobium activated yttrium tantalate samples, the sample prepared with 15 mol% niobium shows the highest emission intensity and is considered as internal standard for the following series of samples. Typical emission spectra, for $YTaO_4$:Nb samples prepared at different temperatures, with Li_2SO_4 as a flux are presented in Fig. 4.20 [262].

In the inside detail picture, the emission and excitation spectra of sample T7 are depicted. The excitation spectrum consists in a band with maximum at around 260 nm, whereas the emission peak is situated at around 390 nm. One notes that the increase of the calcination temperature determines the decrease of the $YTaO_4$:Nb phosphor emission intensity. The influence of the flux nature and firing regime on the PL relative intensity of $YTaO_4$:Nb phosphor samples is presented in Fig. 4.21. The sample prepared at 1200°C using Li_2SO_4 as a flux presents the highest luminescent intensity. When Li_2SO_4–Na_2SO_4 mixture or Na_2SO_4 are used as a flux, a relatively higher temperature (about 1300°C) is needed to obtain good luminescent properties.

The crystalline structure and order degree of phosphor samples were evaluated on the basis of X-ray diffraction patterns and FTIR spectra.

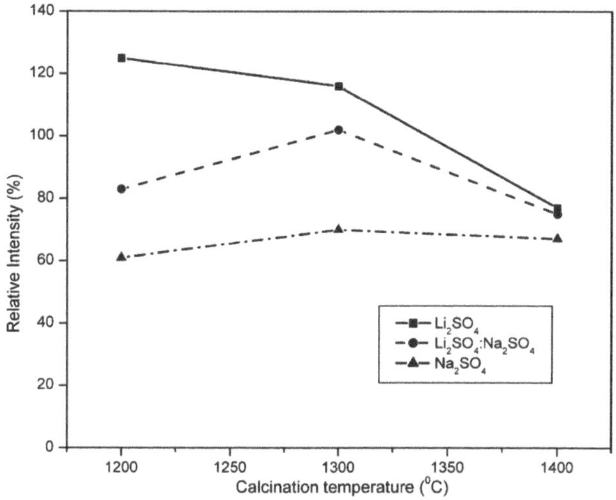

Figure 4.21 Influence of flux nature and firing regime on the PL intensity of some $YTaO_4$:Nb [262].

XRD patterns and FTIR spectra of $YTaO_4$:Nb powder samples prepared at 1200–1400°C with Li_2SO_4 and/or Na_2SO_4, as a flux are presented in Figs. 4.22 and 4.23 [262].

The most homogeneous $YTaO_4$:Nb powder was obtained at 1200°C, using Li_2SO_4 as a flux. Sample T7, $YTaO_4$:Nb (Li_2SO_4; 1200°C) is a single phase material, formed from monoclinic M prime polymorph form (Fig. 4.22a). At the same temperature, but replacing Li_2SO_4 with Na_2SO_4 as a flux (sample T9) additional reflections could be observed that could be ascribed to some intermediate compounds such as the orthorhombic Y_3TaO_7. This shows that, at 1200°C, Na_2SO_4 does not assure the complete conversion of oxides into $YTaO_4$:Nb.

For samples prepared with Li_2SO_4 as a flux, by increasing the calcination temperature, the crystalline host lattice becomes more disordered. XRD patterns illustrate that T11 sample, $YTaO_4$:Nb (Li_2SO_4; 1300°C), is a mixture of M′ and M phases, whereas T15 sample, $YTaO_4$:Nb (Li_2SO_4; 1400°C), is predominantly a mixture of M and T phase (Fig. 4.23a).

The infrared spectra (Figs. 4.22b and 4.23b), confirm that the synthesis conditions, i.e. the flux nature and the firing regime influence the crystalline order degree of phosphors, as already illustrated by

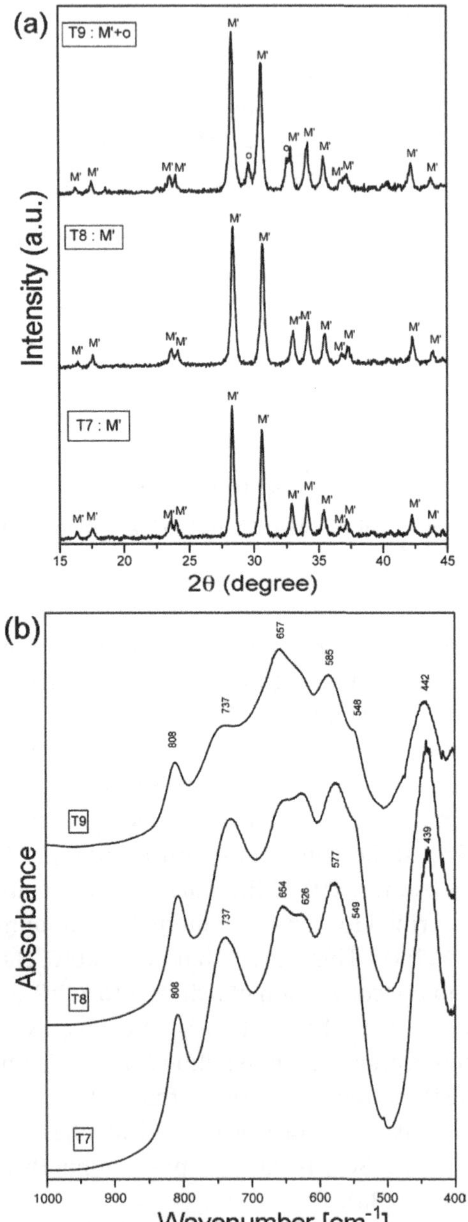

Figure 4.22 XRD patterns (a) and FTIR (b) spectra of some phosphor samples prepared at 1200°C with different fluxes: T7 = $YTaO_4$:Nb (Li_2SO_4); T8 = $YTaO_4$:Nb (Li_2SO_4+Na_2SO_4); T9 = $YTaO_4$:Nb (Na_2SO_4).

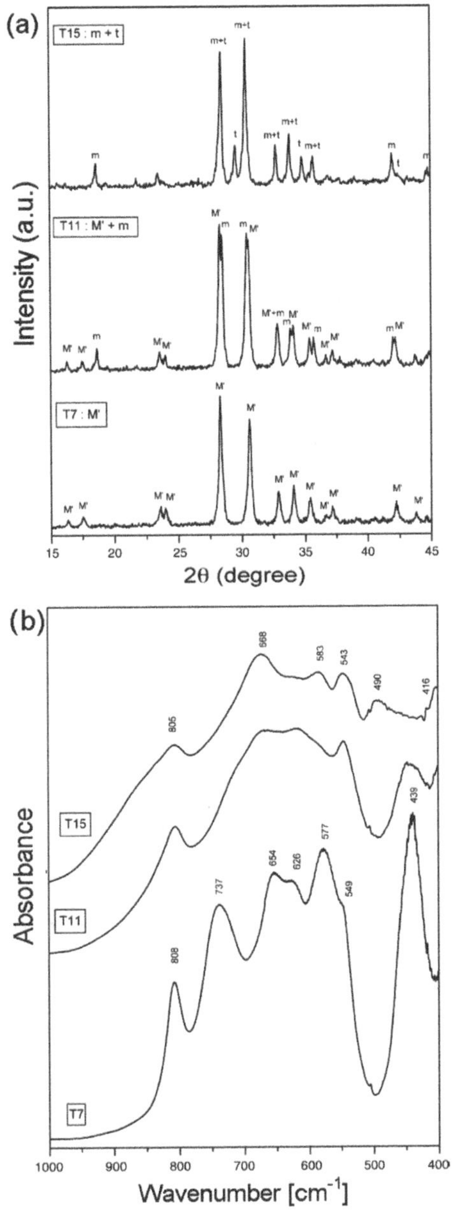

Figure 4.23 XRD patterns (a) and FTIR (b) spectra of some phosphor samples prepared with Li_2SO_4 at different temperatures: T7 = $YTaO_4$:Nb (1200°C); T11 = $YTaO_4$:Nb (1300°C); T15 = $YTaO_4$:Nb (1400°C).

XRD patterns. FTIR spectra, registered in the 1000–400 cm^{-1} range, show a complex pattern, which could be interpreted according to a structure with more or less distorted TaO_6 units, which share edges with one another in the monoclinic modification of $YTaO_4$.

FTIR spectra sustain the XRD data for samples prepared with different fluxes. Sample T7 prepared with Li_2SO_4 possesses a single phase M′ structure, as illustrated by the well-formed 439 and 808 cm^{-1} absorption bands, that are not characteristic to M phase (Fig. 4.22b). In sample T9, obtained with Na_2SO_4 as a flux, the M′ phase still exists, but crystalline structure is more disorganized, as suggested by FTIR spectra. The increase of the calcination temperature induces the conversion of M′ into M phase, process that is shown by the disappearance of the 439 cm^{-1} band and the diminishing of the 808 cm^{-1} band (Fig. 4.23b).

By comparing the XRD data with the luminescence properties, one can observe that the sample prepared at 1200°C using Li_2SO_4 as a flux, presents the highest luminescent intensity that could be associated with the M′-$YTaO_4$ phase. For Y_2O_3–Ta_2O_5–Nb_2O_5 system, Na_2SO_4 shows weaker flux reactivity as compared with Li_2SO_4. This behavior is responsible for the weakest luminescence emission observed in this sample series. Moreover, the heterogeneity of the crystalline phases of T11 and T15 samples could explain the decrease of the emission intensity with the increase of the calcinations temperature. By heating, M′ phase could convert into T phase which, during the cooling stage, could transform into the M phase [12, 246]. The emission intensity of samples prepared at higher temperature is lower due to the beginning of the M′ phase conversion.

The PL intensity is strongly affected by the crystalline homogeneity and particle dimensions of the phosphor powders. A series of SEM images (Fig. 4.24) was made to characterize and compare the particle morphology and sizes for $YTaO_4$:Nb samples prepared in different synthesis conditions.

Using Li_2SO_4 as a flux (sample T7), uniform and non-agglomerated particles are obtained. The most of the polyhedral, elongated crystals are of 2–5 μm in length and about 1μm in thickness. The partial or the total replacement of Li_2SO_4 by Na_2SO_4 (sample T9) generates polyhedral particles with less regulated shapes and also induces the formation of aggregates. In the same time, a decrease of the particle size can be observed.

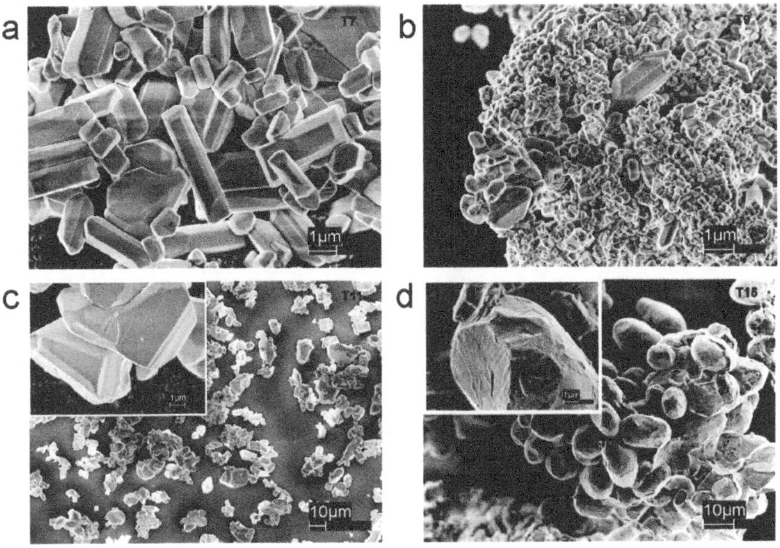

Figure 4.24 SEM images of some niobium activated yttrium tantalate phosphor samples: T7 = $YTaO_4$:Nb (Li_2SO_4; 1200°C); T9 = $YTaO_4$:Nb (Na_2SO_4; 1200°C); T11 = $YTaO_4$:Nb (Li_2SO_4; 1300°C); T15 = $YTaO_4$:Nb (Li_2SO_4; 1400°C) [262].

Samples prepared with Li_2SO_4 as a flux, in different thermal synthesis regime (T7, T11, and T15), present variable morphology, and particle dimensions. By increasing the calcinations temperature (T15 sample), the particles shape becomes nearly spherical and they trend to form clusters. The particle size increases and most of them are of about 10–20 μm.

The most heterogeneous material is the $YTaO_4$:Nb crystalline powder obtained at 1200°C, with Na_2SO_4 as a flux while the most uniform is the crystalline powder prepared at 1200°C, but with Li_2SO_4 as a flux. One can note that, in spite of the fact that Li_2SO_4 (m.p. = 884°C) and Na_2SO_4 (m.p. = 860°C) possess close melting point values, their behavior as a flux is very different due to the large difference between the two cation sizes. Lithium (Li^+ = 0.073 nm) that is a smaller cation than sodium (Na^+ = 0.113 nm), is in the favor of the total conversion of oxides into M-$YTaO_4$ crystalline phase.

The scanning electron microscopy put in evidence the physical differences between M′ and the other crystalline phases formed during the thermal synthesis of niobium tantalate phosphors. Generally, the M′ phase crystals grow in regular, non-agglomerated polyhedral particles.

In conclusion, the flux nature and firing regime are two important factors which influence luminescence properties, structural and morphological characteristics of niobium activated yttrium tantalate phosphor. XRD patterns, sustained by FTIR spectra, put in evidence the crystalline order degree of $YTaO_4$:Nb powders, whereas the SEM images illustrate the physical differences between M′ and the other phases formed during the synthesis stage.

A correlation between the morpho-structural properties, luminescent characteristics and synthesis conditions was established. In shown experimental conditions, non-agglomerated, polyhedral particles of M′-$YTaO_4$:Nb crystalline powder with good photoluminescence properties was obtained.

This compound M′-$YTaO_4$:Nb is versatile because it can also be doped with other elements to make phosphors efficiently emitting different "colors." Tb and Eu activation, as well as double activation by these RE elements is shown in the next paragraphs of this chapter.

4.2.3 *Terbium Activated $Y(Ta,Nb)O_4$ Phosphors*

Terbium activated yttrium tantalate $YTaO_4$:Tb samples were prepared by solid state reaction from homogeneous mixtures of Y_2O_3 (99.95% Sigma Aldrich) and Ta_2O_5 (99.85% Alfa Aesar) for host-lattice formation, Tb_4O_7 (99.9%, Aldrich) or $Tb(NO_3)_3 \cdot 5H_2O$ (99.9%, Aldrich) as an activator source and Li_2SO_4, Na_2SO_4, CsF, $Na_4P_2O_7$,or $Na_2B_4O_7$ (p.a. Chimopar) as a flux. In this respect, stoichiometric amounts of oxides were milled with flux in acetone, calcined at 1200°C for 4 h and finally, water washed, dried and sieved.

Samples with general formula $Y_{0.95}Tb_{0.05}TaO_4$ were prepared and characterized by photoluminescence measurements, X-ray diffraction, FTIR spectroscopy and scanning electron microscopy. Photoluminescence emission and excitation spectra were registered in identical experimental conditions, using Jasco FP-6500 Spectrofluorimeter Wavel equipped with photomultiplier PMT R3377 (Farbglasfilter WG 320; phosphor Zn_2SiO_4:Mn /N40 as control sample). FTIR spectra were registered with Thermo Scientific Nicolet 6700 FTIR Spectrometer (KBr pellets). XRD patterns were taken with Bruker D8 Advance X-Ray Diffractometer (CuKα radiation; alumina/corundum powder as standard for instrument broadening correction). SEM images were taken with Jeol-JSM 5510LV Microscope (Au-coated samples). Several samples series were prepared and investigated as follows:

Series A: $YTaO_4$ (IF42) and $Y_{0.95}Tb_{0.05}TaO_4$ (IF44) samples prepared using high quality Ta_2O_5 (Merck, Optipur), Tb_4O_7 as activator source and Li_2SO_4 (30 wt%) as a flux; sample IF44 is used as internal standard for all the photoluminescence measurements (PL intensity estimated by the 522–576 nm peak area is considered 100%).

Series B: $Y_{0.95}Tb_{0.05}TaO_4$ samples prepared with Tb $(NO_3)_3$ as activator source, and Li_2SO_4, Na_2SO_4, Li_2CO_3, CsF, Na_2B4O_7, or $Na_4P_2O_7$ as a flux material (30 wt%).

Series C: $Y_{0.95}Tb_{0.05}TaO_4$ samples prepared with Tb_4O_7 as activator source, and variable amounts of Li_2SO_4, e.g. 0, 10, 20, 30, 40, 50, 60 wt%.

Series D: $Y_{1-x}Tb_xTaO_4$ samples ($x = 0.005 \ldots 0.15$) prepared with 50 wt% Li_2SO_4 as a flux and variable amounts of Tb_4O_7.

Terbium activated yttrium tantalate $Y_{0.95}Tb_{0.05}TaO_4$ samples were obtained in one-step synthesis, by solid state reaction, using different mineralizing agents or terbium sources. The influence of the flux nature and concentration on the morpho-structural and photoluminescent properties is investigated and demonstrated in numerous figures.

The starting point of this study could be considered to be $YTaO_4$ phosphor in which, 5% of the yttrium ions are isovalent substituted with terbium. The accommodation of Tb^{3+} into the monoclinic $YTaO_4$ crystalline lattice is possible due to the close ionic radii of Tb^{3+} (0.104 nm) and Y^{3+} (0.102 nm), in sixfold coordination. Formation of $Y_{0.95}Tb_{0.05}TaO_4$ (Series A) proceeds in the presence of $Li2SO_4$ that, according to the literature, directly participates in the host lattice formation ($YTaO_4$ matrix) and activator incorporation [246, 262].

Under 254 nm excitation, pure $YTaO_4$ phosphor (IF42) shows a very weak ultraviolet emission. Incorporation of terbium into the tantalate matrix (IF44) brings about the specific luminescence, as illustrated by photoluminescence excitation and emission spectra (Fig. 4.25). The complexity of PL spectra is correlated with the

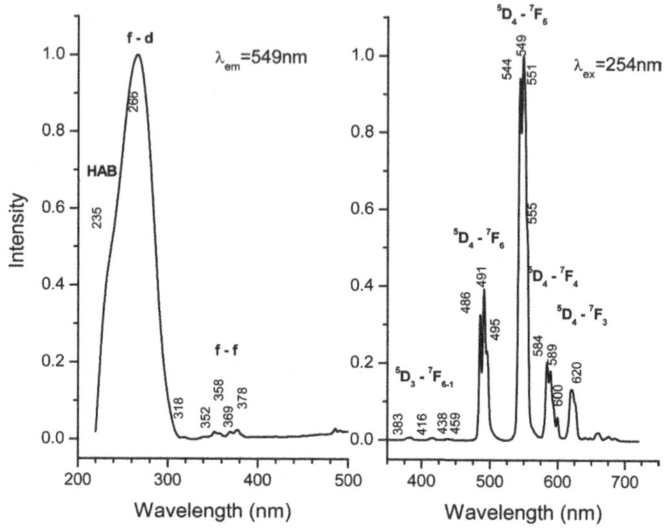

Figure 4.25 Normalized PLE (left) and PL (right) spectra of $YTaO_4$:Tb (5%).

presence of Tb^{3+} – centers that are responsible for the characteristic blue and green luminescence, associated with $^5D_3 \rightarrow {}^7F_J$ and $^5D_4 \rightarrow {}^7F_J$ electronic transitions, respectively [261]. Many narrow emission bands/lines in different ratios can be noticed on the entire visible spectrum. The strongest emission in the green spectral domain is related to the line group associated with $^5D_4 \rightarrow {}^7F_5$ transitions. For Tb concentration in use, i.e. 5 mol Tb/100 mol (Y+Tb), the broad emission band of the host lattice (TaO_4) is not evident and the blue emission bands associated with $^5D_3 \rightarrow {}^7F_J$ transitions are very weak.

PLE spectrum of $YTaO_4$:Tb is rather complicated. Different types of electronic transitions could be identified, such as the host lattice charge transfer (O→Ta) related to the host absorption band (HAB) and inter-configuration ($4f^8 \rightarrow 4f^7 5d^1$) or intra-configuration ($4f^8 \rightarrow 4f^8$) transitions, related to the intra-atomic excitation bands. According to XRD investigation, terbium activated yttrium tantalate phosphor $YTaO_4$:Tb (Li_2SO_4) is a single phase material with M′-$YTaO_4$ monoclinic structure, similar to the non-activated $YTaO_4$ sample.

4.2.3.1 Effect of flux materials

The synthesis of rare earth activated yttrium tantalate phosphor by solid state reaction proceeds slowly because of the two high refractory materials involved in the formation of $YTaO_4$ matrix, i.e. Y_2O_3 (m.p. = 2410°C) and Ta_2O_5 (m.p. = 1880°C). In order to increase the reaction rate (or even the powder reactivity), different flux materials could be used for both matrix formation and activator incorporation. The flux materials, which are alkaline or alkaline earth salts melting below the temperature of phosphor formation and dissolving partially one or both of the reactants, take no part in the solid state reaction and undergo no reaction with the reaction product [52].

In this study, different alkaline salts with melting temperature in 600–900°C range were used as a flux materials, i.e. Li_2SO_4 (859°C), Na_2SO_4(884°C), Li_2CO_3(618°C), CsF(682°C), $Na_2B_4O_7$ (880°C), or $Na_4P_2O_7$ (741°C). The choice of the mineralizing agents was performed on the basis of previous studies referring to the flux-assisted synthesis of different yttrium oxide-based phosphors, i.e. $Y(Ta,Nb)O_4$ (lithium/sodium sulfate) [262], Y_2O_3:Eu (sodium tetraborate/ pyrophosphate) [273], $Y_3Al_5O_{12}$:Ce (alkaline fluoride like CsF) [274] and Gd_2O_2S:Tb (lithium carbonate) [275].

The effect of the flux material on PL and the morpho-structural properties of $YTaO_4$:Tb (5%) samples from Series B was investigated, aiming to find a correlation between the lattice microstructural parameters and the specific luminescent characteristics.

PLE and PL spectra of samples from Series B show similar features, the only differences occurring in the PL brightness. Relative intensity of the most important bands associated with $^5D_4\rightarrow{}^7F_6$ (476–521 nm range), $^5D_4\rightarrow{}^7F_5$ (522–576nm), $^5D_4\rightarrow{}^7F_4$ (577–610 nm), and $^5D_4\rightarrow{}^7F_3$ (611–536 nm) transitions was monitored for samples prepared with different alkaline salts (Fig. 4.26). Mention has to be made that, due to the spectrum complexity, the PL intensity of $YTaO_4$:Tb samples is better evaluated by measuring the peak area in the spectral domains associated with the most relevant electronic transitions.

Lithium sulfate and sodium tetraborate show the best mineralizing action, whereas lithium carbonate seems to hinder the formation of activator–host lattice system and to dramatically damage the luminescence.

According to the literature data, lithium sulfate reacts with the constituent oxides to produce reaction intermediates that are more reactive than the starting oxides, Y_2O_3 and Ta_2O_5 [272]. They combine to give the $YTaO_4$ product and to regenerate the flux. By similitude with the solid state synthesis of non-activated $YTaO_4$ [246], the

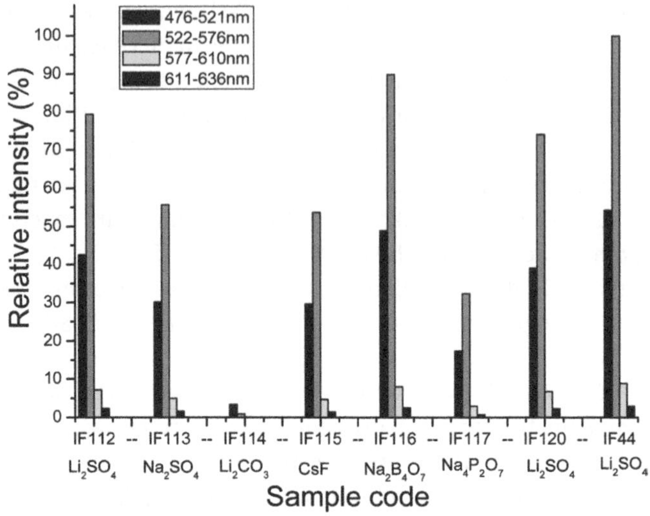

Figure 4.26 Dependence of PL intensity of $YTaO_4$: Tb samples (Series B) on flux materials.

formation of yttrium tantalate doped with terbium (provided by terbium nitrate) can be described by the next equations:

$$(1-x)Y_2O_3 + 2xTb(NO_3)_3 + Li_2SO_4 + Ta_2O_5 \rightarrow 2LiTaO_3 + Y_{2-2x}Tb_{2x}O_2SO_4 + 6NO_2 + 3/2O_2$$

$$Y_{2-2x}Tb_{2x}O_2SO_4 + 2LiTaO_3 \rightarrow Li_2SO_4 + 2Y_{1-x}Tb_xTaO_4$$

Similar equations could be written for the formation of $YTaO_4$:Tb when Tb_4O_7 is used as an activator source. In both cases, lithium sulfate interacts with the raw materials and the oxysulfate and lithium tantalite intermediates produce the terbium–yttrium tantalate luminescent system.

It might also be supposed that sodium tetraborate acts in a similar manner, taking into account its known reactivity against yttrium oxide. Therefore, one can conclude that both flux materials are favorable to the formation of a more ordered activator-host lattice system.

Lithium carbonate shows an unexpected low mineralizing action, suggesting the disturbance of the synthesis mechanism. It is obvious that, Li_2CO_3 flux do not allow the formation of terbium–yttrium tantalate luminescent system.

Samples $YTaO_4$:Tb from Series B prepared from Ta_2O_5 (Alpha Aesar) and $Tb(NO_3)_3$ as an activator source are compared with two powders prepared also with Li_2SO_4 as a flux, but with different raw materials i.e. sample IF44 (internal standard) obtained from Ta_2O_5 (Optipur, Merck) and Tb_4O_7 as well as sample IF120 prepared from Ta_2O_5 (Alpha Aesar) and Tb_4O_7 as terbium source. It can be noted that the synthesis of terbium activated yttrium tantalate is sensitive to the quality of tantalum oxide and terbium oxide seems to be a better source for activator ions than terbium nitrate.

XRD patterns of $YTaO_4$:Tb samples indicate the formation of a single phase materials with pure monoclinic M′ structure (PDF 24-1425) when lithium sulfate (IF112), sodium sulfate (IF113) and sodium tetraborate (IF116) were used as a flux (Fig. 4.27).

Besides the monoclinic M′-YTaO4, different other crystalline phases were identified in Series B, e.g. cubic Y_2O_3 (noted YO/PDF 43-0661), tetragonal YPO4 (YP/ PDF 74-2429), monoclinic Ta_2O_5 (TO/ PDF 33-1391), monoclinic Li_3TaO_4 (LT/PDF 76-2274), and cubic YOF (YOF/PDF 75-0051).

With cesium fluoride (IF115) and sodium pyrophosphate (IF117) as a flux, the secondary products yttrium oxy-fluoride and yttrium phosphate are formed beside the M′-$YTaO_4$ phase. The use of lithium carbonate almost totally destroys the $YTaO_4$ matrix. In this case (IF114), the XRD pattern is dominated by Li_3TaO_4 and Y_2O_3 reflections.

The flux-assisted synthesis of $YTaO_4$ –based phosphors can be monitored by infrared absorption spectroscopy [262]. FTIR spectra sustain the XRD observations referring to the degree of crystalline order in powders prepared with different alkaline salts as a flux (Fig. 4.28).

It is well known that $YTaO_4$ crystallize in two monoclinic polymorphs, that differ from each other by the crystal space group and the cell unit dimensions i.e. M′-$YTaO_4$ [PDF 24-1425; SG *P* 2/*a* (13)] and M-$YTaO_4$ (PDF 24-1415; SG I2/*a* (15)] [185, 269]. FTIR spectroscopy can be used to illustrate the crystalline arrangement in $YTaO_4$ host lattice, based on the characteristic feature of spectra in the 900–400 cm^{-1} range (finger print domain) related with the

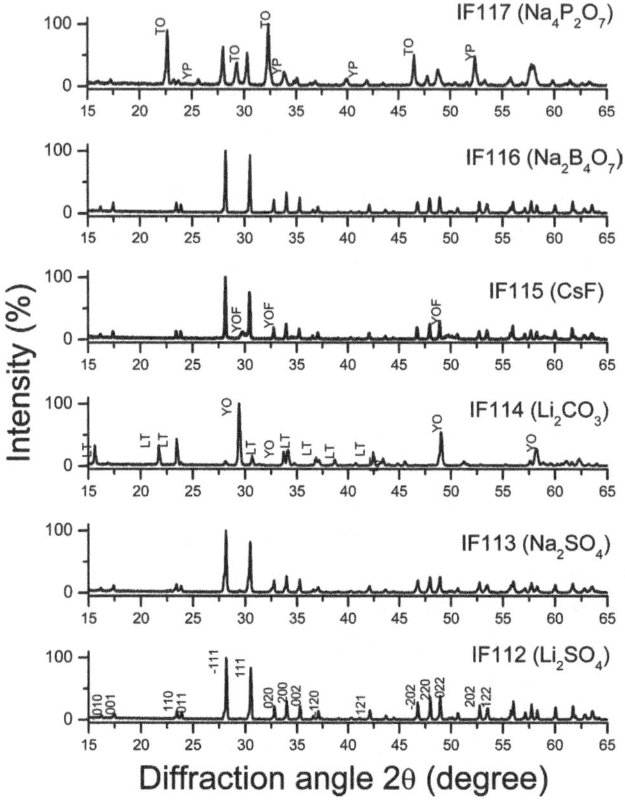

Figure 4.27 XRD patterns of $YTaO_4$: Tb samples prepared with different flux materials.

specific local symmetry. FTIR spectra of the two polymorph differ mainly by bands assigned to ν (Ta–O) and ν (Ta–O–Ta) vibrations that appear at ~812–806 cm^{-1} (medium), ~660–620 cm^{-1} (strong), and ~445–440 cm^{-1} (very strong) in M′ monoclinic phase and 805–800 cm^{-1}(medium), 670–680 cm^{-1} (very strong), and 470–480 (medium) cm^{-1} in M monoclinic arrangement [264]. The complexity of the FTIR patterns is associated with the atomic arrangement formed from more or less distorted TaO_6 units, which build the $YTaO_4$ monoclinic modifications. In M′-type structure, tantalum atoms are in distorted octahedron coordination (six Ta–O bonds), whereas in monoclinic M-type $YTaO_4$ tantalum atoms are in tetrahedral coordination (four Ta-O or bonds) [179].

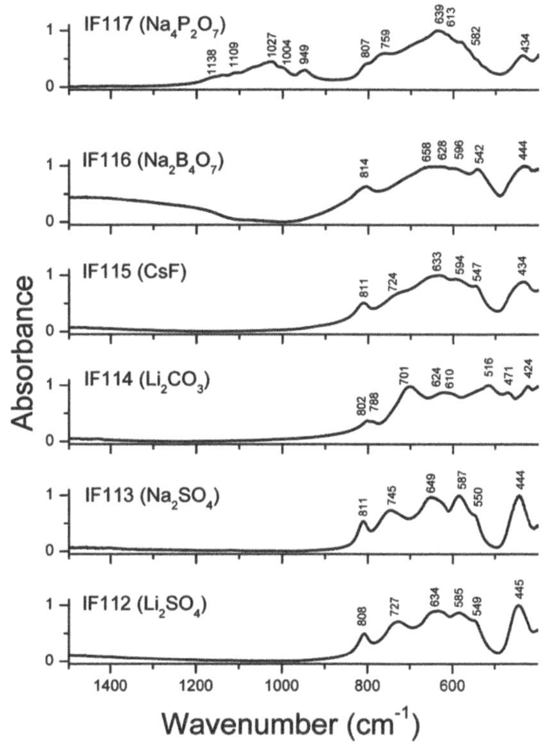

Figure 4.28 FTIR spectra of $YTaO_4$: Tb samples obtained with different flux materials.

Powders prepared with Li_2SO_4 (IF 112) and Na_2SO_4 (IF113) are well-crystallized materials, with M′-structure, whereas $YTaO_4$ matrix is almost completely destroyed with Li_2CO_3 (IF 114) as a flux. Surprisingly, the spectrum of $YTaO_4$:Tb ($Na_2B_4O_7$) is not well resolved, the specific TaO_4 vibration bands being "washed." In this case, it is possible that the powder particles are covered/mixed with vitreous material (residual flux), unobserved in the XRD patterns. FTIR spectra revealed the presence of phosphate impurity (PO_4^{3-} vibrations in the range 1200–900 cm^{-1}) in $YTaO_4$:Tb ($Na_4P_2O_7$), the PO_4 groups being responsible for the modification of $YTaO_4$ finger print domain.

Note also that FTIR spectra contain no specific vibrations of SO_4 structural unit, i.e. the stretching (1140–1080 cm^{-1}) and bending

(680–610 cm^{-1}) vibrations; sulfate intermediates were totally removed during the wash stage.

FTIR spectroscopy completes the structural information obtained from XRD investigation i.e. homogeneous crystalline powders are obtained with Li_2SO_4, Na_2SO_4, or $Na_2B_4O_7$, but a heterogeneous material was isolated with $Na_4P_2O_7$ or Li_2CO_3. In the latter case, no yttrium tantalate was formed, as illustrated by XRD pattern and FTIR spectrum.

Microstructural information was obtained by X-ray profile Fourier analysis. The single [111], [200], [220], and [222] profiles were analyzed by the Warren–Averbach method [276] and processed by XRLINE program [277]. The effective crystallite mean size (D_{eff}) and the root mean square (r.m.s.) of the microstrains $<\varepsilon^2>^{1/2}_{hkl}$ averaged along the [hkl] direction were determined. The unit cell parameters were calculated by Rietveld refinement, using the Powder Cell software [278].

Micro-structural parameters, i.e. the effective crystallite mean size and the root mean square of the microstrains, illustrate that Li_2SO_4, Na_2SO_4, and $Na_2B_4O_7$ prove to be very efficient flux materials, whereas $Na_4P_2O_7$, CsF, and Li_2CO_3 show medium or low mineralizing action (Table 4.5).

The microcrystalline strain is stronger for $YTaO_4$:Tb (Li_2CO_3) and $YTaO_4$:Tb ($Na_4P_2O_7$) samples that are based on high disordered host lattice (due to the secondary phases), constituted from very small crystallites (10–14 nm). In contrast, there are relatively few stacking faults in $YTaO_4$:Tb (Li_2SO_4) and $YTaO_4$:Tb ($Na_2B_4O_7$) crystals that are built from relatively larger crystallites (73–75 nm).

SEM images provide evidence for the efficiency of different mineralizing agents (Fig. 4.29).

Different particle dimensions and shapes are formed when using different flux materials. Irregular polyhedral crystals of 0.5–10 μm are obtained with Li_2SO_4, whereas with sodium tetraborate $Na_2B_4O_7$, regular prismatic particles of ~10 μm length are formed. In addition, small particles of amorphous borate (secondary product) could be noticed. SEM images confirm the supposition (suggested by FTIR spectra) that $YTaO_4$:Tb ($Na_2B_4O_7$) powder contains traces of vitreous material (borax glass). As a matter

Table 4.5 The unit cell and microstructural parameters calculated for M′-$YTaO_4$ phase of powders obtained with variable flux material

Samples	Flux	a [nm]	b [nm]	c [nm]	beta [deg.]	*V* $[nm]^3$	D_{eff} [nm]	r.m.s × 10^3
IF112	Li_2SO_4	0.53019	0.54561	0.51148	96.46	0.14702	73.5	0.576
IF113	Na_2SO9	0.52981	0.54549	0.51115	96.41	0.14680	58.1	0.648
IF114	Li_2CO_3	0.53118	0.54640	0.51289	96.56	0.14791	14.1	4.291
IF115	CsF	0.52958	0.54495	0.51102	96.41	0.14656	68.9	0.862
IF116	$Na_2B_4O_7$	0.53015	0.54582	0.51135	96.45	0.14703	75.1	0.510
IF117	$Na_4P_2O_7$	0.53039	0.55101	0.50936	96.46	0.14791	10.6	4.882

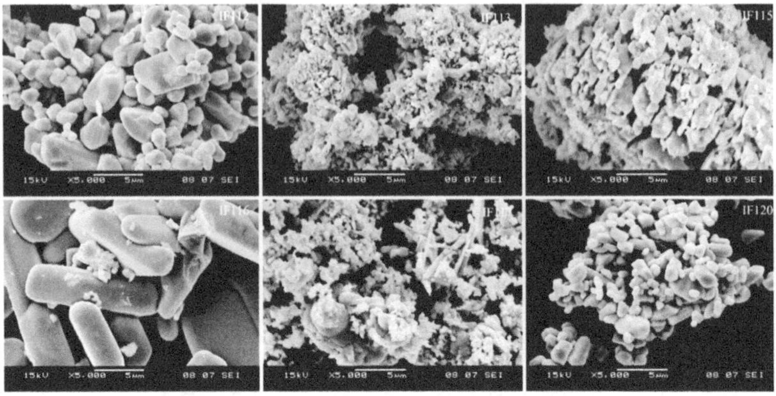

(**Figure 4.29** SEM images of YTaO4:Tb powders (Series B) obtained with different Tb-source and flux materials: IF112 ($Tb(NO_3)_3$, Li_2SO_4); IF113 ($Tb(NO_3)_3$,Na_2SO_4); IF115 ($Tb(NO_3)_3$,CsF); IF116 ($Tb(NO_3)_3$; $Na_2B_4O_7$); IF117 ($Tb(NO_3)_3$;$Na_4P_2O_7$); IF120 (Tb_4O_7; Li_2SO_4).

of fact, the powder quality could be improved by an additional conditioning treatment (ultrasonication in acidic medium).

In spite of the fact that Li_2SO_4, (m.p. = 859°C) and Na_2SO_4, (m.p. = 884°C) possess close melting point values, it can be noted that their behavior as flux is different due to the large difference between the two cation sizes. Lithium (Li^+ = 0.073 nm), that is a smaller cation than sodium (Na^+ = 0.113 nm), is in the favor of the total conversion of oxides into M'-$YTaO_4$ crystalline phase.

XRD patterns, FTIR spectra and SEM images put in evidence that lithium sulfate, sodium sulfate, and sodium tetraborate are efficient mineralizing agents, whereas sodium pyrophosphate, cesium fluoride, and lithium carbonate lead to heterogeneous powders with medium and low PL intensity, respectively.

4.2.3.2 Effect of flux concentration

Lithium sulfate is a reactive flux that directly participates in the solid state reaction. For this reason, Li_2SO_4 amount has to play an important role, not only for activator incorporation, but also for host lattice preparation and crystallization. In order to study the influence of flux amount on morpho-structural and luminescence properties, $Y_{0.95}Tb_{0.05}TaO_4$ samples with 0–60 wt% Li_2SO_4 (Series C) were prepared and investigated.

Although the feature of PL and PLE spectra is not influenced by the flux amount, the photoluminescence intensity strongly depends on Li_2SO_4 concentration. The relative PL intensity (estimated by peak area of band in 522–576 nm range) increases with flux amount to a maximum value (at 50%) and then starts to decrease (Fig. 4.30a).

As expected, the no-flux and 10% flux $Y_{0.95}Tb_{0.05}TaO_4$ samples show relative low green luminescence, probably due to an incomplete formation of the host lattice. High PL intensity is observed for powders prepared with 40–60% Li_2SO_4.

XRD patterns show that, with the exception of the sample without flux, all materials are single phase monoclinic M′-$YTaO_4$:Tb powders. On the basis of XRD data, the microstructural parameters were calculated on two crystallographic directions, namely [111] and [200] planes (Figs. 4.30b and 4.30c).

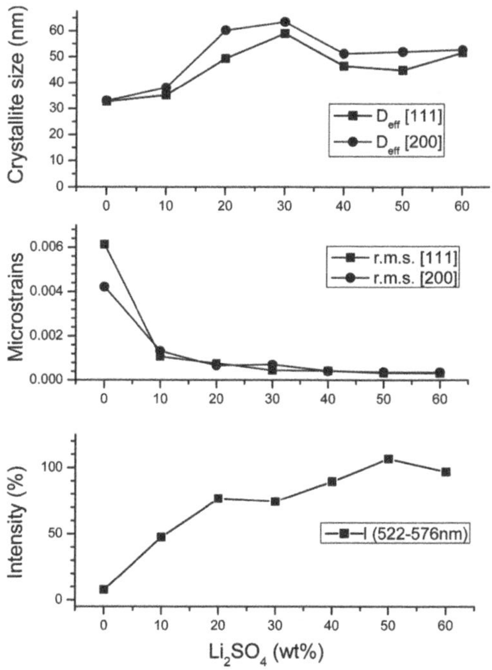

Figure 4.30 Dependence of PL intensity of $YTaO_4$: Tb samples (Series C) and microstructural parameters, effective crystallite size (D_{eff}) and root mean square of the microstrains (r.m.s.), on flux concentration.

Li_2SO_4 concentration determines both the effective crystallite mean size (D_{eff}) and the root mean square (r.m.s.) of the microstrains $<\varepsilon^2>^{1/2}_{hkl}$. The r.m.s. value decreases almost monotonically with Li_2SO_4 amount, whereas the effective crystallite size increases up to 30% flux and then decreases. The $YTaO_4$ matrix of powders prepared with no flux or with 10% Li_2SO_4 show the lowest degree of crystalline order. Moreover, the D_{eff} and r.m.s. values for the same flux concentration, calculated on [111] and [200] reflections, are different.

One can also note that, up to 30% Li_2SO_4, there is a parallelism between the PL intensity and the crystalline order. The increase of the flux amount determines the increase of the crystallite dimensions and the decrease of the lattice microstrain, in parallel with the improvement of the photoluminescence performances. No correlation between the microstructural parameters and PL intensity could be envisaged, over 30% Li_2SO_4.

It is obvious that the crystallization of $YTaO_4$ matrix does not proceed equally alongside the two crystallographic directions. In order to provide supplementary evidence, the crystallite size distribution functions along [111] and [200] crystallographic directions were determined from the second derivative of the strain corrected Fourier coefficients [277].

The crystallite size distribution illustrates the anisotropic growth of $YTaO_4$:Tb crystals curves along the two crystallographic directions (Fig. 4.31). In addition, all distribution curves are multimodal suggesting a complex growth mechanism.

For samples prepared without flux, two distribution maxima could be noticed at 29 and 41 nm (D1 curves) and 33 nm and 45 nm (D2 curves). The use of small Li_2SO_4 amount (10%) shifts the distribution maxima toward larger values; the shift is higher alongside [200] crystallographic direction. Crystallization in the presence of 20–30% Li_2SO_4 proceeds with a considerable increase of the crystallite dimensions (relative to sample with 10% flux) that finally determines a new maximum to appear. With 40–60% Li_2SO_4, almost identical trimodal crystallite distributions are obtained. Alongside [111] direction, distribution curve has three maxima at about 32 nm, 45 nm, and 60 nm, the highest probability being in the small crystallite domain. Similar results are obtained alongside

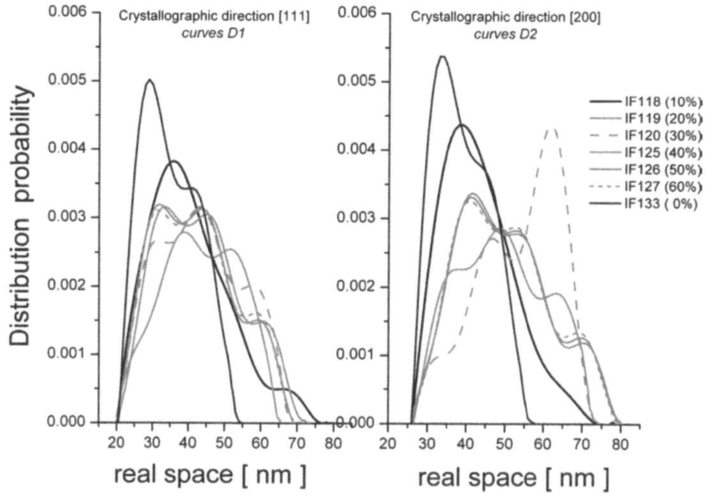

Figure 4.31 Crystallite size distribution calculated along [111] and [200] crystallographic directions for $YTaO_4$: Tb powders (Series C).

[200] direction, but the probability maxima are situated at larger crystallite dimensions (41 nm, 53 nm, and 70 nm).

The crystallite distribution curves sustain the supposition that, during the Li_2SO_4-assisted synthesis, the crystal growth proceeds in an anisotropic manner.

The effect of lithium sulfate amount on phosphor formation is illustrated also by FTIR spectra. In samples without or with 10% flux, the raw material conversion is incomplete and the monoclinic $YTaO_4$ matrix is partially formed (Fig. 4.32). The powder crystallinity is improved by higher flux amount (20%). Similar to the crystallite distribution curves, FTIR spectra allow no differentiation between the samples with 40–60% Li_2SO_4.

The heterogeneous powder of $YTaO_4$:Tb obtained without flux consists from irregular conglomerates of particles that probably incorporate unreacted raw materials (Fig. 4.33).

Yttrium tantalate matrix is already formed with 10% Li_2SO_4, but the particles are irregular in shape. The powder becomes homogeneous with 20–30% Li_2SO_4 when $YTaO_4$:Tb sample consists mostly in small polyhedral particles (1–3 μm). Higher Li_2SO_4 amount determines the increase of particle dimensions (2–5 μm), with a loss in homogeneity. $YTaO_4$:Tb (40–60% Li_2SO_4) samples are a mixture

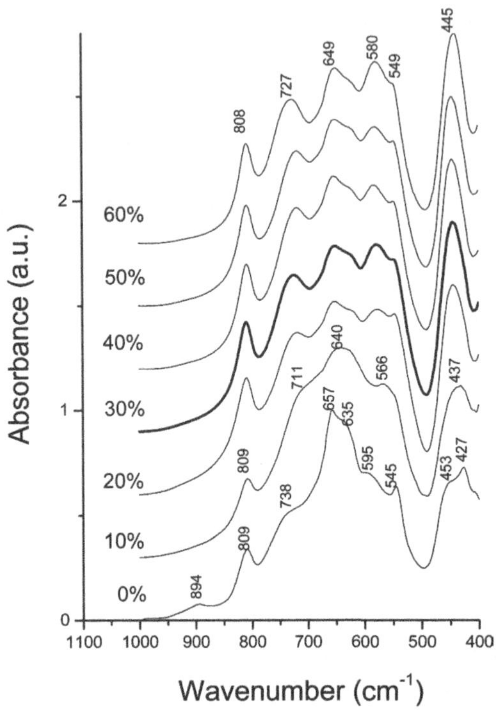

Figure 4.32 FTIR spectra of $YTaO_4$:Tb powders (Series C) obtained with variable Li_2SO_4 amount.

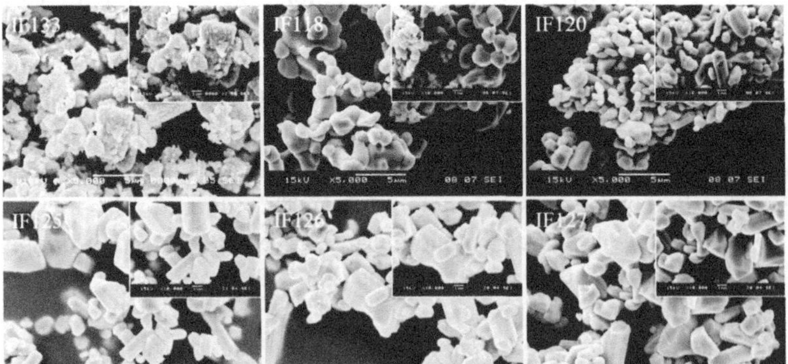

Figure 4.33 SEM images of $YTaO_4$:Tb powders (Series C) obtained with variable amounts of Li_2SO_4: IF133 – no flux; IF118 – 10%; IF 120 – 30%; IF 125 – 40%; IF126 – 50% ;IF 127– 60%.

of small polyhedral particles with elongated crystals. It seems that the anisotropic growth of the crystallites induces a tendency of similar growth at crystals level. At over 30% flux, larger particles are built from smaller crystallites, with relatively smaller number of microstrains, alongside both directions.

4.2.3.3 Effect of terbium concentration

Gradual substitution of yttrium with terbium into the monoclinic crystalline lattice of $YTaO_4$ can determine a more or less monotonous variation of phosphor properties. In order to study the influence of activator amount on luminescence and morpho-structural properties, $Y_{1-x}Tb_xTaO_4$ samples with 0.5–15 mol% terbium (Series D) were prepared and investigated (Fig. 4.34).

Although the feature of PL and PLE spectra is not influenced by the activator amount, the photoluminescence intensity strongly depends on terbium concentration. As expected, in parallel with the increase of the number of emission centers, the PL intensity of the emission bands (estimated by the peak area in the well-delimited spectral range) tends to increase with Tb amount to a maximum value and then to decrease slowly (concentration quenching). However, an abnormal dependence of PL intensity on Tb concentration can be noticed for some of the different PL spectral bands (Fig. 4.34a).

It is well known that the luminescence color of some terbium doped X-ray phosphors (i.e. Y_2O_2S:Tb, Gd_2O_2S:Tb, La_2O_2S:Tb) can be tuned from blue to green by using small or relatively high Tb amounts, respectively [279].

In our case, the balance between the UV-blue (365–475 nm) and the green-orange bands (476–630 nm) determines that all $YTaO_4$:Tb samples are green emitting powders.

The most intense blue contribution is observed when 0.5% from yttrium positions are substituted by terbium. But even at this favorable UV-blue/green-yellow ratio, the blue emission is 1/10 of the specific green terbium band (522–576 nm). In the 365–475 nm domain, the PL dependence is normal, the maximum blue band value being at 0.5 mol% terbium.

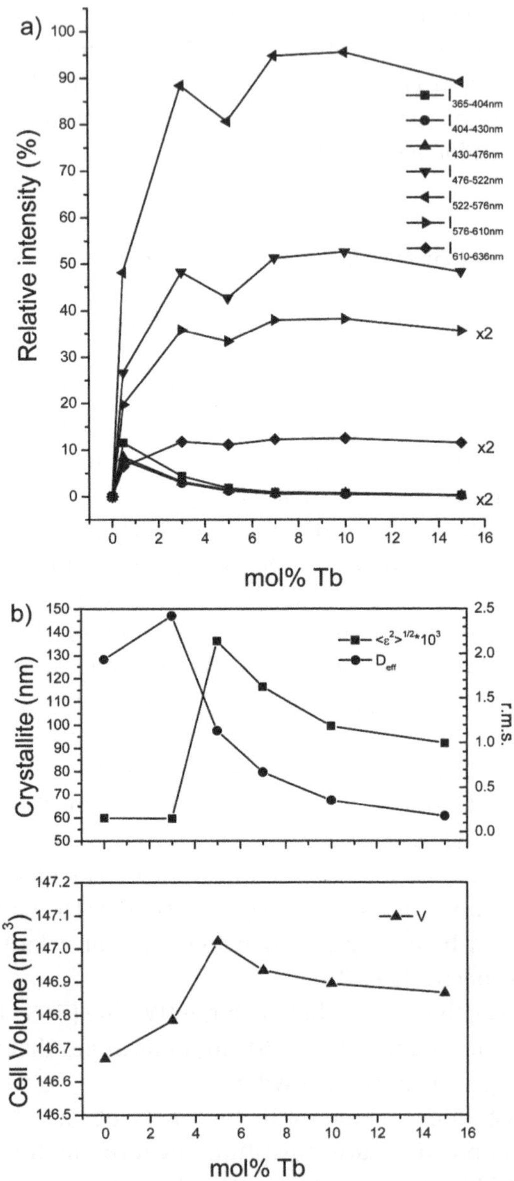

Figure 4.34 (a) Dependence of PL intensity and (b) microstructural and unit cell parameters of $YTaO_4$:Tb(x%) samples on terbium concentration.

The abnormal growth of PL intensity with Tb concentration could be noticed for PL bands situated at relatively higher wavelengths (>476 nm) where the 5 mol% Tb seems to be a critical activator concentration. In this domain, the most intense green PL is generated at 7–10 mol% Tb.

The observed anomaly suggests an unhomogeneous incorporation of terbium into the monoclinic $YTaO_4$ crystalline lattice. Mention has to be made that, the effective Tb concentration was determined in all $Y_{1-x}Tb_xTaO_4$ powders (using ICP optical emission spectroscopy), and no composition deviation was observed in this series of samples.

XRD patterns show that all powders possess the monoclinic $YTaO_4$:Tb structure (PDF 24-1425). The microstructural and the unit cell parameters were calculated (as already described above) and they were depicted versus terbium concentration (Fig. 4.34b).

The accommodation of Tb^{3+} into the monoclinic $YTaO_4$ crystalline lattice brings about a small increase of the unit cell volume from 146.67 nm^3 (no Tb) to the maximum value 147.02 nm^3 (5 mol% Tb) and then a monotonous decrease down to 146.87 nm^3 (15 mol% Tb). This non-monotonous variation indicates that, in spite of the relatively larger ionic radii of Tb^{3+}, i.e. 0.104 nm compared with 0.102 nm of Y^{3+}, the unit cell volume becomes smaller over 5 mol% terbium, thus suggesting a more disordered atom packing in terbium-doped $YTaO_4$:Tb lattice.

The microstructural parameters, i.e. the effective crystallite mean size (D_{eff}) and the root mean square (r.m.s.) of the microstrains $<\varepsilon^2>^{1/2}_{hkl}$ sustain this supposition.

The effective crystallite size slowly increases from ~130 nm (no Tb) up to ~150 nm (3 mol% Tb), and then decreases monotonically down to ~ 60 nm (15 mol% Tb). It seems that, during the thermal synthesis stage, Tb_4O_7 acts either as a weak mineralizing agent (in small amount), or as a dispersing agent (in relatively higher amounts).

From the r.m.s. point of view, the best terbium accommodation proceeds for 3 mol% Tb where the $YTaO_4$ unit cell remains not disturbed ($<\varepsilon^2>^{1/2}_{hkl}$ = 0.00015). The r.m.s. value increases up to 0.0021 (5 mol% Tb) and then decreases almost monotonically down to 0.0099 (15 mol% Tb).

When introduced as Tb_4O_7, terbium amount determines the effective crystallite mean size and the microstrains concentration as well as the unit cell volume, with consequences on the PL performances. The anomaly on the PL versus mol% Tb curve could be explained by a relatively higher microstrains amount at 5 mol% activator.

Structural investigations pointed up that Li_2SO_4, Na_2SO_4, and $Na_2B_4O_7$ flux materials produce homogeneous crystalline powders, whereas $Na_4P_2O_7$ and CsF form heterogeneous ones. The most effective mineralizing agents are Li_2SO_4 and $Na_2B_4O_7$ that lead to single phase powders with monoclinic M'$YTaO_4$:Tb structure and superior PL properties.

Photoluminescence and morpho-structural properties of $YTaO_4$:Tb strongly depend on flux concentration and terbium amount. The dependence of microstructural parameters on Li_2SO_4 amount provides evidence that flux determines an anisotropic crystal growth and influences not only the crystals microstructure, but also the concentration of Tb centers. As expected, the PL intensity tends to increase with terbium amount to a maximum value (in parallel with the number of emission centers) and then to decrease slowly (concentration quenching). However, a small disturbance was observed at 5 mol% Tb that could be explained with the local structural disorder, as indicated by the micro-structural parameters. The maximum PL intensity is shown by $YTaO_4$:10%Tb powder prepared with 50% Li_2SO_4 as a flux.

One can conclude that, using well-defined amounts of Tb_4O_7 as activator source and Li_2SO_4 or $Na_2B_4O_7$ as a flux, homogeneous powders of green emitting $YTaO_4$:Tb can be obtained. Flux materials participate in many competitive processes such as formation of $YTaO_4$ matrix, organization of the crystalline lattice or activator incorporation. Therefore, supplementary work has to be done referring to incorporation of terbium ions into the yttrium tantalate crystalline lattice.

4.2.3.4 Effect of niobium addition

Terbium activated yttrium niobium–tantalate phosphors general formula $Y_{0.95}Tb_{0.05}Ta_{1-x}Nb_xO_4$ ($x = 0 \div 1.00$) were prepared by solid state reaction from homogeneous mixtures of Y_2O_3 (99.95% Alfa Aesar), Ta_2O_5 (Optipur, Merck), and/or Nb_2O_5 (Optipur, Merck), for

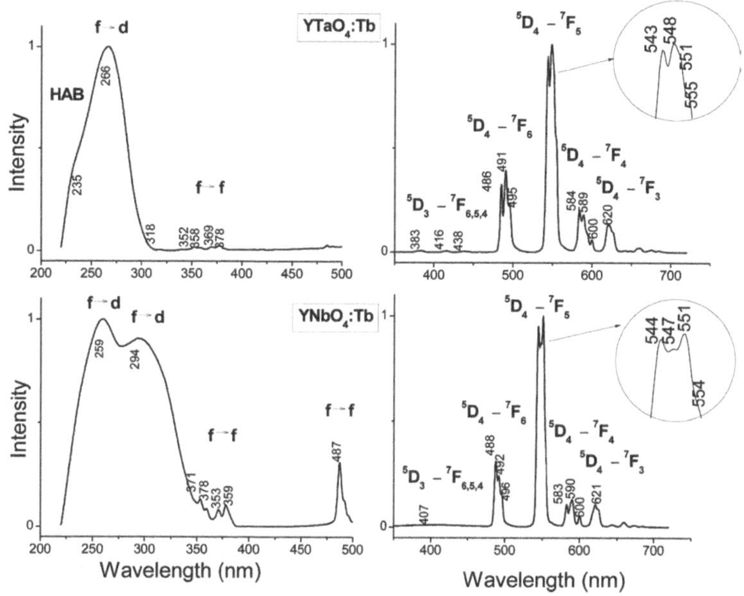

Figure 4.35 Normalized PLE (left; λ_{em}= 549 nm) and PL (right; λ_{ex}= 254 nm) spectra of $YTaO_4$:Tb and $YNbO_4$:Tb samples.

host lattice formation, Tb_4O_7 (99.9%, Aldrich) as activator and Li_2SO_4 as a flux. Stoichiometric amounts of oxides and flux were ball-milled, annealed at 1200°C and, finally, water washed, dried, and sieved. Formation of Y(Ta,Nb)O_4:Tb proceeds by isovalent substitution of tantalum with niobium, and yttrium with terbium during the thermal synthesis that determines the luminescence and structural properties.

Photoluminescence excitation and emission spectra of $YTaO_4$:Tb and $YNbO_4$:Tb are depicted in Fig. 4.35. PL spectra of Y(Ta,Nb)O_4:Tb samples are very similar, illustrating the presence of Tb^{3+} centers that are responsible for the characteristic green luminescence, associated with $^5D_4 \rightarrow {}^7F_J$ electronic transitions [261]. Multiple narrow bands could be noticed in the green region. The strongest peaks from the $^5D_4 \rightarrow {}^7F_5$ group show different ratios for $YTaO_4$:Tb and $YNbO_4$:Tb (Fig. 4.35).

For the Tb concentration in use, the broad emission band of the host lattice (TaO_4/NbO_4) is not so evident. Nevertheless, a small

PL contribution of the matrix can be noticed in the blue domain, for $YNbO_4$:Tb. Moreover, the blue emission bands associated with $^5D_3 \rightarrow ^7F_J$ transitions are very weak.

PLE spectra of $YTaO_4$:Tb and $YNbO_4$:Tb are different and rather complicated (Fig. 4.35). For $YTaO_4$:Tb, the various electronic transitions can be identified such as host-charge transfer (O→Ta) related to the host absorption band, inter-configuration ($4f^8 \rightarrow 4f^7 5d^1$) and intra-configuration ($4f^8 \rightarrow 4f^8$) transitions, associated with the intra-atomic excitation bands. In $YNbO_4$:Tb, the contribution of the host lattice is very weak and Tb is directly excited.

Niobium addition induces a non-monotonic change of spectral features and a strong PL decrease (Fig. 4.36); 30–40% Nb content seems to be the PL transition point.

The relative intensity of PL emission versus Nb amount is depicted in Fig. 4.37. Emission was monitored for the band groups associated with $^5D_4 \rightarrow ^7F_6$ (474–520 nm), $^5D_4 \rightarrow ^7F_5$ (527–576 nm), $^5D_4 \rightarrow ^7F_4$ (577–605 nm) and $^5D_4 \rightarrow ^7F_3$ (613–635 nm) transitions. Excitation was monitored for two representative bands in the 220–350 nm and 475–500 nm domains, associated with HAB and $f \rightarrow d$ and $f \rightarrow f$ transitions. The intensity of all PL bands decreases with Nb increase toward a minimum value (at 30–40% Nb) and then it slightly increases, in parallel with the PLE bands that are directly

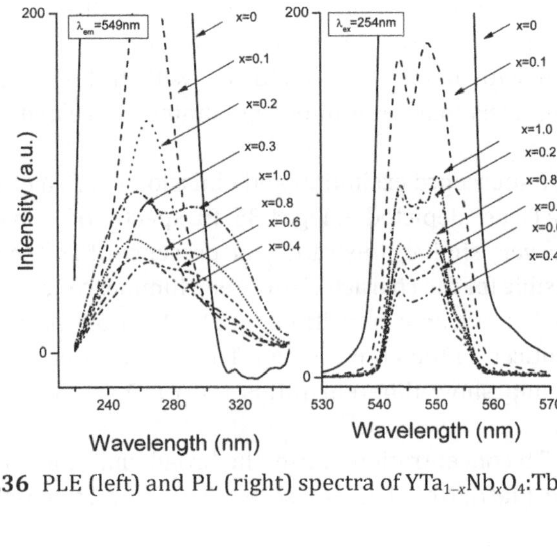

Figure 4.36 PLE (left) and PL (right) spectra of $YTa_{1-x}Nb_xO_4$:Tb samples.

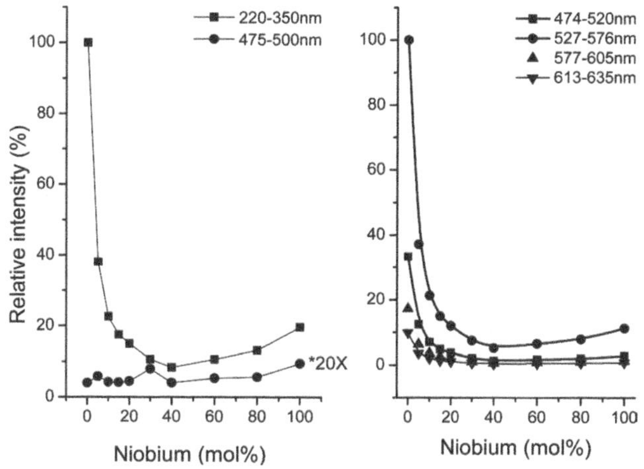

Figure 4.37 Relative intensity (peak area) of the most relevant PLE (left) and PL (right) bands versus niobium amount.

(HAB) or indirectly (*f*→*d*) associated with Tb surroundings. Intra-ion excitation (*f*→*f*) increases slowly at 30 mol% Nb.

UV excitation of Tb ion is efficient only if the activator is incorporated into the $YTaO_4$ crystalline lattice, when the energy transfer from the host lattice to the activator is very active (Fig. 4.37). Quite to the the contrary, $YNbO_4$ is not a good host lattice for Tb ions.

Crystalline structure could explain the PL properties of $Y_{0.95}Tb_{0.05}Ta_{1-x}Nb_{s}O_4$ phosphors. The phase compositions as well as the microstructural parameters were calculated on the basis of XRD patterns. All samples are powders with a monoclinic structure, in agreement with previous works [262].

Unit cell parameters were calculated through Rietveld refinement using the Powder Cell software that inputs structural data, creates models, and calculates diffraction patterns [278]. Structural parameters are fitted until the calculated patterns match the experimental ones (interactive refinement).

Previous works showed that $YTaO_4$ (Li_2SO_4; 1200°C) with no activator possesses M′ monoclinic structure, whereas $YNbO_4$ (Li_2SO_4;1200°C) have the fergusonite M monoclinic structure [262]. Therefore, for investigated series, the monoclinic M′-$YTaO_4$ (PDF 24-1425) and monoclinic M-$YNbO_4$ (PDF 83-1319) were used for input data.

Table 4.6 Structural parameters of $Y_{0.95}Tb_{0.05}Ta_{1-x}Nb_xO_4$ samples.

		Cell unit				Microstructure	
Nb x	Phase	A (nm)	b (nm)	c (nm)	β (°)	D_{eff} (nm)	r.m.s. (×1000)
0	M′	0.5302	0.5455	0.5115	96.47	124.5	0.367
0.05	M′	0.5302	0.5453	0.5114	96.45	73.87	0.825
0.10	M′	0.5301	0.5450	0.5112	96.44	65.62	0.263
0.15	M′	0.5303	0.5451	0.5114	96.43	28.23	1.786
0.20	M′	0.5301	0.5450	0.5112	96.43	65.84	0.616
0.30	M	0.7676	1.0940	0.5324	138.98	43.55	0.876
0.40	M	0.7663	1.0929	0.5316	138.94	34.44	1.551
0.60	M	0.7656	1.0942	0.5317	138.83	28.19	3.398
0.80	M	0.7637	1.0946	0.5306	138.62	57.48	0.778
1.00	M	0.7616	1.0942	0.5296	138.44	41.98	1.530

All phosphors are homogeneous, single phase materials with M′ (0–20% Nb) or M (30–100%Nb) monoclinic structure (Table 4.6). This is contrary to the literature data reported for no-activator $Y(Ta,Nb)O_4$ phosphors, when the mixed M + M′ structure was identified [185].

In above mentioned conditions, two structural groups were obtained, the M′-to-M transition occurring at approximately 30% Nb. The cell unit parameters are almost equal inside the groups, due to the similar Ta and Nb dimensions (0.78 nm, for six-fold coordination). Samples with 0.05–20% Nb could be considered as Nb-rich M′-$YTaO_4$ (unit cell volume $V = 0.147$ nm^3), whereas samples with 30–80% Nb are Ta-rich M-$YNbO_4$ phase (unit cell volume $V = 0.293$ nm^3).

The microstructural information was obtained by X-ray profile Fourier analysis. The single [111], [200], [220], and [222] profiles were analyzed by the Warren–Averbach method [276] and processed by XRLINE program [277]. The effective crystallite mean size, D_{eff} (nm) and the root mean square (r.m.s.) of the microstrains averaged along the [hkl] direction, $<\varepsilon^2>^{1/2}_{hkl}$ were calculated.

The addition of only 5 mol% Nb produces the decrease with 50% of particle dimensions and, correspondingly, the increase of the microstructural stress. No regularity in the variation of the microstructure parameters with niobium amount could be defined.

The major difference between the two monoclinic phases arises in the coordination of Y and Ta atoms. Tantalum atoms are situated in a distorted octahedral coordination, in M′-$YTaO_4$ and in tetrahedral coordination, in M-$YTaO_4$. Moreover, the unit cell volume of M′ phase is approximately half of the M phase and the average Ta–O distances are correspondingly smaller. As a result, the crystalline field in M′-$YTaO_4$ crystals is in the favor of a more efficient energy transfer that determines superior luminescence. The two monoclinic M and M′ phases are organized from similar building blocks, stacked in a different sequence [185].

The slight modification of the crystal field strength at ~30% Nb (the point for the disruptive structural change) could explain the PL spectra feature. Moreover, the M′ and M transformation requires the motion of large blocks during the high-temperature re-crystallization that accompanies the phosphor formation. The apparent irregular microstructural parameters could be the result of these competitive processes.

Morphology and particle size of Y(Ta,Nb)O_4:Tb powders were investigated by scanning electronic microscopy. SEM pictures provide evidence of the disruptive change of the crystalline arrangement at about 30% Nb (Fig. 4.38). Sample $YTaO_4$:Tb (IF 44) is a mixture of small, irregular polyhedral particles (1–4 μm) with larger prismatic ones (5/10 μm). This morphology is preserved up to 20% Nb (IF40). A drastic change is observed in sample with 30% Nb (IF 56) where the majority of particles are multifaceted, with larger dimensions (2–10 μm). Sample $YNbO_4$:Tb (IF47) is the most homogeneous material, being formed from small, regular, almost spherical particles of 4–6 μm. The increase of Nb amount improves the powder homogeneity.

In conclusion, it should be noted that the color characteristics and the relative ratio between the multiple PL peaks are determined by the host lattice composition, degree of crystalline order and particle

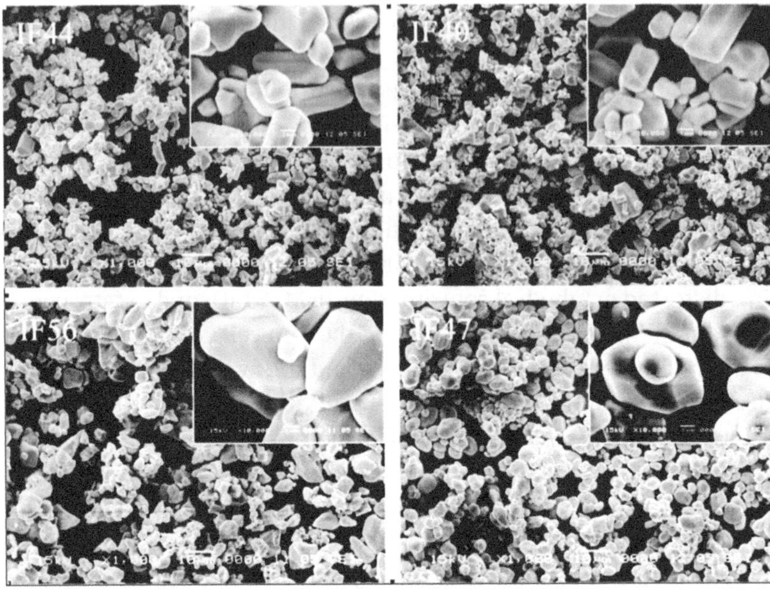

Figure 4.38 SEM Images of some $Y(Ta,Nb)O_4$:Tb samples with 0% Nb (IF44), 20% Nb (IF40), 30% Nb (IF56). and 100% Nb (IF47).

morphology and size. Nb addition determines the strong decrease of $YTaO_4$:Tb photoluminescence. The $Y(Ta,Nb)O_4$:Tb phosphors are single phase materials with M′ and M structure, respectively. A dramatic change of almost all properties is noticed at 30% Nb (the approximate transition point). The phosphors form two groups with different structural parameters and particles morphologies and sizes.

Due to their high X-ray absorption capability, efficient green luminescence and good powder properties, $YTaO_4$:Tb phosphor is a promising scintillating material.

4.2.4 *Europium Activated $Y(Ta,Nb)O_4$ Phosphors*

The general luminescence of Eu^{3+} in tantalate–niobate phosphors has been reported previously [12, 20, 250–252]. Some recent publications with activation by Eu^{3+} have appeared in the last years [179, 256–258, 260, 262, 280, 281]. To the best of our knowledge, no systematic investigation about the influence of the yttrium tantalate–niobate matrix on both the luminescent and structural properties of Eu-doped phosphors has been performed so far. The main goal of this contribution is to elucidate some physical and chemical aspects related to the influence of the host lattice composition and crystalline order on the luminescence of Eu activated Y(Ta,Nb)O_4 phosphors. Photoluminescence spectroscopy, X-ray diffraction, FTIR spectroscopy, and SEM have been used to investigate the structural and luminescent properties of Y(Ta,Nb)O_4:Eu^{3+} phosphors. The correlation between the PL performances and the powder morphological and structural characteristics is attempted in order to evaluate the transfer and transport of energy through the $YTa_{1-x}Nb_xO_4$ lattice. The interaction between the tantalate–niobate matrix and the Eu-emission centers is considered in correlation with the PL characteristics.

Sample preparation was performed by ceramic method. Stoichiometric mixtures of Y_2O_3 (99.99% SIGMA-ALDRICH), Ta_2O_5 (OPTIPUR Merck) and Nb_2O_5 (OPTIPUR Merck) as host lattice generator, Eu_2O_3 (99.99% SIGMA-ALDRICH) as activator source (activated with 5 mol%) and Na_2SO_4 or Li_2SO_4 as a flux were ball-milled with acetone. After drying and firing at 1200°C for 4 h, the powders were water-washed, dried, and sieved.

Samples characterization was performed by photoluminescence measurements (JASCO FP-6500 Spectrofluorimeter Wavel; Glass filter WG 320; commercial phosphors as calibration standard), X-ray diffraction (D8 Advanced Brucker Diffractometer, Cu-Kα radiation; Ni filtered), FTIR spectroscopy (JASCO 610 Spectrometer; KBr pellets), and scanning electron microscopy (JEOL–JSM 5510LV Microscope; Au coated samples). Photoluminescence performances were compared to Y_2O_3:Eu^{3+} (KEMIRA) for Eu-doped powders. XRD quantitative phase analysis was performed using the Powder Cell program for Windows. The crystalline phases were retrieved after a peak search semi-automatic routine using the data base PDF-2.

Two samples series were prepared by solid state reaction, namely $YTaO_4$, $YTa_{0.85}Nb_{0.15}O_4$, and $YNbO_4$ phosphors, with Na_2SO_4 as a flux, and $YTa_{1-x}Nb_xO_4$ phosphors with Li_2SO_4 as a flux (where $x = 0.05–1$). Flux allows the control of particle size and assists the formation of the yttrium tantalate-based phosphors. During the thermal synthesis stage, the lithium/sulfate flux reacts with the oxide mixture and produces intermediate compounds that are more reactive than the starting oxides. Formation of $YTaO_4$ based phosphors takes place by the isovalent substitution of tantalum and yttrium with niobium and europium, respectively.

$$(1-x)Y_2O_3 + xEu_2O_3 + (1-y)Ta_2O_5 + yNb_2O_5 \rightarrow \frac{Na_2SO_4}{Li_2SO_4}$$
$$\rightarrow 2\ Y_{1-x}Eu_xTa_{1-y}Nb_yO_4$$

The influence of the host lattice composition and flux on photoluminescence properties and morphological and structural characteristics is investigated.

4.2.4.1 Photoluminescence properties

The UV-blue luminescence of tantalate-type phosphors is due to a charge transfer transition into the TaO_4 and/or NbO_4 group. Under UV excitation, $YTaO_4$ with no activator shows weak luminescence at about 330–350 nm [20, 246]. When tantalum ions are substituted by niobium, the sensitivity of the host lattice to UV excitation increases and the luminescence becomes significant in the blue spectral region. When Eu^{3+} is used to partially replace the yttrium ions into the host crystalline lattice, the luminescence is shifted toward longer wavelengths. The Eu centers are responsible for the characteristic red luminescence, associated with the specific electronic transitions $^5D_0 \rightarrow {^7F_J}$. The excitation takes place essentially by charge transfer from O^{2-} to Eu^{3+} and to a less extent by direct excitation of Eu^{3+} species. The photoluminescence emission spectrum shows the characteristic Eu-red band centered at about 613 nm. The photoluminescence excitation spectrum monitored for 613 nm emission contains the characteristic charge-transfer band located at about 256 nm beside the less intense excitation bands associated with $^7F_0 \rightarrow {^5D_J}$ transitions (Fig. 4.39.)

Moreover, UV excitation of Eu ions is efficient when they are incorporated into a well-organized crystalline lattice (Fig. 4.40.).

The energy transfer from the host lattice to activator is very efficient when the thermal synthesis is performed with Li_2SO_4 as a flux, as illustrated by the higher (~20%) intensity values. This is in

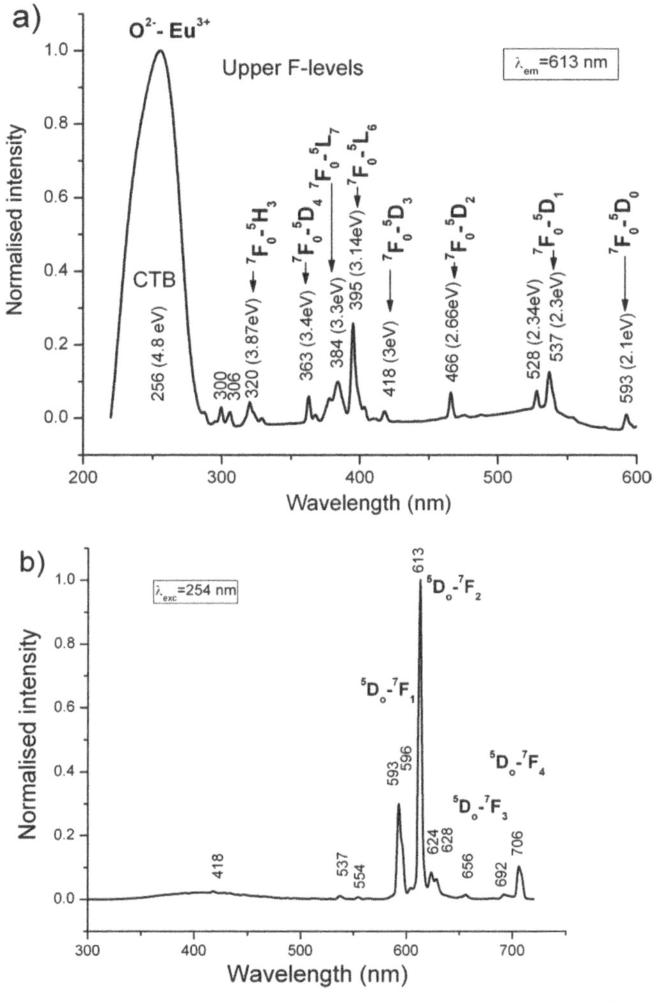

Figure 4.39 PLE (a) and PL (b) spectra of $YTa_{0.85}Nb_{0.15}O_4$: Eu^{3+} (Li_2SO_4) sample.

agreement with other works that shows that Li_2SO_4 is a more efficient mineralizing agent than the corresponding sodium salt [262].

Composition of the host lattice influences the ratio between the different Eu-emission bands (Fig. 4.41).

In the blue domain, the weak Eu bands correlated with the higher level emission $^5D_J \rightarrow {}^7F_J$ ($J > 0$) are superimposed over the broad emission band of the host lattice. In this spectral domain, there is a

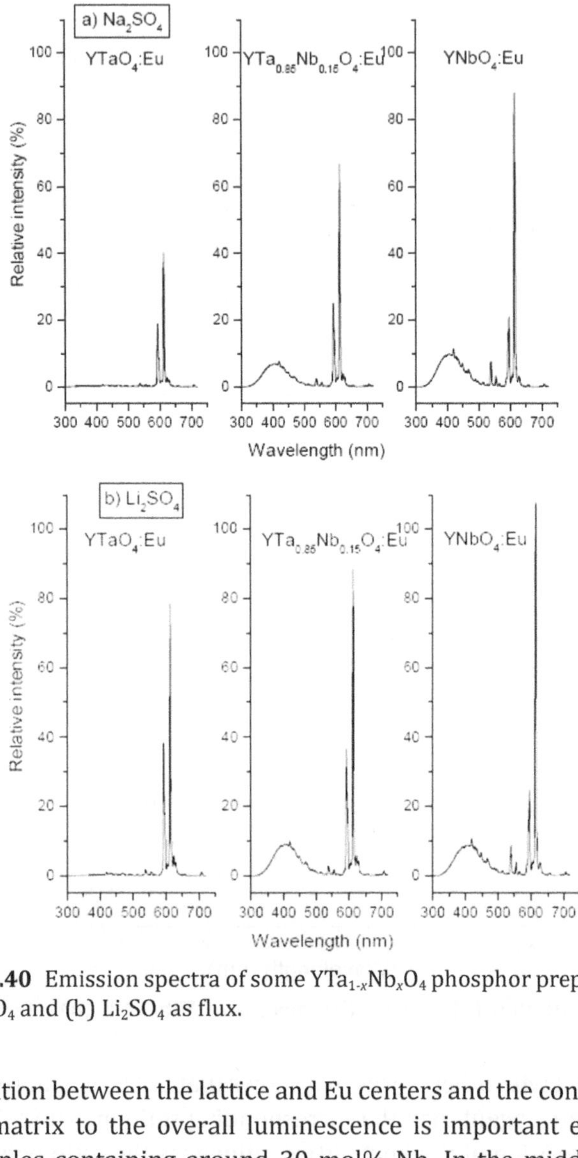

Figure 4.40 Emission spectra of some $YTa_{1-x}Nb_xO_4$ phosphor prepared with (a) Na_2SO_4 and (b) Li_2SO_4 as flux.

competition between the lattice and Eu centers and the contribution of the matrix to the overall luminescence is important especially for samples containing around 30 mol% Nb. In the middle of the $YTaO_4$–$YNbO_4$ series, the energy migration through the lattice seems to be disturbed by the disordered matrix. The most important lattice effect is observed in the orange-red spectral domain, particularly for the specific Eu-bands associated with $^5D_0 \rightarrow {}^7F_1$ or $^5D_0 \rightarrow {}^7F_2$ transitions

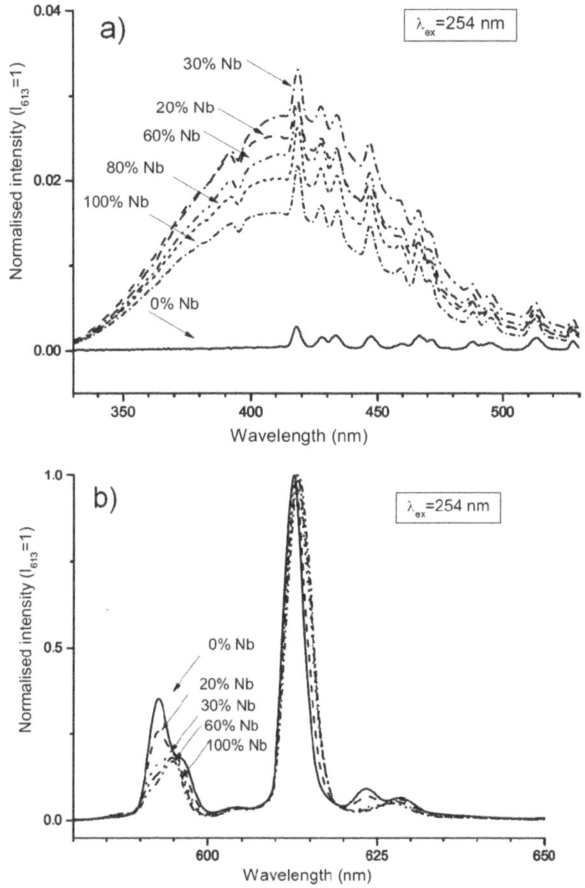

Figure 4.41 Blue (a) and red-orange domain (b) of the PL spectra of some $YTa_{1-x}Nb_xO_4$:Eu^{3+} (Li_2SO_4) samples.

(Fig. 4.41). This is the reason why, under UV exposure, the PL color of Eu-doped samples is so different along this series, varying from yellow to orange to pink.

The dependence of the relative intensity of the PL main bands in Y(TaNb)O$_4$:Eu sample series is illustrated in Fig. 4.42.

Alongside the Y(TaNb)O$_4$:Eu^{3+} sample series, the ratio between 593 and 596 nm bands as well as between 623 and 628 nm is suddenly changing at about 25 mol% Nb. The intensity of the characteristic Eu band at 613 nm increases with the niobium concentration, thus

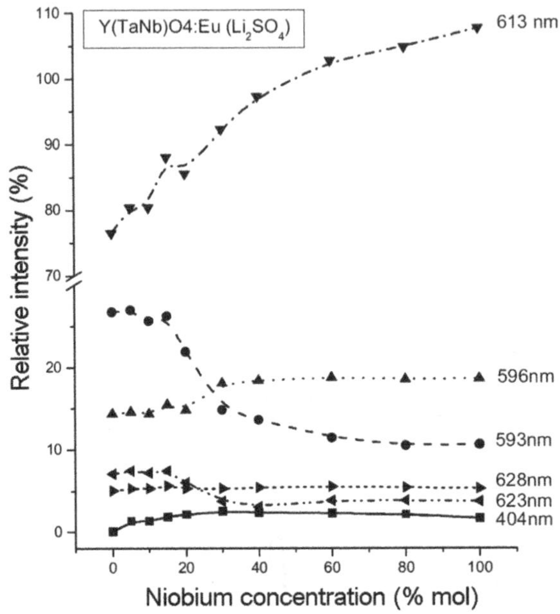

Figure 4.42 Relative intensity of Eu bands versus Nb amount.

illustrating a more efficient energy transfer from NbO_4 groups. Moreover, the change of the ratio between the emission bands corresponding to $^5D_0 \rightarrow ^7F_1$ magnetic-dipole transition and $^5D_0 \rightarrow ^7F_2$ electric-dipole transition illustrates the decrease of the symmetry of the Eu^{3+} surrounding, in parallel with the increasing of the niobium content, probable related with the crystalline phase composition [258].

4.2.4.2 Crystalline structure

XRD patterns support the idea that Li_2SO_4 is more efficient as a flux as the corresponding sodium salt. $YTaO_4$ (Na_2SO_4) phosphor is a multiple phase material containing monoclinic M′-type crystals and orthorhombic O-type Y_3TaO_7 phase, whereas $YTaO_4$ (Li_2SO_4) is a single M′-phase powder (Fig. 4.43). As for the $YNbO_4$ samples, they are single phase with monoclinic M-type crystalline structure; the lithium salt improves the powder's crystallinity.

It is well known that $YTaO_4$ crystallize in two monoclinic polymorphs M and M′ that differ from each other by the cell unit

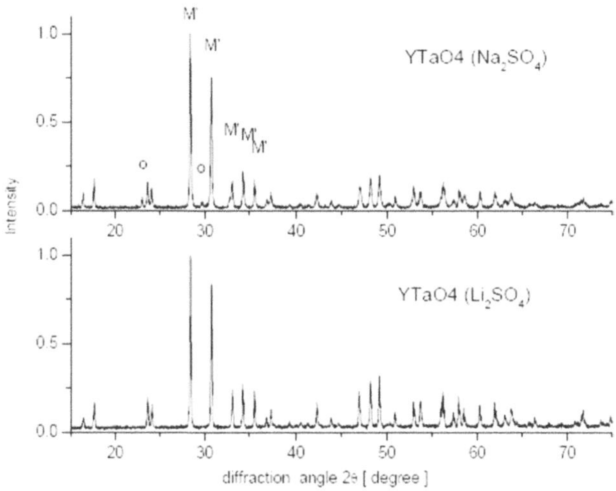

Figure 4.43 X-ray patterns of some $YTaO_4$ samples.

dimensions while $YNbO_4$ adopts only the monoclinic M type. XRD analysis shows that almost all phosphors from the $YTaO_4$–$YNbO_4$ system are M and M′ polyphasic powders. Actually, only small amounts of niobate (<15%) are accommodated into the M′-$YNbO_4$ matrix and vice versa (Table 4.7).

According to the XRD quantitative phase analysis, the M$YNbO_4$ phase is well observed in $YTa_{0.85}Nb_{0.15}O_4$ powder (14.6 vol%) and becomes predominant in $YTa_{0.60}Nb_{0.40}O_4$ powder (55.8 vol%). The phosphor heterogeneity disturbs the energy transfer and migration through the matrix and acts on the PL performances.

FTIR spectra confirm that the crystalline arrangement of YTa_{1-x}, Nb_xO_4 samples is different. The complexity of the FTIR patterns is associated with the atomic arrangement formed from more or less distorted TaO_6 units, which build the $YTaO_4$ monoclinic modifications. In M′-type structure, tantalum atoms are in distorted octahedron coordination (six Ta–O bonds), whereas in monoclinic M-type $YTaO_4$ or YNbO4, tantalum/niobium atoms are in tetrahedral coordination (four Ta–O or Nb–O bonds) [179].

The shape of FTIR spectra in the domain of MO_4 group vibrations illustrates the change of the number of metaloxygen bonds with the increase of Nb concentration, in parallel with the structural conversion from M′-$YTaO_4$ to M$YNbO_4$ (Fig. 4.44). The change of the spectra

Table 4.7 Crystalline phase composition (%vol) and cell unit parameters for $YTa_{1-x}Nb_xO_4$ (Li_2SO_4) phosphors

Code	Formula	Cell unit parameters					Phase composition
		a (nm)	b (nm)	c (nm)	beta [degree]	V (nm3)	
	$YTaO_4$ (theor)	0.5303	0.5456	0.5116	96.45	0.1471	M′-$YTaO_4$; PDF 24-1425
IF42	$YTaO_4$	0.5304	0.5457	0.5117	96.45	0.14719	100% M′ phase
IF28	$YTa_{0.95}Nb_{0.05}O_4$	0.5304	0.5454	0.5116	96.45	0.14706	100% M′ phase
IF29	$YTa_{0.90}Nb_{0.1}O_4$	0.5303	0.5423	0.5116	96.45	0.14699	100% M′ phase
IF30	$YTa_{0.85}Nb_{0.15}O_4$	0.5302	0.5451	0.5114	96.45	0.14688	85.35% M′ phase; 14.65% M′ phase
IF31	$YTa_{0.80}Nb_{0.2}O_4$	0.5300	0.5457	0.5102	96.45	0.14673	67.32% M′ phase; 32.38% M′ phase
IF48	$YTa_{0.70}Nb_{0.3}O_4$	0.7664	0.1096	0.5315	138.95	0.29305	30.84% M′ phase; 69.16% M′ phase
IF49	$YTa_{0.60}Nb_{0.4}O_4$	0.7657	0.1995	0.5313	138.91	0.29285	55.82% M′ phase; 44.18% M′ phase
IF50	$YTa_{0.40}Nb_{0.6}O_4$	0.7638	0.1094	0.5311	138.75	0.29260	65.92% M′ phase; 34.08% M′ phase
IF51	$YTa_{0.20}Nb_{0.8}O_4$	0.7627	0.1094	0.5302	138.62	290.23	74.12% M′ phase; 25.88% M′ phase
IF45	$YNbO_4$	0.7624	0.1094	0.5301	138.62	0.29215	100% M′ phase;
	$YNbO_4$ (theor)	*0.7645*	*0.1100*	*0.5317*	*138.40*	*0.29686*	*M′-$YNbO_4$; PDF* 83-1319

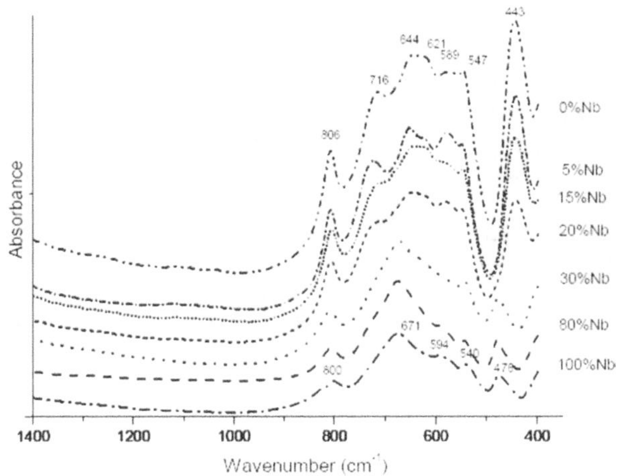

Figure 4.44 FTIR spectra of $YTa_{1-x}Nb_xO_4$ (Li_2SO_4) samples.

feature is well observed for sample $YTa_{0.60}Nb_{0.40}O_4$ phosphor that contains ~56 vol% M-$YNbO_4$ phase and ~44 vol% M'-$YTaO_4$ phase.

4.2.4.3 Morphology and particle size

The mineralizing agent influences the particle dimensions and morphology, as expected (Fig. 4.45). SEM images sustain the assertion that Li_2SO_4 is a more effective mineralizing agent than Na_2SO_4. In this case, PL performances are improved by the large and relatively regular shape particles. It is worth noting the mineralizing effect of Nb_2O_5, illustrated by the more dense powders formed from round-shaped particles. In reaction with yttrium oxide, Nb_2O_5 (m.p. 1380°C) shows a higher reactivity than Ta_2O_5 (m.p. 1872°C).

4.2.4.4 Vibrational spectra

To investigate the vibrational spectra in Y(Ta,Nb)O_4:Eu^{3+} phosphors Raman spectroscopy was used.

a) $YNbO_4$:Eu^{3+}: Some additional series of yttrium niobate phosphors activated by Eu^{3+} were prepared by solid state reaction method from homogeneous mixture consisting of Y_2O_3 (99.9%), Nb_2O_5 (99%) or Ta_2O_5 (Optipur) and Eu_2O_3. The amount of Eu_2O_3 was in the range of 0–20 mol%. Na_2SO_4 (99%) and Li_2SO_4 (99%) as a flux were used

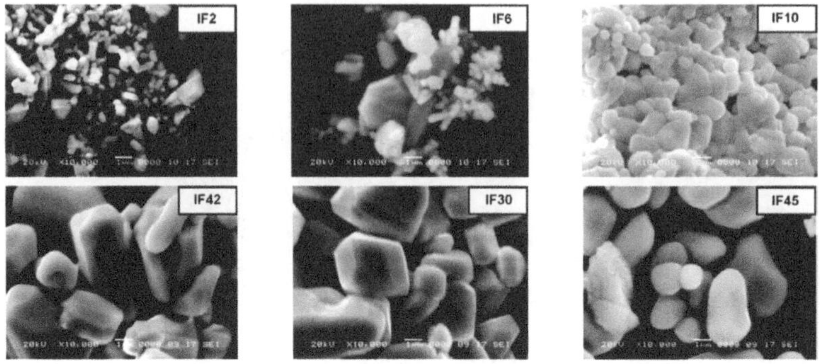

Figure 4.45 SEM images of $YTa0_4$ (IF2), $YTaNb0_4$(IF6), $YNb0_4$(IF10) phosphors prepared with Na_2S0_4 (up) and $YTa0_4$ (IF42), $YTaNb0_4$(IF30), $YNb0_4$(IF45) phosphors obtained with Li_2S0_4(down).

in the first series [262]. The mixtures were homogenized with a ball-mill in acetone medium, and dried at 70°C. The phosphor samples were baked at 1200°C for 4 h and slowly cooled to room temperature. Finally, the samples were water washed, dried and then sieved.

In the second series $YNb0_4:Eu^{3+}$ phosphor powders were fabricated by firing the stoichiometric mixtures of Y_20_3, Nb_20_5, and Eu_20_3 powders. LiCl (7 mol%) was added as a flux to promote a solid state reaction and the luminescent properties by accelerating the kinetics for the compound formation due to the increased diffusion coefficients. The amount of Eu_20_3 was in the range of 0–45 mol%. After ball-milling the mixtures for 24 h, they were fired at 1300°C for 12 h in an electric tube furnace flowing N_2 gas.

Raman spectra were obtained in the spectral range 100–1000 cm^{-1} using the 632.8 nm excitation wavelength from a continuous wave He–Ne laser at room temperature. Incident light is focused on a sample through optical microscope with a spatial resolution of <2 μm in the backscattering geometry. Scattered light was dispersed using triple monochromator system (TRIVISTA, ACTON) (see Fig. 3.30). Two of three monochromators in the triple monochromator system were used for laser line rejection to filter out Rayleigh scattering from the sample upon laser excitation. In subtractive mode, the Raman scattering signature as close as 20 cm^{-1} to the laser excitation line could be observable. The signal was recorded using a thermoelectrically cooled charge-coupled device detector.

In the Raman spectra, more associated with lattice vibrations and symmetrical bonds, small amount of activator can be registered in additional peaks. Some characteristic vibration modes ν_1, ν_2, ν_3, and ν_4 from the host lattice $YNbO_4$ were observed in spectrum 1 in Fig. 4.46). These modes correspond to a regular NbO_4 tetrahedron with no interactions and distortions at 815 (ν_1), 650 (ν_3), 420 (ν9), and 340 cm^{-1} (ν_2) predicted by Blasse [267]. Raman bands appearing at 815 and 340 cm^{-1} are due to Nb–O symmetric modes of the NbO_4 tetrahedral structure, and Raman bands appearing at 650 and 420 cm^{-1} are due to Nb–O antisymmetric modes of the NbO_4 tetrahedral structure. The phonon energies below 300 cm^{-1} are assigned to external vibrations.

One can see four new peaks around 446, 503, 524, and 589 cm^{-1} in the spectrum 2 and 3 from the activated $YNbO_4$ by 2.5 mol% Eu^{3+} and 5 mol%, correspondingly (Fig. 4.46–2,3). The detailed analysis shows that no additional peak is observed when yttrium niobate is doped by terbium and the Raman spectra of $YNbO_4$:Tb^{3+} is identical with spectra $YNbO_4$ (Fig. 4.46–1). It is reasonable to assume that these new peaks around 446, 503, 524, and 589 cm^{-1} (Fig. 4.46–2,3) are related to europium activation. Moreover, the intensity of these peaks is proportional to europium concentration, while the intensity

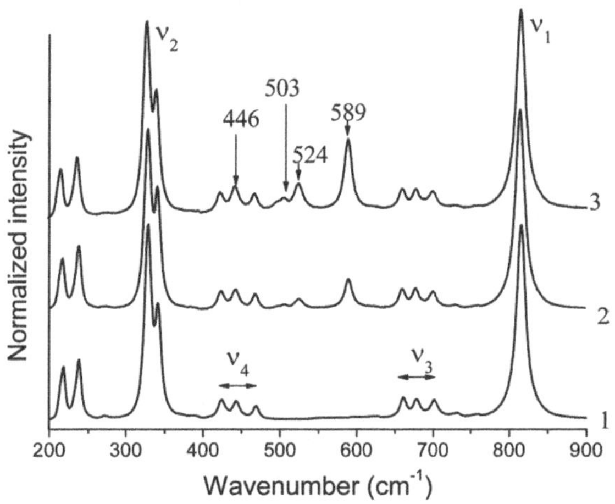

Figure 4.46 Raman spectra of the $YNbO_4$ (1), $YNbO_4$:Eu^{3+} (2.5 mol%) (2), and $YNbO_4$:Eu^{3+} (5 mol%) (3).

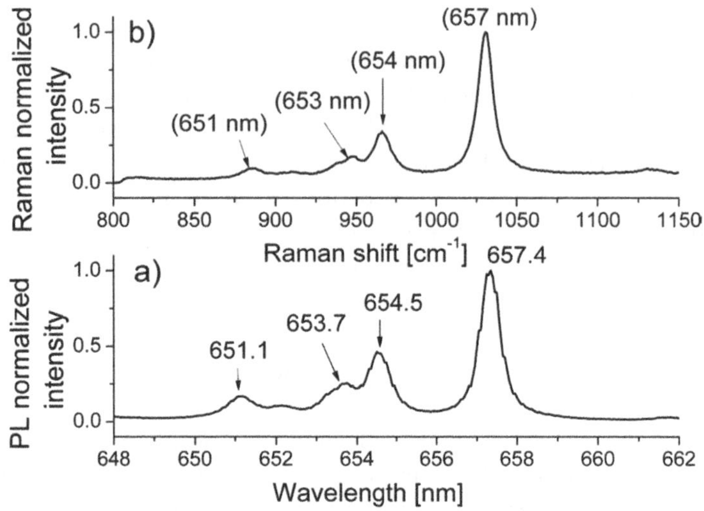

Figure 4.47 (a) PL emission spectrum of $^5D_0$–$^7F_3$ transition in $YNbO_4:Eu^{3+}$ under X-ray excitation. (b) Fragment of $YNbO_4:Eu^{3+}$ spectra in Raman spectroscopy under 615.57 nm excitation which shows the shift of Eu^{3+} luminescence peaks from 589 cm^{-1} (under 632.8 nm excitation) to 1030 cm^{-1} (615.57 nm excitation). Recalculated in (nm) peak positions (b) correspond to luminescent peaks (a).

of vibration modes from the host lattice $YNbO_4$ at 815 cm^{-1} (ν_1) does not change. The nature of this phenomenon is discussed.

According to Blasse [267], ν_1 is related to symmetrical stretching vibrations of NbO_4 near 815 cm^{-1}. It is proposed that new peaks are related to europium activation because their shape and position correspond to luminescence spectra. At room temperature, He–Ne laser (632.8 nm) can excite Eu^{3+} ions to $^5D_0$ energy level. The luminescence from $^5D_0$ energy level to $^7F_3$ band gets to 650–660 nm wavelength regions. The PL emission spectrum under X-ray excitation is shown in Fig. 4.47a.

It can be supposed that in Raman spectroscopy the registered spectrum at 632.8 nm excitation is a mixture of spontaneous Raman spectrum of $YNbO_4$ host crystal and luminescence lines of Eu^{3+} since these spectra overlaps in this area. When the host lattice is excited by 615.57 nm these four peaks (Fig. 4.46–3) shifted from 450–600 cm^{-1} to 850–1100 cm^{-1} (this area is shown in Fig. 4.47b) and no additional peaks were observed between antisymmetric modes ν_3 and ν_4 of the

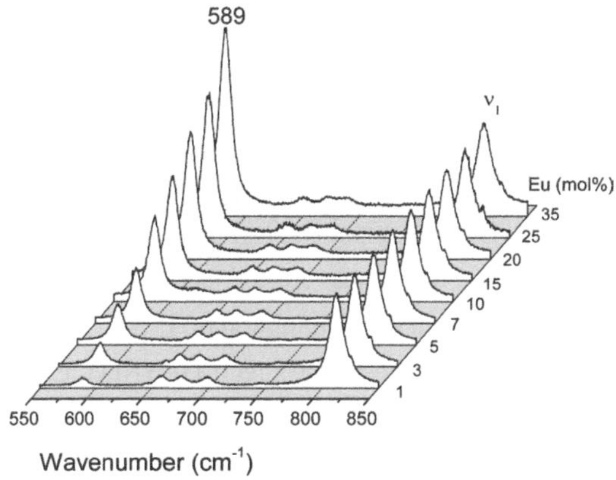

Figure 4.48 Fragments of Raman and luminescence spectra of the $YNbO_4$:Eu^{3+} with different Eu concentration and LiCl flux.

doped phosphors. Moreover, the shape and peak positions in Raman spectroscopy (Fig. 4.47b) are identical to PL emission spectrum (Fig. 4.47a).

This information is very useful for practical application. The dependence of intensity in these peaks as a function of the Eu^{3+} content in $YNbO_4$ samples was investigated. The samples were prepared with different fluxes: LiCl (Eu concentration up to 35 mol%, Fig. 4.48) and Li_2SO_4 with Eu concentration up to 20 mol%, Fig. 4.47). As experimental results show the tendency of Raman and luminescence spectra of $YNbO_4$:Eu^{3+} phosphor practically does not depend on the flux.

All observed vibration modes ν_1, ν_2, ν_3, and ν_4 in Raman spectra don't depend on the Eu^{3+} concentration. Only four luminescent peaks at 446, 503, 524, and 589 cm^{-1} show strong linear dependence on the Eu concentration (Figs. 4.46 and 4.49).

Because the excitation energy is too weak to excite the charge transfer band or upper F-levels, no saturation or concentration quenching up to 35 mol% was observed. With this energy only one excitation mechanism is possible: the direct excitation of $^5D_0$ level.

The liner dependence allows us to estimate the europium concentration in yttrium niobate host lattice. The calculation of relative intensity of two peaks at 589 cm^{-1} and 815 cm^{-1} (ν_1) were performed and the ratio of peak intensities I_{589}/I_{815} versus europium

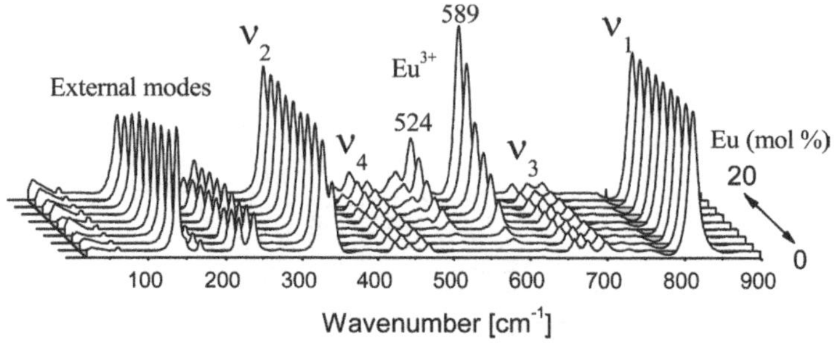

Figure 4.49 Raman and luminescence spectra of $YNbO_4:Eu^{3+}$ with different Eu^{3+} concentration and Li_2SO_4 flux.

concentration is presented in Fig. 4.50–2. Same linear dependence but with a little lower intensity was obtained for the samples from series (1) prepared with LiCl flux and it is presented in Fig. 4.50–1.

This phenomenon can be explained by different activator incorporation degree (AID) [226] in synthesis process, if we have a single phase powder (XRD data) and all row materials disappeared. The term AID is used to define the ability of europium to efficiently substitute yttrium from the host lattice. Activator substitution in general is a function of concentration, annealing temperature, flux and other synthesis parameters. The real Eu concentration in powders synthesized with different fluxes was measured. Quantitative non-destructive elemental analyses were performed with electron probe

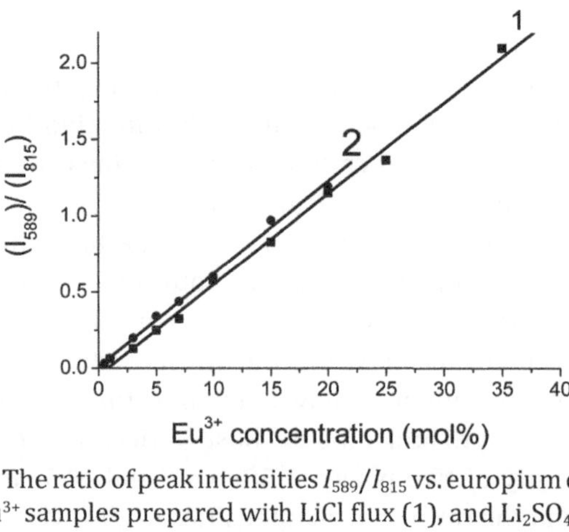

Figure 4.50 The ratio of peak intensities I_{589}/I_{815} vs. europium concentration for $YNbO_4:Eu^{3+}$ samples prepared with LiCl flux (1), and Li_2SO_4 flux (2).

Table 4.8 Synthesis and real Eu mol% concentration with different fluxes

Eu mol% concentration (synthesis)	Flux LiCl		Flux Na_2SO_4		Flux Li_2SO_4	
	Real Eu%	AID (%)	Real Eu%	AID (%)	Real Eu%	AID (%)
0.5					0.45	90
1					0.9	90
2.5			2.12	85		
3	2.44	81			2.64	88
5	3.60	72	4.25	85	4.6	92
7	5.58	79			6.3	90
10	9.10	91			8.9	89
15	12.51	83			13.5	90
20	16.11	81			18.2	91
35	27.95	80				
Average AID (%) →		81		85		90

microanalysis machine (SX-100, the electron acceleration 15 kV, beam current 10 nA and the diameter of electronic beam was 1 μm). The detailed explication of microanalysis was done in section 3.3 and the final data are summarized in Table 4.8.

From the Table 4.8 we can conclude that the most important parameter is not the concentration of activator proposed during the synthesis, but the real concentration in prepared samples. The best results can be achieved with Li_2SO_4 as a flux when the AID is about 90%.

b) $YTaO_4$:Eu^{3+}: The Raman spectrum of non-activated M′-$YTaO_4$ in comparison with M-$YNbO_4$ is shown in Fig. 4.51.

We can see from this figure that the Raman spectra from tantalate and niobate differ significantly. M-$YNbO_4$ and M′-$YTaO_4$ compounds has a unique Raman spectrum that is related to the symmetry and bond order of its structure. Only symmetric modes ν_1 in Raman spectra of M-$YNbO_4$ coincide with M′-$YTaO_4$ and peaking

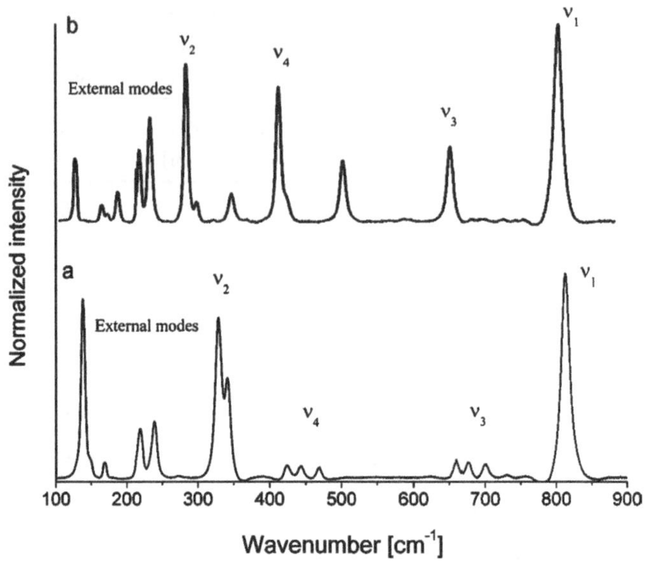

Figure 4.51 Raman spectra of the M-$YNbO_4$ (a) and M′-$YTaO_4$ (b).

both at 818 cm^{-1}. The main differences are due to different Nb–O antisymmetric modes of the NbO_4 tetrahedral structure (ν_3 and ν_4 vibration modes) and Ta–O antisymmetric modes in distorted octahedral TaO_4 structure. Moreover, the Ta–O symmetric mode in Raman spectra ν_2 is shifted to the short wavenumber area at about 285 cm^{-1}, while the Nb–O symmetric mode ν_2 is at 340 cm^{-1}. The similar tendency was observed by Brixner between M′ and M structure in $YTaO_4$ [185].

It is interesting to check the influence of activators on vibrational spectra in tantalate system. The results of activation M′-$YTaO_4$ by Tb^{3+} and Eu^{3+} are presented in Fig. 4.52.

Experimental data show that Tb^{3+} activation does not change the vibrational spectra, while activation by Eu^{3+} results in appearing four new peaks at 446, 484, 542, and 592 cm^{-1} (Fig. 4.52). The intensity of these peaks strongly depends on Eu^{3+} concentration (Fig. 4.53).

Both peaks at 446 and 484 cm^{-1} (small peak at 484 cm^{-1} is not seen in this picture) have a weak dependence on Eu and can be described by linear function that is clear seen from Fig. 4.54. Two other peaks at 542 and 592 cm^{-1} show linear dependence only at

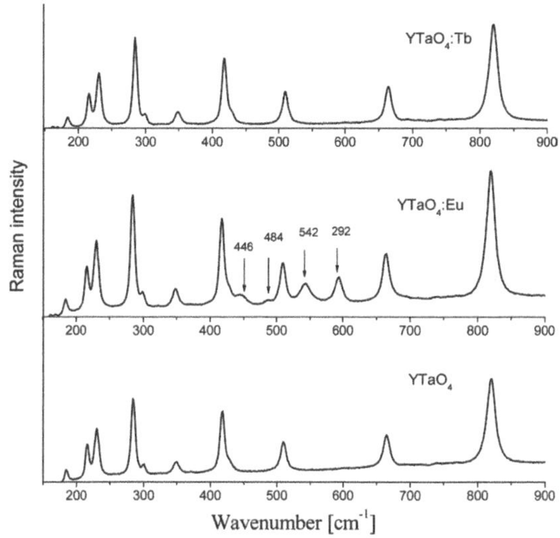

Figure 4.52 Raman spectra of the M′-YTaO$_4$, M′-YTaO$_4$:Eu^{3+} (5 mol%), and M′-YTaO$_4$:Tb^{3+} (5 mol%).

low Eu concentrations (less than 10 mol%) and can be described by exponential functions.

The partial substitution of the Y atoms by the rare earth ions Eu^{3+} into M′-YTaO$_4$ and M-YNbO$_4$ structures does not change significantly the basic structure and the results do not differ from those reported before for the host lattices [185, 246]. However, the incorporation of rare earth ions to the host lattices may change the luminescent properties and overlap the Raman spectra of the investigated

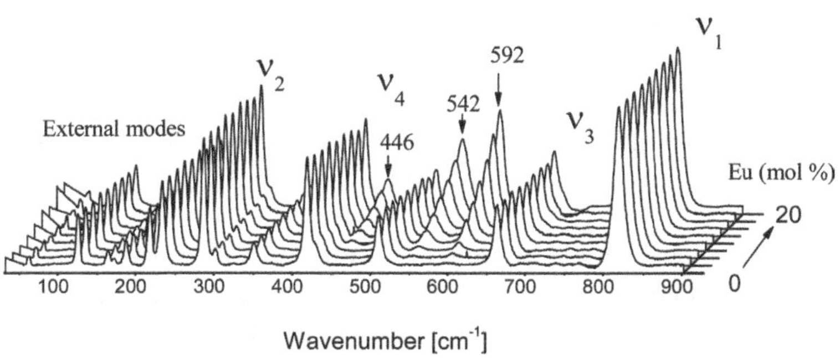

Figure 4.53 Series of M′-YTaO$_4$:Eu^{3+} Raman spectra with different Eu^{3+} concentration.

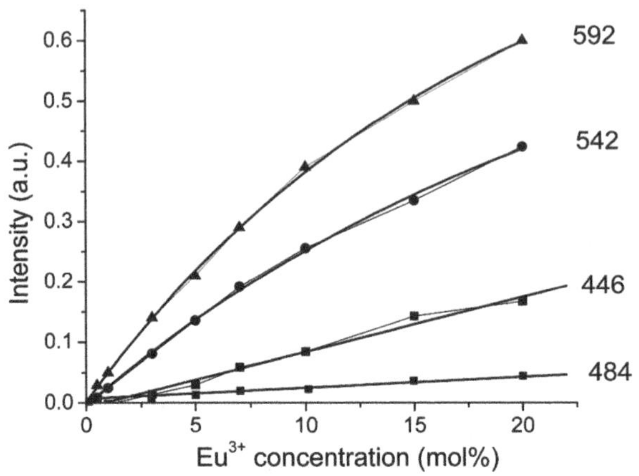

Figure 4.54 Intensities of Raman peaks at 446, 484, 542, and 592 cm^{-1} (normalized on intensity of ν_1 at 818 cm^{-1}) vs. europium concentration.

phosphors. We suppose that some additional peaks in Raman spectra related with Eu activation can be explained by this phenomenon.

We compared these data with Blasse [267] who studied non-activated M-$YNbO_4$ and M-$YTaO_4$ compounds by Raman spectroscopy and predicted the vibrational modes of a regular NbO_4 tetrahedron with no interactions and distortions to occur at 816 (ν_l), 650 (ν_3), 420 (ν_4), and 340 cm^{-1} (ν_2). The real experimental data [282] are a little different from the theory and shown in Table 4.9.

For the same host lattices (M-$YNbO_4$), Blasse data [267] and presented here experimental results are very similar, but for different structures (M′ and M-$YTaO_4$) the results differ. It is no wonder, because the tantalum atoms in these structures are in a distorted octahedral and tetrahedral coordination, respectively. The multiple Raman bands appearing in the 400–800 cm^{-1} region for the $YNbO_4$ compound, reported by Blasse [267], are probably due to the simultaneous presence of NbO_4 tetrahedra and NbO_6 octahedra in the structure.

In summary should be noted that Eu activated phosphors show four additional luminescence peaks in yttrium tantalate and niobate host lattices. Intensities of these peaks depend on Eu concentration. The Raman method can be recommended as an accurate and quantitative express analysis of europium activated phosphors.

Table 4.9 Raman spectra of the $YTaO_4$ and $YNbO_4$ groups in fergusonite. M is monoclinic fergusonite structure, M′ is modified monoclinic structure, (R) means Raman-active, s strong, m medium, w weak, and sh shoulder.

Free NbO_4 ion	NbO_4 ion in $YNbO_4$	Experimental (values in cm^{-1})						Assignment
		$YNbO_4$		$YNbO_4$:Eu	$YTaO_4$		$YTaO_4$:Eu	
		M	M	M	M	M′	M′	
(T_d)	(C_2)	Ref. [267]	Ref. [282]	Ref. [282]	Ref. [267]	Ref. [282]	Ref. [282]	
ν_1:A_1(R) 860	A	832 (s)	815 (s)	815 (s)	825 (s)	818 (s)	818 (s)	Symmetric Nb–O stretching vibrations [267, 283]
		–	–	783 (sh)	–	–	–	
ν_3:T_2(R) 650	A+2B	715 (w)	758 (w)	763 (w)	720 (w)	–	–	Antisymmetric Nb–O stretching vibrations [267]
		695 (w)	730 (w)	743 (w)	705 (m)	–	–	
		675 (w)	700 (w)	720 (w)	670 (m)	662 (m)	662 (m)	
		650 (w)	701 (w)	700 (w)	655 (w)	–	–	
		–	678 (w)	676 (w)	–	–	–	
		560 (w)	660 (w)	660 (w)	–			
				589 (*)			592 (*)	(*) Eu^{3+} luminescence peaks [282].
				524 (*)			542 (*)	
				503 (*)			484 (*)	
				446 (*)			446 (*)	
ν_4:T_2(R) 420	A+2B	480 (w)	469 (w)	468 (w)	480 (w)	508 (m)	508 (m)	Nb–O antisymmetric modes of the NbO_4 tetrahedral structure [267].
		455 (w)	445 (w)	443 (w)	450 (w)	–	–	
		435 (w)	427 (w)	424 (w)	–	417(s)	417 (s)	
		–	403(w)	403(w)	–	–	–	
		385(w)	370(w)	372(w)	375(w)	–	–	
		–	356(w)	355(w)	–	349(w)	349(w)	
ν_2:E(R)	2A	350(s)	340(s)	340(s)	345(s)	300(w)	300(w)	Symmetric stretching vibrations, ν_s (NbO_4) [267, 283]
		340(s)	328(s)	328(s)	320(s)	285(s)	285(s)	
External modes		245(m)	238(m)	238(m)		233(m)	233(m)	Lattice vibrations [284]
		225(m)	218(m)	218(m)		217(m)	217(m)	
		170(w)	168(w)	168(w)	215(s)	185(w)	185(w)	
		–	147(w)	147 (w)	120(m)	162(w)	162(w)	
		–	138(s)	138(s)		124(m)	124(m)	

Raman and luminescence experiments show that the best results can be achieved in synthesis of high-efficiency phosphors using Li_2SO_4 as a flux.

4.2.5 *Double Activated Y(Ta,Nb)O_4:Eu^{3+},Tb^{3+}*

In this section yttrium tantalate ($YTaO_4$), yttrium niobium–tantalate ($YTaNbO_4$) and yttrium niobate ($YNbO_4$) doubly doped by Eu^{3+} and Tb^{3+}, were investigated using X-ray diffraction and X-ray excitation luminescence in order to study their structural and luminescent properties. By means of X-ray diffraction, the crystallographic data for $YTaO_4$ and $YNbO_4$ with double activation by Eu^{3+} and Tb^{3+} was first calculated. Under X-ray excitation luminescence, the rare earth emission centers contribute to the overall luminescence. Due to their various luminescence chromaticities, the proposed rare earth activated phosphors are promising materials for optoelectronics as well as for X-ray intensifying screens for medical diagnosis providing the broad variation of visible photoluminescence from blue to red.

4.2.5.1 Sample preparation

Yttrium tantalate, yttrium niobium–tantalate and yttrium niobate phosphors activated by Eu^{3+} and Tb^{3+} were prepared by solid state reaction method from homogeneous mixture consisting of Y_2O_3 (99.9%), Ta_2O_5 (Optipur), and Nb_2O_5 (99%). The oxide precursors for the host lattice, Eu_2O_3 and/or Tb_4O_7, are used in the activator system and Na_2SO_4 (99%) as a flux. The mixtures were homogenized with a ball-mill, in acetone medium, and dried at 70°C. The phosphor samples were baked at 1200°C for 4 h and slowly cooled to room temperature. Finally, the samples were water-washed, dried, and then sieved.

4.2.5.2 Particle morphology and sizes

To have an efficient phosphor for fluorescent lamps, electroluminescence, or X-ray screens, it is necessary to have well-crystallized, preferably polyhedral particles of approximately 3–10 μm in diameter. Larger sizes would cause difficulty in screen making, and some efficiency losses are experienced via internal scattering as the particle size gets below 3 μm. A series of scanning electron microscopy (Hitachi S-4700) images were made to characterize and compare the particle morphology and sizes for $YTaO_4$ and $YNbO_4$ samples doubly activated by Eu^{3+} and Tb^{3+}. The Na_2SO_4 flux used in preparing the powder samples helped to generate mainly polyhedral particles for all samples. The crystalline particle of M′-$YTaO_4$:Eu^{3+},Tb^{3+} has an elongated polyhedral form of 3 to 5 μm in length and about 1–2 μm in thickness. The grain size increases for the M-$YNbO_4$:Eu^{3+},Tb^{3+} structure to around 8 to 10 μm in dimension, having a more round shape (Fig. 4.55).

4.2.5.3 X-ray diffraction measurement

The structural properties of fabricated phosphor samples were investigated by performing X-ray diffraction experiments. Cu-Kα radiation (λ = 1.54 Å) source generated from a Rigaku rotating anode generator was used for operating at 40 kV and 40 mA. Basically, the powder diffraction profiles (θ–2θ scan) were measured in the range from 10° to 80° covering the range between 10° and 80°.

Figure 4.55 SEM images of phosphor samples: (a) $YTaO_4$:Eu^{3+},Tb^{3+} and (b) $YNbO_4$:Eu^{3+},Tb^{3+}.

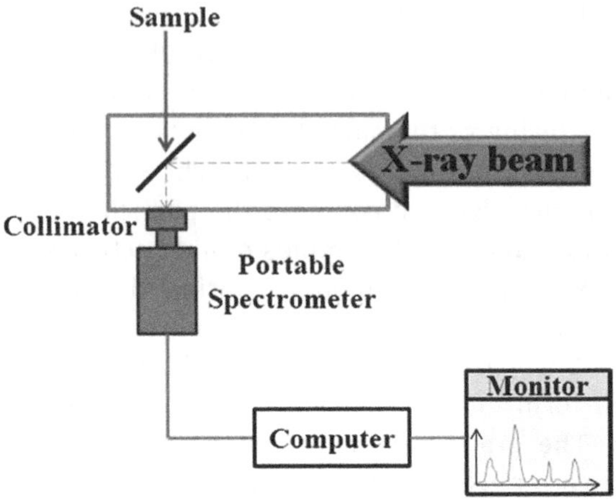

Figure 4.56 Schematic diagram of the experimental configuration to study the luminescence of phosphors under X-ray excitation.

4.2.5.4 X-ray luminescent measurement

To study phosphor luminescence under X-ray excitation, the Bremsstrahlung radiation from a Rigaku rotating anode generator operating in the range from 20 to 50 kV and 100 mA was used. Powder phosphors were illuminated into the broadband X-ray radiation. As illustrated in Fig. 4.56, samples were placed at a position close to exit of Be X-ray windows to minimize the loss of X-ray flux by air scattering, and rotated by 45° from the incident X-rays. Since the luminescence signals were irradiated into all directions, a collimating lens was used to collect the light effectively. The emission spectrum was recorded with a spectrometer (SM240, Spectral Products) in the range of ~200–1200 nm.

Three types of host lattices with double activation by rare earth elements Eu^{3+} and Tb^{3+} were prepared. The general formula of the phosphors is $Y_{1-x-y}Eu_xTb_yTa_{1-z}Nb_zO_4$, denoted as $Y(Ta,Nb)O_4$:Eu^{3+},Tb^{3+}. The next parameters were used: $x = 0.025$, $y = 0.025$, and $z = 0$; 0.15 or 1. The chemical compositions of prepared phosphor samples are listed in Table 4.10. In this study, equal amounts of europium and terbium were used at a small doping level, as to better put in evidence the mutual interaction between the rare earth ions and host lattice.

Table 4.10 Composition of yttrium tantalate/niobate phosphors with double activation by Eu^{3+} and Tb^{3+}

		Hostel lattice (mol %)		RE–activator (mol %)	
Phosphor type	Phosphor formula	$YTaO_4$	$YNbO_4$	Eu^{3+}	Tb^{3+}
$YTaO_4$:Eu, Tb	$Y_{0.95}Eu_{0.025}Tb_{0.025}TaO_4$	100	–	2.5	2.5
Y(TaNb) O_4;Eu,Tb	$Y_{0.95}Eu_{0.025}Tb_{0.025}Ta_{0.85}Nb_{0.15}O_4$	85	15	2.5	2.5
$YNbO_4$;Eu,Tb	$Y_{0.95}Eu_{0.025}Tb_{0.025}NbO_4$	–	100	2.5	2.5

4.2.5.5 X-ray diffraction characterization

The X-ray diffraction spectra of two different host lattices M′- $YTaO_4$ and M-$YNbO_4$ doubly activated by Eu^{3+} and Tb^{3+} are shown in Fig. 4.57.

Figure 4.57a shows M′-$YTaO_4$ crystal structure. The X-ray diffraction spectrum of $YNbO_4$:Eu^{3+},Tb^{3+} in Fig. 4.57b shows the evidence of the monoclinic M-$YNbO_4$ fergusonite crystal structure.

In the Fig. 4.57, the black points (b) are the measured data and the solid lines (a) are the calculated fitting of Rietveld analysis, which was developed by Rietveld for structure profile refinement of X-ray powder diffraction data [285]. The measured peak positions are in good agreement with Rietveld fitting and data given in Powder Diffraction Files: JCPDS 72-2018 (for $YTaO_4$) and JCPDS 72-2077 (for $YNbO_4$) from International Center for Diffraction (Fig. 4.57c).

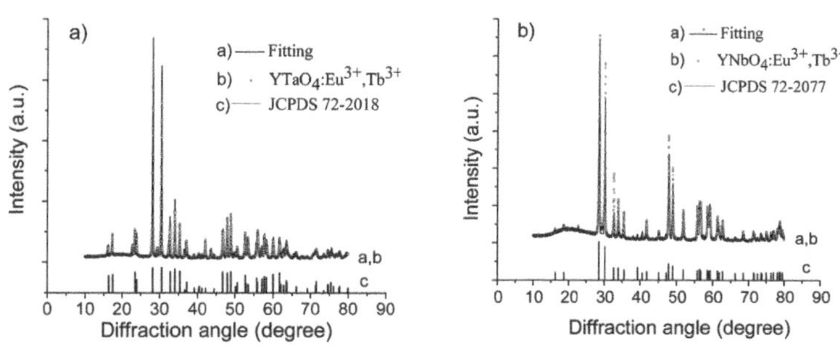

Figure 4.57 XRD patterns of $YTaO_4$:Eu^{3+},Tb^{3+} (a) and $YNbO_4$:Eu^{3+},Tb^{3+} (b), and their fitting. Expected peak postitions from JCPDS card are indicated.

Partial substitution of the tantalum ions by 15 mol% niobium ions ($YTaNbO_4$) did not change the M′ crystal structure [262]. We also found that the incorporation of the rare earth ions (5 mol% concentration of Eu^{3+} and Tb^{3+}) into the M′-$YTaO_4$ structure and into the M-$YNbO_4$ structure did not change the basic structure significantly, but increased the unit cell volume according to Vegard's law. The amounts of the unit cell volume increment were 0.28% for $YTaO_4$:Eu^{3+},Tb^{3+} and 0.15% for $YNbO_4$:Eu^{3+},Tb^{3+} over the volume of the original host lattices. The difference in the ionic radii of yttrium, terbium, and europium (1.02 Å, 1.04 Å, and 1.06 Å, respectively) may account for these increments.

The Rietveld calculation provides the crystallographic information by comparing the model profile with X-ray or neutron curves using the least squares method. The Rietveld analysis is generally used to get the lattice parameters, atomic positions, and atomic distances. The peak positions of X-ray diffraction curve are related to the unit cell lattice constants of the crystal structure and the peak intensities are affected by various parameters such as atomic position, atomic occupancy, and thermal effect. No work has been published using Rietveld analysis for these phosphors doubly activated. The detailed crystallographic information of the measured phosphors from Rietveld analysis is listed in Tables 4.11 and 4.12. From the occupation factor *F* (Table 4.11), one can see that F = 1, which means the site is fully occupied by an atom.

Using the data in Tables 4.11 and 4.12 the structures M′-$YTaO_4$ and M-$YNbO_4$ doubly activated by Eu^{3+} and Tb^{3+} were calculated and presented in Fig. 4.58. Figure 4.58 illustrates the process of rare earth activators Eu^{3+} and Tb^{3+} substituting yttrium atoms from the host lattice.

In M′-$YTaO_4$:$Eu^{3+}Tb^{3+}$, the Y atoms (Fig. 4.58a) are surrounded by 8-coordinated oxygen atoms forming a distorted cube. From mathematical calculations the Y–O averaged distance is 2.37 Å. The Ta atoms are in a distorted octahedral coordination with four shorter Ta–O bonds at 1.96 Å and 1.87 Å and two longer at 2.23 Å. The total amount of atoms inside of the M′-$YTaO_4$ structure is 2 yttrium atoms, 2 tantalum, and 8 oxygen atoms.

Following the same procedure, the crystallographic structure of $YNbO_4$:Eu^{3+},Tb^{3+} was calculated and presented in Fig. 4.58b. The Y

Table 4.11 Atomic positions and lattice parameters for different host lattices

M′-YTaO$_4$:Eu^{3+},Tb^{3+}	*Atom*	*x*	*y*	*z*	*F*
a = 5.298 Å b = 5.461 Å c = 5.112 Å β = 96.37	Y	0.25	0.7605	0	1
	Ta	0.25	0.3111	0.5	1
	O(1)	0.4192	0.4033	0.3079	1
	O(2)	0.1230	0.1117	0.2824	1
M′-Y(TaNb)O$_4$:Eu^{3+},Tb^{3+}	*Atom*	*x*	*y*	*z*	*F*
a = 5.299 Å b = 5.476 Å c = 5.072 Å β = 94.51	Y	0.25	0.726	0	1
	Ta	0.25	0.2611	0.5	1
	O(1)	0.5251	0.4236	0.2746	1
	O(2)	0.0857	0.0810	0.2140	1
M-YNbO$_4$:Eu^{3+},Tb^{3+}	*Atom*	*x*	*y*	*z*	*F*
a = 7.617 Å b = 10.95 Å c = 5.298 Å β = 138.4 2	Y	0	0.3791	0.25	1
	Nb	0	0.8551	0.25	1
	O(1)	0.2286	0.79	0.2422	1
	O(2)	0.2519	0.9576	0.6818	1

atoms are also surrounded by 8-coordinated oxygen atoms forming a distorted cube. The Y–O averaged distance is 2.35 Å. The Nb atoms are in a distorted octahedral coordination, with four shorter Nb–O at 1.94 Å and 1.83 Å and two longer at 2.43 Å.

As it was shown in Fig. 4.57, the partially substitution of the Y atoms by the rare earth ions Eu^{3+} and Tb^{3+} into M′-YTaO$_4$ and M-YNbO$_4$ structures does not change significantly the basic structure and the results do not differ from those reported before for the host lattices [185, 246]. However, as we will see, the incorporation of rare earth ions to the host lattices may change the luminescent properties of the investigated phosphors.

Table 4.12 Interatomic distances between anions and cations of different host lattices

Structure	*Y–O (Å)*	*Ta–O (Å)*	*Nb–O (Å)*
M'-$YTaO_4$:Eu^{3+},Tb^{3+}	O(1)2x 2.524	O(1)2x 1.957	
	O(2)2x 2.348	O(1)2x 2.230	
	O(1)2x 2.322	O(2)2x 1.870	
	O(2)2x2.296		
M'- Y(TaNb) O_4:Eu^{3+},Tb^{3+}	O(1)2x 2.545	O(1)2x 1.926	
	O(2)2x 2.331	O(1)2x 2.251	
	O(1)2x 2.289	O(2) 2x 1.874	
	O(2)2x 2.325		
M-$YNbO_4$:Eu^{3+},Tb^{3+}	O(1)2x 2.391		O(1)2x1.944
	O(2)2x 2.331		O(1)2x2.426
	O(1)2x 2.304		O(2)2x1.830
	O(2)2x 2.387		

4.2.5.6 Cathodoluminescence

Local cathodoluminescence makes it possible to thoroughly characterize the luminescence properties of phosphor materials. The

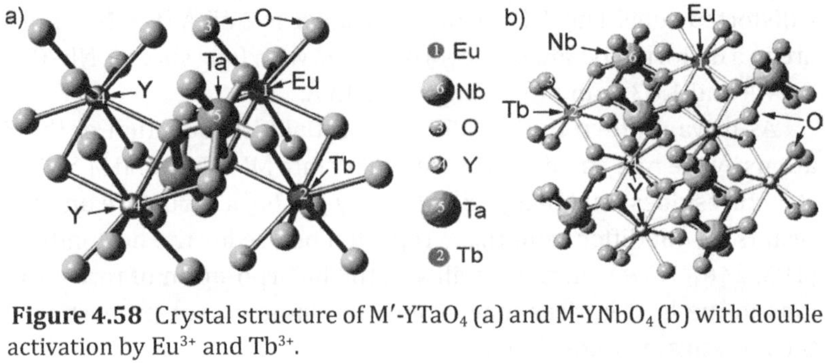

Figure 4.58 Crystal structure of M'-$YTaO_4$ (a) and M-$YNbO_4$ (b) with double activation by Eu^{3+} and Tb^{3+}.

main advantages of this method are follows: (1) locality providing a spatial resolution of several micrometers; (2) high excitation energy allowing observation of optical transitions whose excitation is complicated or impossible in the optical wavelength range; (3) the possibility of varying the excitation energy density in a wide range and observation of optical transition saturation and other non-linear effects; (4) real color images of investigated samples under defocused electron beam excitation.

Sample characterizations were performed with a special CL spectrometer combined with a standard "Camebax" (Fig. 3.8). The CL was studied with the same type of spectrometer used previously [177]. The CL spectrometer was installed on the optical microscope port of "Camebax" electron microprobe. The CL spectra were acquired under accelerating voltage of 15 kV at room temperature. The diameter of beam on the sample was about few micrometers and the beam current 1.5 nA. The color images in real colors were made with defocused electron beam about 200 μm. The combination of CL imaging and CL spectroscopy allowed us to get spatially resolved secondary electron images to characterize relatively large parts of the object under study, and to record CL spectra in specific areas (Fig. 4.59). We compare here color cathodoluminescence images and cathodoluminescence spectra of non-activated, Eu^{3+} and Tb^{3+} activated tantalates with double activated phosphors.

The photoluminescence variations of the host lattice $YTaO_4$ due to the incorporation of europium and/or terbium activators and their respective CL spectra are shown in Fig. 4.59. The color cathodoluminescence image of $YTaO_4$ host lattice without activators is shown in Fig. 4.59a (left). Visual CL observation through the optical microscope binocular of the "Camebax" EPMA showed that the non-activated $YTaO_4$ sample is visually perceived as violet. A weak blue-violet color is seen due to the small tail of spectrum after 400 nm in the CL spectra (Fig. 4.59a, right). Some green points are also seen in this structure, which confirms the presence of small amount Tb^{3+} admixtures into the host lattice. These admixtures can be identified on CL emission spectrum as $^5D_4 \rightarrow {}^7F_6$ transition at 485 nm. When the material is heterogeneous, the local cathodoluminescence method can be applied to avoid these admixtures. Choosing one or more spots in the sample and averaging them, it is possible to elude extra peaks, in our case the $^5D_4 \rightarrow {}^7F_5$ and $^5D_4 \rightarrow {}^7F_6$ transitions (Fig. 4.60).

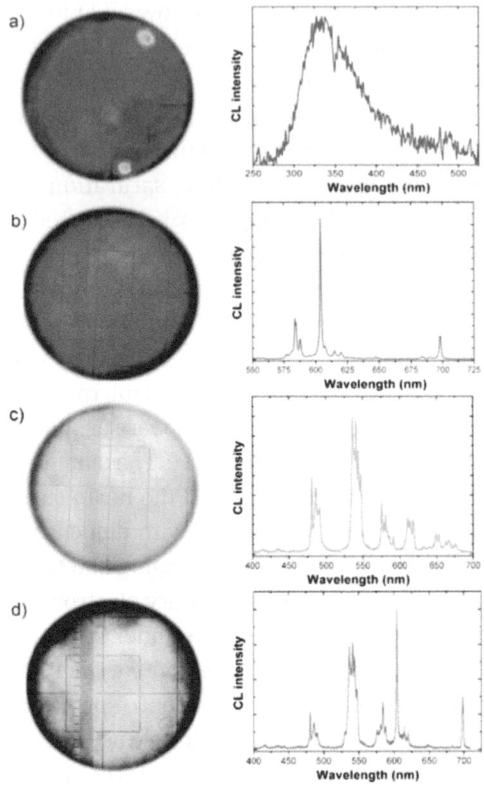

Figure 4.59 Color cathodoluminescence images and cathodoluminescence spectra: (a) $YTaO_4$ (original blue); (b) $YTaO_4:Eu^{3+}$ (original red); (c) $YTaO_4:Tb^{3+}$ (original green); (d) $YTaO_4:Eu^{3+},Tb^{3+}$ (original yellow).

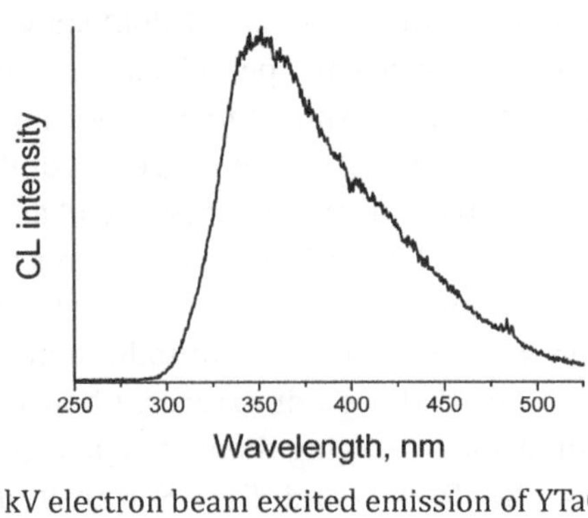

Figure 4.60 15 kV electron beam excited emission of $YTaO_4$.

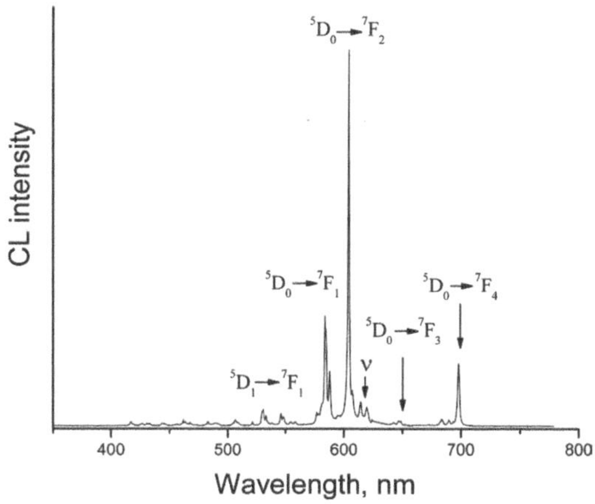

Figure 4.61 Emission CL spectra of M′-type $Y_{0.95}Eu_{0.05}TaO_4$. (ν-vibronic transitions).

The CL emission spectrum in Fig. 4.60 shows a broad band with a peak at 350 nm wavelength and can be ascribed to the emission from the charge transfer transition of the TaO_4^{3-} group [250]. This CL spectrum correlates well with X-ray excited emission spectrum of $YTaO_4$ given in Fig. 4.19

In Fig. 4.59b a bright red color of Eu^{3+} luminescence is presented. For the Eu^{3+}-doped M′-type $YTaO_4$, the CL spectra (Fig. 4.61) differ from UV PL spectra (Fig. 4.40).

The intensity of the charge transfer transition emission of TaO_4^{3-} decreases completely, and the strong emission of the Eu^{3+} $f \rightarrow f$ transition appears. It means that the $TaO4^{3-}$ group really absorbs excitation energy through $O^{2-} \rightarrow Ta^{5+}$ charge transfer transition and transfers the energy to Eu^{3+} luminescent centers, which give rise to the corresponding characteristic emission. The strongest emission peak is at 612 nm, which can be ascribed to the Eu^{3+} $^5D_0 \rightarrow ^7F_2$ transition emission. It shows Eu^{3+} situates on the Y^{3+} site of C_2 asymmetry. We could not observe all allowed transitions to the ground state ^7F_J ($J = 5$, 6, 7) under CL excitation because the intensities of these transitions are very weak. However, in any case, the intensity of some transitions, for example, the ratio $(^5D_0 \rightarrow ^7F_4)/(^5D_0 \rightarrow ^7F_2)$ under CL, is at 12 times greater in comparison with the PL. It is expected that the empty levels and traps are readily filled under electron beam excitation and the intensity of such weak transitions increases.

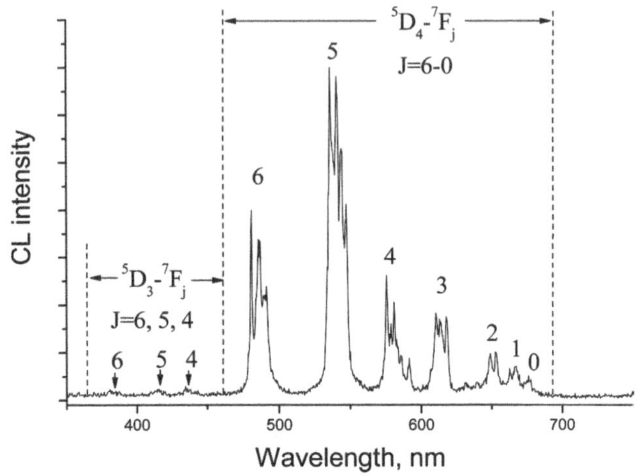

Figure 4.62 Emission spectra of M′-type $Y_{0.95}Tb_{0.05}TaO_4$ under electron beam excitation.

Eu^{3+}-doped $YTaO_4$ shows a strong red emission under electron beam excitation. This proves this material to be efficient phosphors for CL applications.

In Fig. 4.59c a bright green color of Tb^{3+} luminescence is shown. One more advantage of CL investigation is illustrated in the emission spectrum of Tb^{3+} luminescence in Fig. 4.62.

Under UV and partially under VUV excitation, the $^5D_4 \rightarrow {}^7F_J$ (J = 0, 1, 2) emissions are mostly trapped by the empty upper levels. In contrast, it is more likely that such empty levels can be readily filled under electron beam excitation. This model explains the appearance of relatively weaker emissions from $^5D_4 \rightarrow {}^7F_J$ (J = 0, 1, 2) transitions and the increase of the cathodoluminescence intensity in comparison with UV excitation (Fig. 4.25). These results are in good agreement with selection rules for Tb^{3+} ($4f^8$ shell) and splitting of $^{2S+1}D_J$ multiplets for Tb^{3+} in the crystal field at the C_2 symmetry site. All allowed transitions between Stark components of terbium $^5D_4 \rightarrow {}^7F_J$ (J = 6, 5, 4, 3, 2, 1, 0) were observed under electron beam excitation and presented in Fig. 4.62.

When Eu^{3+} and Tb^{3+} ions are incorporated simultaneously in the host lattice with the same concentration (Eu^{3+} = 2.5 mol%; Tb^{3+} = 2.5 mol%), the luminescence of phosphor exhibits a yellowish color

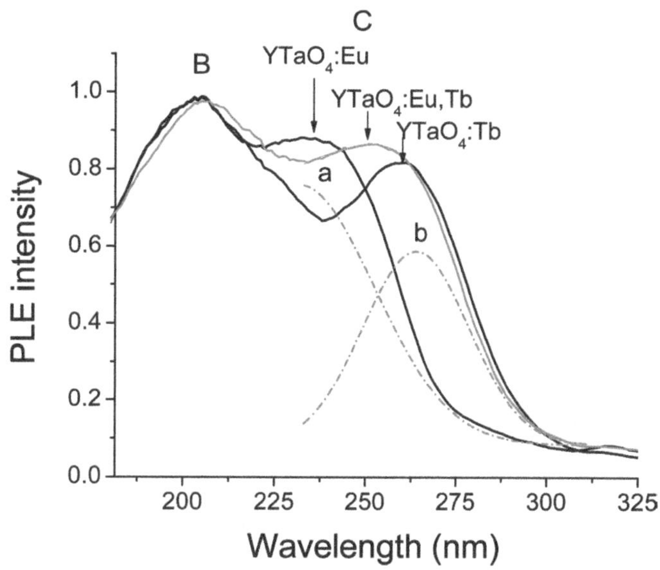

Figure 4.63 Excitation spectra of $YTaO_4:Eu^{3+}$, $YTaO_4:Tb^{3+}$, and $YTaO_4$: Eu^{3+}, Tb^{3+}.

(CCL in Fig. 4.59d). Varying the proportion of Eu^{3+} and Tb^{3+} we can get a wide range of different colors from blue to red.

4.2.5.7 VUV excitation

The PLE spectra from $YTaO_4:Eu^{3+}$, $YTaO_4:Tb^{3+}$, and $YTaO_4$: Eu^{3+}, Tb^{3+} were measured in the range of 100–350 nm using sodium salicylate powder as a reference and shown in Fig. 4.63.

Two broad bands B and C are observed in this figure. The band B in Fig. 4.63 from 180 to 220 nm is the same for all the samples and does not depend on the activator. This band is associated with the absorption of the host lattices. The TaO_4^{3-} group can absorb the excitation energy through $O^{2-}\rightarrow Ta^{5+}$ charge transfer transition and transfer the energy to the rare earth luminescent centers.

Band *C* is shifted from 240 nm for $YTaO_4:Eu^{3+}$ to 260 nm for $YTaO_4:Tb^{3+}$. The nature of the band *C* is different for Tb^{3+} and Eu^{3+} activation. An absorption band peaked at 260 nm in the PLE spectrum of the Tb^{3+} emission belongs to the Tb^{3+} center itself. This

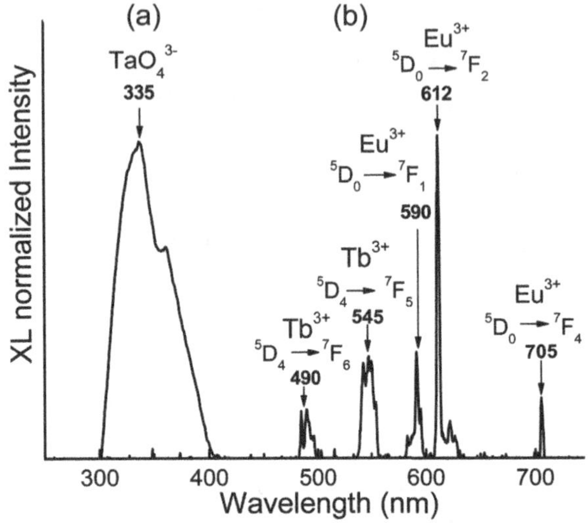

Figure 4.64 X-ray emission spectra of $YTaO_4$ (a) and $YTaO_4$:Eu^{3+},Tb^{3+} (b).

may be either a 4*f*→5*d* band [253, 286] or a Tb^{3+}-tantalate charge-transfer band of O^{2-}→Tb^{3+} in $YTaO_4$ host lattices [250]. Namely, the delocalized electrons move from the filled 2*p* shell of O^{2-} to the partially filled 4*f* shell of Tb^{3+}. In the recent years this conclusion was revised and many authors assign this band peaked at 260 nm to the ^9D level of the $4f^8 \rightarrow 4f^7 5d^1$ transition of Tb^{3+} within its $4f^8$ configuration [256, 287, 288]. The absorption band *C* for Eu^{3+} peaked at 240 nm associated with charge-transfer band (CTB) of O^{2-}→Eu^{3+} [289, 290]. For double activated $YTaO_4$:Eu^{3+}, Tb^{3+} this band is placed at 250 nm. It is broader than such bands for Eu^{3+} or Tb^{3+} single activation. We believe that this band is composite and it can be decomposed on two bands (a) and (b) belonging to the Eu^{3+} CTB and Tb^{3+} 4*f*→5*d* transition, correspondingly.

The VUV emission peaks of double activated $YTaO_4$: Eu^{3+}, Tb^{3+} (not shown here) are similar to CL spectra (Fig. 4.59d) and XL spectra (Fig. 4.64b) and also combine Eu^{3+} and Tb^{3+} lines.

4.2.5.8 X-ray luminescence characterization

Yttrium tantalate phosphor is an ultraviolet emitter (λ_{em} = 335 nm) whose luminescence corresponds to a charge-transfer transition

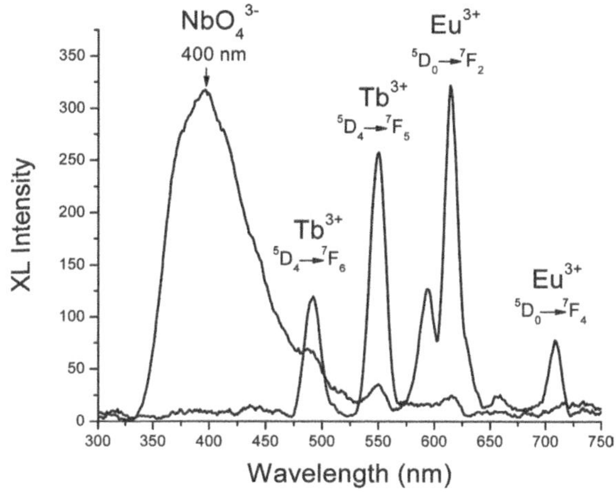

Figure 4.65 X-ray emission spectra of non-activated Y(Ta,Nb)O_4 (left) and Y(Ta,Nb)O_4:Eu^{3+},Tb^{3+} (right).

involving TaO_4 groups (Fig. 4.64). Replacing part of the tantalum ions with niobium ions, NbO_4 centers are formed that generate a blue-shifted luminescence emission (385–395 nm) (Fig. 4.65). Under X-ray excitation in $YTa_{0.85}Nb_{0.15}O_4$, only the niobate group and not the host lattice, is emitted because the energy transfer occurs from tantalate to niobate. (Only for very low niobate concentration this transfer is not efficient enough to quench the emission of all the tantalate groups, so that the emission still contains a band due to the tantalate itself [250]).

Figure 4.66 shows the luminescence spectra of M′-$YTaO_4$, M′-Y(Ta,Nb)O_4 and M-$YNbO_4$ activated by Eu^{3+} and Tb^{3+} when they were exposed into the X-ray beam. The linear dependency of luminescence intensity under different applied voltage was found.

When rare earth ions such as Eu^{3+} and Tb^{3+} are incorporated simultaneously to partially substitute the yttrium ions from the host crystalline lattice, Eu^{3+} and Tb^{3+} emission centers are created. In this case, the luminescence can be redshifted toward longer wavelengths, and both emission centers might contribute to the overall luminescence. The main peaks were found in the range from 450 to 725 nm. In Fig. 4.66 the first two peaks at 490 nm and at 545 nm correspond to terbium, and the peaks at 590 nm, 612 nm and 705 nm are the europium contribution to the luminescence. The

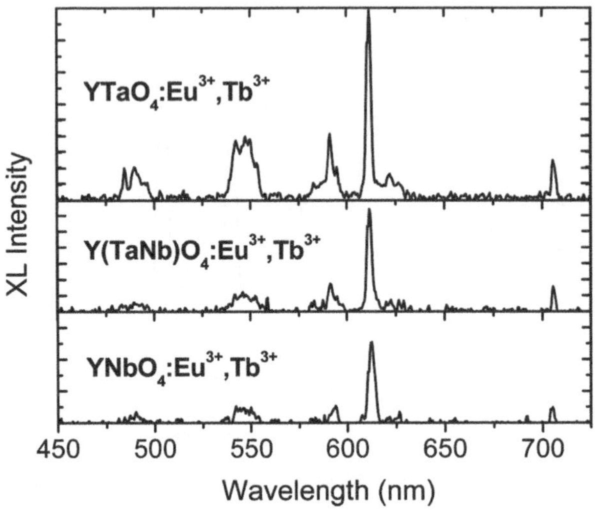

Figure 4.66 X-ray emission spectra of $YTaO_4$, $Y(TaNb)O_4$, and $YNbO_4$ host lattices doubly activated Eu^{3+} and Tb^{3+}.

doubly activated M′-$YTaO_4$ structure shows the better luminescence compared with M-$YNbO_4$. The incorporation of these two activators into the host crystalline lattice substituting the yttrium ions seems to efficiently enhance the charge-transfer process for the M′-$YTaO_4$ structure, where the average Ta–O distances are smaller in M′ structure than Nb–O distances in M structure. Under X-ray excitation, it is quite reasonable to assume that the excitation energy is absorbed first by the host lattice, which involves the transition between 4*d*-like states of Y and 2*p*-like states of O. The absorbed energy may then be transferred to TaO_4^{-3} and NbO_4^{-3} groups and at last transferred to the Eu^{3+} and Tb^{3+} emission centers.

Under X-ray excitation the $^5D_0 \to {}^7F_4$ transition intensity is appeared. Usually, as was mentioned above, this emission is very weak and could not be seen under UV excitation [265]. Furthermore, comparing the luminescence under UV (254 nm) VUV (147 nm), CL and the broad X-ray excitation, we found that the relative intensity of $^5D_0 \to {}^7F_j$ ($j = 1, 2, 4$) and $^5D_4 \to {}^7F_j$ ($j = 5, 6$) emission peaks of Eu^{3+} and Tb^{3+}, respectively, so-called branch ratios, varies with not only the host lattices, but also with the incorporation of the rare earth activators and the activation

Table 4.13 Branching ratios of $^5D_0 \rightarrow {}^7F_j$ and $^5D_4 \rightarrow {}^7F_j$ transitions (Eu^{3+} and Tb^{3+} concentration = 5 mol%)

Transition	$YTaO_4Eu^{3+},Tb^{3+}$				$YTaNbO_4Eu^{3+},Tb^{3+}$				$YNbO_4Eu^{3+},Tb^{3+}$			
	UV	VUV	XL	CL	UV	VUV	XL	CL	UV	VUV	XL	CL
$^5D_4 \rightarrow {}^7F_6$	57	46	51	38	59	84	49	47	51	48	67	42
$^5D_4 \rightarrow {}^7F_5$	100	100	100	100	100	100	100	100	100	100	100	100
$^5D_0 \rightarrow {}^7F_1$	53	40	35	30	40	46	27	25	22.5	24	21	22
$^5D_0 \rightarrow {}^7F_2$	100	100	100	100	100	100	100	100	100	100	100	100
$^5D_0 \rightarrow {}^7F_4$	1.8	25	21	16	1.8	35	24	12.5	0.8	26	19	9

excitation wavelength. The detailed branching ratios of $^5D_0 \rightarrow {}^7F_4$ and $^5D_4 \rightarrow {}^7F_j$ transitions are presented in Table 4.13.

Under UV excitation, the $^5D_0 \rightarrow {}^7F_4$ emission is mostly trapped by the empty upper levels. On the contrary, it is more likely that such empty levels can be readily filled under VUV, CL and X-ray excitation. This explains the appearance of visible $^5D_0 \rightarrow {}^7F_4$ transition and the increment of the luminescence intensity of Eu^{3+} and Tb^{3+} emission centers under high energy excitation.

Mention has to be made that, usually, for the same host-lattice, the optimal activator concentration for photoluminescent and X-ray phosphors is quite different, due to the variable PL and XL excitation mechanism in relation with the energy of UV and X-ray exciting photons. For this reason, no comparison between the efficiency of Y(Ta,Nb)O$_4$: Eu^{3+},Tb^{3+} samples under UV and X-ray excitation could be made.

Moreover, the performances of yttrium tantalate based phosphors strongly depend on the activator-flux system in association with the thermal synthesis regime.

4.2.5.9 Luminescence mechanism in double activated Y(Ta,Nb)O$_4$:$Eu^{3+}Tb^{3+}$

The general luminescence in tantalates and niobates consists of luminescence of the host lattice and luminescence of activators.

The luminescence in the host lattice $YTaO_4$ or $YNbO_4$ is ascribed to the ligand-metal charge-transfer transition of tetrahedral $[TaO_4]^{3-}$ or $[NbO_4]^{3-}$ molecular ions and showed a wide emission bands [291]. For the TaO_4 group the charge-transfer transition will require a higher energy than in the case of the niobate. This is due to the lower 5th ionization potential (I_5 = 45 eV for Ta and 52 eV for Nb) [250]. The excitation and emission spectra of the tantalate are, therefore, shifted to higher energies relative to the niobate. Because for TaO_4, this charge transfer transition requires considerably higher energy than for the NbO_4 case, it is the reason why $YTaO_4$ cannot be excited with 254 nm radiation. The absorption edge is situated at about 225 nm, so that excitation is only possible with very short-wavelength ultraviolet radiation (or e.g. cathode rays). Blasse [250] reported emission of ~330 nm for M-$YTaO_4$ via 229 nm excitation that is very close to the emission observed under X-ray excitation. The speed difference for the M and M′ $YTaO_4$ structures as reported by Brixner and Chen is, therefore, ascribed to a more efficient charge-transfer process for the M′ structure, where the average Ta–O distances are closer than in the M structure [185]. For $YNbO_4$ the band position is peaked around 400 nm under UV excitation at 254 nm [185, 251], CL with 0.5–1 keV [292] or X-ray excitation (Mo 30 kVp (peak kilovoltage)) [12, 185].

The emission of low concentrations of NbO_4 groups in tantalates was investigated by Blasse and Bril and they showed that the NbO_4 group is an efficient luminescent center in the $YTaO_4$ host [250]. Only the niobate group in $Y(Ta,Nb)O_4$ and not the host lattice itself is then excited under 254 nm excitation. All the properties are nearly the same as in $YNbO_4$: excitation and emission spectrum, temperature dependence and efficiency. The decay time of $Y(Ta,Nb)O_4$ is a little shorter than that of $YNbO_4$. All this indicates that the interaction between the NbO_4 groups in $YNbO_4$ is negligible, as argued before [251]. The situation is comparable to that of the WO_4 group in $CaSO_4$ and $CaWO_4$ [293]. For excitation into the tantalate group (either by 229 nm radiation or by X-ray or cathode-ray excitation) of $Y(Ta,Nb)O_4$ the situation is somewhat different: the energy transfer occurs from tantalate to niobate. For low niobate concentration this transfer is not efficient enough to quench the emission of all the tantalate groups, so that the emission still contains a band due to the tantalate itself [250].

The luminescence of Eu^{3+} and Tb^{3+} in lanthanide niobates has been described previously [12, 20, 250–252]. Tb^{3+}-doped $YTaO_4$ shows a strong green emission while Eu^{3+}-doped $YTaO_4$ and $YNbO_4$ shows a red emission under UV, VUV, CL, and X-ray excitation, and both phosphors have proved to be efficient phosphors for different applications.

The activators can be excited directly by excitation sours or by energy transfer transition from the host lattice. In the lattice of rare earth doped yttrium tantalate or niobate the doped luminescent centers, such as Eu^{3+} and Tb^{3+} occupy the Y sites. The TaO_4^{3-} and NbO_4^{3-} groups can absorb excitation energy through $O^{2-}\rightarrow Ta^{5+}$ and $O^{2-}\rightarrow Nb^{5+}$ charge transfer transition, respectively, and transfer the energy to rare earth luminescent centers, which give rise to the corresponding characteristic emission. The system $Y_{1-x}Eu_xNbO_4$ was studied more thoroughly [251] because it illustrates clearly the importance of excitation energy transport through the lattice in order to obtain efficient phosphors. Brixner [249] mentions the intensity and color of the luminescence of a number of fergusonites. Later on this author reported on the luminescence of the Eu^{3+} ion in niobates and tantalates with fergusonite structure [294]. That study contains two remarkable conclusions, firstly that concentration quenching occurs at an extremely high Eu concentration, viz., at about 50%, and secondly that the efficiency of the tantalate phosphors relative to the niobate phosphors is very low.

Some recent publications with separated activation by Eu^{3+} or Tb^{3+} yttrium niobate systems [256–258, 280, 281, 295] and yttrium tantalate systems [179, 259–261] have been appeared in the last years. However, no work has been reported using simultaneous doubly activation (Eu^{3+} and Tb^{3+}) for Y(Ta,Nb)O_4 phosphors. The simultaneous incorporation of Eu^{3+} and Tb^{3+} ions may allow us to obtain different luminescence colors on the entire visible spectrum. It is also possible that, in certain energy conditions a process of energy transfer from Tb to Eu appears, because of the proximity of the Tb^{3+} ($^5D_4$) and Eu^{3+} ($^5D_1$) energy levels. In this case, the Eu–Tb activator couple could function as a donor-acceptor pair [222, 296, 297].

The work presented in this section continuer describes an investigation of the luminescence in tantalate and niobate systems

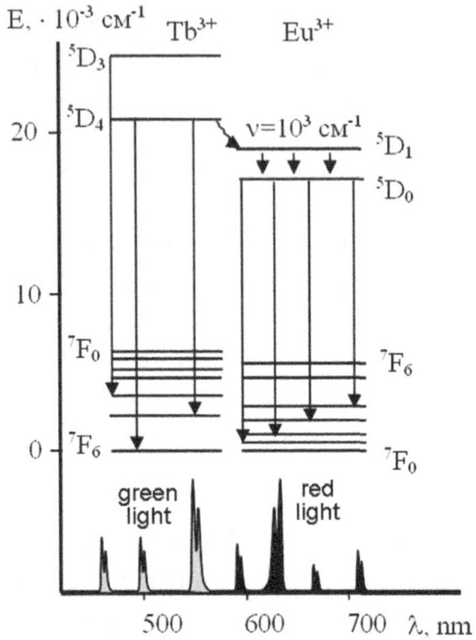

Figure 4.67 Schematic representation of the energy transfer in $CaWO_4$, double activated with Tb^{3+} and Eu^{3+} ions.

and takes into account simultaneous double activation of these compounds by Eu^{3+} and Tb^{3+}.

Two- and *three-level* models for the energy transfer between the activators, proposed first for CL investigation [222] (Fig. 4.5), are developed here for the photoluminescence and other different kinds of excitation. It is possible and reasonable to enlarge this theory because all the spectra PL, CL, and XL have the same luminescent profile, with similar peaks and the relative intensity, which means no matter what the excitation is, the mechanism of energy transfer of $Y(Ta,Nb)O_4$:Eu^{3+},Tb^{3+} is the same (Fig. 4.67). In contrast to cathodoluminescence, in PL we can excite directly the lowest energy levels $^5D_4$ for Tb^{3+} or $^5D_0$ for Eu^{3+} ions. If the photo-excitation is carried out with higher energy, the non-radiative transitions on the corresponding $^5D_4$ and $^5D_0$ levels take place and then radiative transitions on the ground states occurred.

For *the two-level system* (one excited state and one ground state), the UV emission intensity from the excited level (*e*) is determined by the population of this level. According to the two-level luminescent system, the population of the excited level decreases as shown in Eq. (4.18):

$$\frac{dn_e}{dt} = -Cn_e + XIn_g - XIn_e \tag{4.18}$$

where n_e and n_g are the number of luminescent ions in the excited (e) and ground (g) state, respectively, N is the total number of luminescent ions $N = n_e + n_g$, t is the time, C is the spontaneous radiation coefficient, X is the absorption coefficient (simplified representation for excited and ground states), and I is the light yield of UV incident. When it reaches the stationary equilibrium state, $dn_e/dt = 0$. Eq. (4.18) becomes

$$n_e = \frac{XIN}{C+2XI} \tag{4.19}$$

The expression (4.19) explains the linear dependence of luminescence intensity on the activator concentration when the concentration is low and the interaction between ions is not to be considered.

When Tb^{3+} and Eu^{3+} ions are presented at relative high activator concentration, the interaction between activators should be taken into account and the more complicated *three-level model* (with excited states, ground states and one intermediate state $^5D_1$ for Eu^{3+}) is proposed. Two kinds of non-radiative relaxation transitions have to be considered. One of them is from the $^5D_4$ level of Tb^{3+} to $^5D_1$ level of Eu^{3+}, and in this case Tb^{3+} serves as a donor while Eu^{3+} serves as an acceptor. And the other one is from $^5D_1$ level to $^5D_0$ level inside the Eu^{3+} ion with $4f^6$ configuration. One additional channel for population of acceptor level appears. Eu^{3+} and Tb^{3+} concentration determine the probability of energy transfer. The most probable energy transfer corresponds to the case when the activators concentrations are equal. One can assume that the probability of this process could be given by the next equation:

$$K = \frac{N_{\mathrm{Tb}} N_{\mathrm{Eu}}}{N^2} \tag{4.20}$$

where N is the sum of Eu^{3+} and Tb^{3+} concentrations.

The additional energy transfer from the excited $^5D_4$ level of Tb^{3+} reduces the life time of this excited state. For the stationary condition of excitation, the following expression is applied.

$$\left(\frac{dn_e}{dt}\right)_{Tb} = -Cn_e + XIn_g - XIn_e - Kn_e \qquad (4.21)$$

,

where n_e and n_g are populations of the excited and ground state of Tb ions, respectively. Considering that $n_e + n_g = N_{Tb}$ and $N_{Tb} + N_{Eu} = N$, the expression for Tb^{3+} excited level is given by Eq. (4.22):

$$I_{Tb} \sim n_e = \frac{XIN_{Tb}}{2XI + C + (1/N^2)N_{Tb}(N - N_{Tb})} \qquad (4.22)$$

For Eu^{3+} ions, the energy transfer from Tb^{3+} means that one additional channel becomes available. When the stationary conditions of excitation are applied, the expression for Eu^{3+} is given by

$$\left(\frac{dn_e}{dt}\right)_{Eu} = -Cn_e + XIn_g - XIn_e - Kn_e, \qquad (4.23)$$

where n_e and n_g are populations of the excited and ground state of Eu ions, respectively. Considering $n_e + n_g = N_{Eu}$ and $N_{Tb} + N_{Eu} = N$, the expression for Eu^{3+} excited level is given by

$$I_{Eu} \sim n_e = \frac{XIN_{Eu}}{2XI + C - (1/N^2)N_{Eu}(N - N_{Eu})} \qquad (4.24)$$

Comparing Eqs. (4.22) and (4.24), one can conclude that when the interaction between Tb^{3+} and Eu^{3+} is taken into consideration, the luminescence intensity from Eu centers is higher than that from Tb centers.

When $C >> 2XI - (1/N^2)N_{Eu}(N - N_{Eu})$, namely the intensity of incident light, is much more than light yield of Eu^{3+}, the luminescence intensity could be described as

$$I_{Eu} \sim n_e = \frac{XIN_{Eu}}{C} \qquad (4.25)$$

According to the above-mentioned model, the Tb^{3+} emission intensity decreases non-linearly, whereas the Eu^{3+} emission intensity increases linearly with activator concentration.

The proposed models are confirmed further by numerous measurements with different excitations such as high- and low-voltage CL (see section 4.1), PL including UV and VUV, and X-ray luminescence. This model is available in the most matrices, such as yttrium tantalate, yttrium niobate, calcium tungstate [222, 298], zink silicate [296, 299], porous silicon [300], porous phosphate [301], zeolite-Y [288], calcium molibdate [302], and others.

4.2.6 *Stark Levels and Selection Rules*

4.2.6.1 Stark levels and selection rules for the transitions in Tb^{3+} and Eu^{3+} ($4f^8$ shell)

The analysis of the optical lines is based on the calculation of the selection rules for the electronic transitions. First, let us consider the allowed dipole type transitions. In the Russel–Saunders coupling scheme, the electronic states of a free rare earth ion can be labeled as *LSJ* as the conventional notations are used. These states of the f^8-shell of the Tb^{3+} free ion belong to the even irreducible representations (irreps) D_J^+ of the full rotation symmetry group. The pattern of the optical lines in emission and absorption of Tb^{3+} in a crystal is usually related to the following set of the transitions within the low lying levels of the free ion:

$$^5D_4, {}^5D_3 \rightarrow {}^7F_6, {}^7F_5, {}^7F_4, {}^7F_3, {}^7F_2, {}^7F_1, {}^7F_0.$$

The energy levels of a free ion are split in the crystal field produced by the host lattice. Our aim is to qualitatively analyze the crystal field splitting and find the selection rules and polarization of the lines in the crystal field (C_2 point group) of the host lattice for the Tb^{3+} transitions. In order to classify the energy levels of Tb^{3+} in the crystal field, the irreps D_J^+ should be decomposed in the point group C_2. According to the set of the irreps of C_2 resulting from D_J^+ [3], each *LSJ* multiplet is split into the Stark components designated accordingly to the irreps of C_2 as A, B (one-dimensional irreps of C_2)

Table 4.14 Assignment of crystal field multiplets for the Tb^{3+} ion in C_2 point group

LSJ terms	$^5D_4$	$^5D_3$	$^7F_6$	$^7F_5$	$^7F_4$
Crystal field terms	$5A + 4B$	$3A + 4B$	$7A + 6B$	$5A + 6B$	$5A + 4B$
LSJ terms	$^7F_3$	$^7F_2$	$^7F_1$	$^7F_0$	
Crystal field terms	$3A + 4B$	$3A + 2B$	$A + 2B$	A	

and E (two-dimensional irrep) [231]. Using the well-known group-theoretical procedure, one can find the following results for the decomposition $D_J^+ \rightarrow$ irreps of C_2:

$$D_6^+ \rightarrow 7A + 6B,\ D_5^+ \rightarrow 5A + 6B,\ D_4^+ \rightarrow 5A + 4B,$$

$$D_3^+ \rightarrow 3A + 4B,\ D_2^+ \rightarrow 3A + 2B,\ D_1^+ \rightarrow A + 2B,\ D_0^+ \rightarrow A$$

Following these results, one can find out the assignment of crystal field multiplets for the Tb^{3+} ion in C_2 point group (Table 4.14).

The crystal field splitting is schematically illustrated in Fig. 4.68. It should be noted that for the Tb^{3+} ion, the ground term is $^7F_6$ ($J = 6$).

To discuss the pattern of the optical lines, the selection rules for the transitions are to be found. Because the point symmetry does not contain the inversion center, the allowed dipole type transitions play a dominant role. The components of the dipole moment correspond to A and B irreps: $d_z \rightarrow A$, $d_x, d_y = B$. The selection rules for the allowed transitions and their polarizations are determined by the direct products of irreps of C_2 for the two levels participating in the optical transition [3]; $A \times A = A$, $B \times B = B$, $A \times B$. Therefore, one can find the transitions in z-polarization (along C_2) and in x, y-polarization (in plane perpendicular to C_2). They will be referred as "parallel" (||) and "perpendicular"(⊥) polarizations, respectively, according to the active components of the dipole moment. The results are summarized in Table 4.15.

The transition scheme (Fig. 4.68) provides only a possible number of the lines in || and ⊥ polarizations and a qualitative picture of the crystal field splitting. The energy gap intervals between the

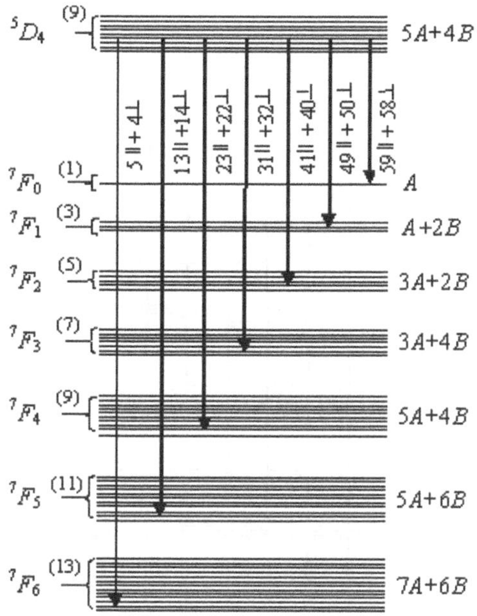

Figure 4.68 Splitting of $^{2S+1}D_J$ multiplets for Tb^{3+} in the crystal field of the C_2 symmetry site.

crystal field levels cannot be found on the basis of symmetry only and require also some additional information including the relative intensities. In particular, some lines are formally allowed but their intensities could be small. Some useful conclusions can be drawn if one assumes that the ground crystal field level is only populated in $^5D_4$. This can be either *A* or *B* at very low temperatures. For example, if the ground sublevel is *A*, we have only the line $A \leftrightarrow A$ (that is ||) in the region of $^5D_4 \rightarrow {}^7F_0$.

The scheme of the levels for the Eu^{3+} ion ($4f^6$) is shown in Fig. 4.69.

Table 4.15 Selection rules and polarizations

Transitions	Polarizations
A↔A	‖
B↔B	‖
A↔B	⊥

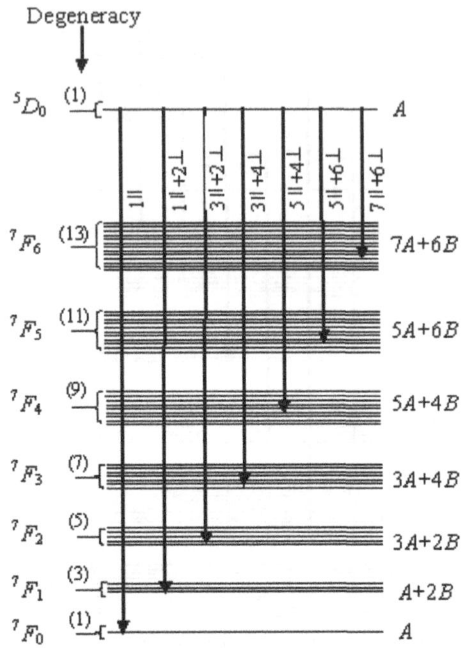

Figure 4.69 Splitting of $^{2S+1}D_J$ multiplets for Eu^{3+} in the crystal field of the C_2 symmetry site.

One can see that the order of the low lying levels is reversed with respect to that in Tb^{3+} ($4f^8$). Now, the ground state is $^7F_0$ ($J = 0$), which gives rise to a singlet A in the crystal field. The assignment of the Stark levels to the *LSJ* multiplets is similar to that previously considered (Table 4.14). Applying the selection rules so far described, one can find for the Eu^{3+} ion a scheme of transitions depicted in Fig. 4.69. It should be noted that in high-energy region of the spectrum one can observe the only weak line ($A \rightarrow A$) polarized along C_2 axis.

4.2.6.2 Experimental measurements

UV excitation: Typical photoluminescence excitation and photoluminescence emission spectra of $Y(Ta,Nb)O_4$:Tb^{3+} under 254 nm excitation are presented in Figs. 4.35 and 4.36. For $Y(Ta,Nb)O_4$:Eu^{3+} the similar spectra are shown in Figs. 4.39 and 4.40.

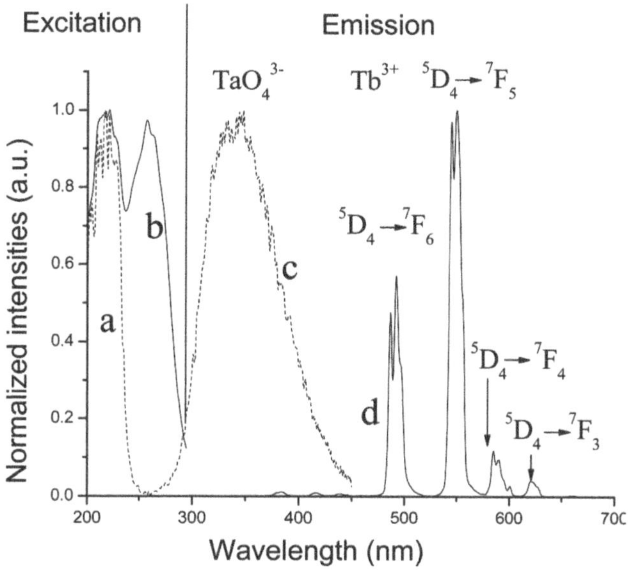

Figure 4.70 PLE (a,b) and PL (c,d) spectra of $YTaO_4$ (a,c) and $YTaO_4$:Tb^{3+} (b,d).

Here we compare excitation and emission spectra of non-activated host lattice $YTaO_4$ with Tb^{3+} activation (Fig. 4.70).

Only one band located at about 227 nm exists in PLE spectrum of the host material (a) and two obvious bands (227 and 260 nm) appeared in Tb^{3+} activated yttrium tantalate (b). The absorption edge of $YTaO_4$ is situated at about 225 nm, so that excitation is possible only with very short-wavelength ultraviolet radiation or X-rays. The dashed line in Fig. 4.70 indicates the PLE spectrum (a) (λ_{em} = 330 nm) and emission spectrum (c) of undoped M′-type $YTaO_4$ under 227 nm excitation. The emission spectrum shows a broad band with a peak at 330 nm wavelength and can be ascribed to the emission from the charge transfer transition of the TaO_4^{3-} group [250]. Blasse and Bril [250] ascribed the broad band (a) at 208–229 nm to the absorption of the host lattice (TaO_4 group). The absorption band (b) peaked at 260 nm belongs to the Tb^{3+} center itself and as it was mentioned above, this band belongs to the ^9D level of the $4f^8 \rightarrow 4f^7 5d^1$ transition of Tb^{3+} within its $4f^8$ configuration. The TaO_4^{3-} group can absorb the excitation energy through $O^{2-} \rightarrow Ta^{5+}$ charge transfer transition and transfer the energy to the rare earth luminescent centers, which give a rise to the corresponding characteristic emission. The luminescence of Tb^{3+} (d in Fig. 4.70) is composed of typical $^5D_4\rightarrow^7F_J$ ($J = 3, 4, 5, 6$) transitions, and the main emission line is $^5D_4\rightarrow^7F_5$ transition at 547 nm. Other transitions from $^5D_4$ excited level to the ground state ^7F_J ($J = 0, 1, 2$) were not observed.

From these spectra (Figs.4.35, 4.36, 4.39, 4.40, and 4.70) one can see that only some of allowed and predicted from the theory transitions in Tb^{3+} and Eu^{3+} can be visualized under UV excitation at room temperature. The luminescence of activators not only depends on the host lattice but also strongly depends on the excitation energy and temperature. VUV ($\lambda_{exc} = 147$ nm) and high-energy X-ray excitation, as well as low-temperature measurements was used to reveal all allowed transitions between Stark components of Tb^{3+} and Eu^{3+} in symmetry of the actual crystal field.

4.2.6.3 VUV excitation

The VUV excited luminescence of $YTaO_4:Tb^{3+}$ is shown in Fig. 4.71.

The excitation spectrum was measured in the range from 100 to 350 nm using sodium salicylate powder as a reference. The excitation spectra were obtained by observing all emission light and calibrated by that of sodium salicylate, which has constant quantum efficiency in 115–300 nm range. It is clearly seen from Fig. 4.71 that there are three obvious bands *A*, *B* and *C* in the VUV PLE spectra. In the VUV PLE spectra of the rare earth ions activated luminescent materials, five physical processes have been reported: (1) $f\rightarrow f$ transition of the rare earth ions, (2) $f\rightarrow d$ transition of the rare earth ions, (3) charge transfer band from coordination anions to the rare earth ions, (4) the absorption of the host lattice, (5) electronic excitation of the host such that the high-energy photons of VUV excitation are absorbed and create charge carriers like free electrons in conduction band and holes in valence band. In other words, excitation in the VUV range produces electron–hole pairs that undergo recombination at activator ions.

The three bands observed in the VUV range in $YTaO_4:Tb^{3+}$ are related to the following processes. The absorption band *C* peaked at 260 nm belongs to the Tb^{3+} center itself and as it was mentioned

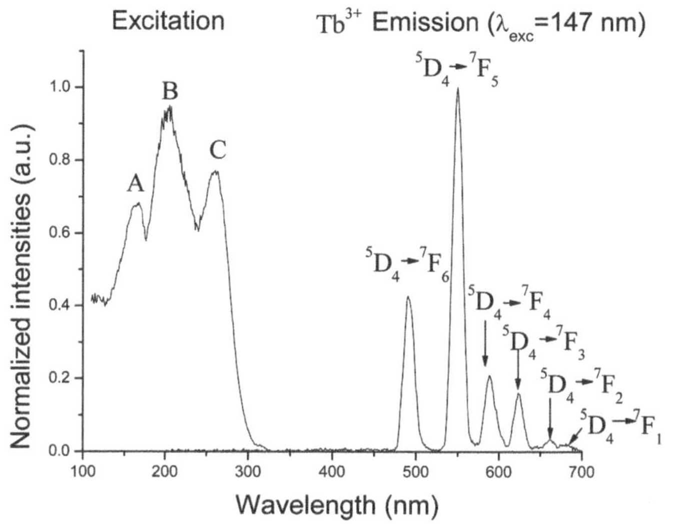

Figure 4.71 PLE and emission spectra from $YTaO_4$:Tb^{3+} under VUV excitation.

above, this band belongs to the ^9D level of the $4f^8 \rightarrow 4f^7 5d^1$ transition of Tb^{3+} within its $4f^8$ configuration.

The band *B* in Fig. 4.71 from 180 to 220 nm is the same as the band (a) in Fig. 4.70, which is associated with the absorption of the host lattices. The TaO_4^{3-} group can absorb the excitation energy through $O^{2-} \rightarrow Ta^{5+}$ charge transfer transition and transfer the energy to the rare earth luminescent centers.

The new band *A* peaked around 160 nm, which was not registered under the normal UV PLE, is probably related to the host absorption. The Y–O bonds are excited and the energy is transferred to TaO_4^{3-} group or to the activators through the host charge-transfer transition (HCTT). The position of this band *A* for Y^{3+} ions can be calculated with the help of an empirical formula given by Jorgensen [303]:

$$E_{CT} = [(X) - (M)] \times 30{,}000 \text{ cm}^{-1}$$

Here, E_{CT} gives the position of the HCTT in cm^{-1}, (X) the optical electronegativity of the anion, and (M) that of the central metal ion. Using Pauling scale for electronegativity [304], namely (O)=3.44, (Y)=1.22, the HCTT of Y–O can be estimated near 67,000 cm^{-1}, or around 150–160 nm. The result is close to the O^{2-}–Tb^{3+} CT bands

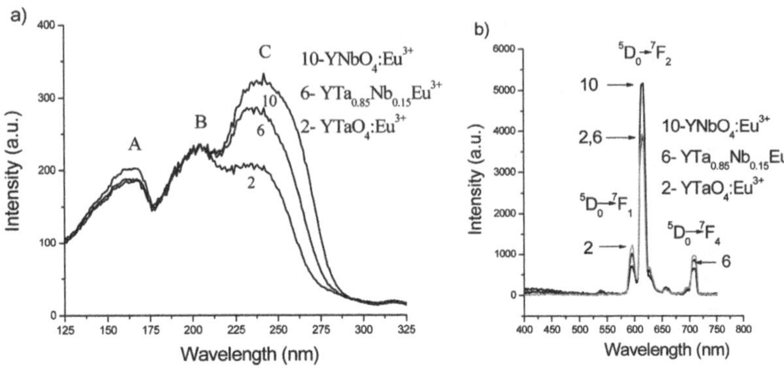

Figure 4.72 Excitation (a) and emission (b) spectra of tantalate-niobate systems activated by Eu^{3+}.

at 68,100 cm^{-1} (RE = Y) and at 63,500 cm^{-1} (RE = La) reported by Gerard et al. [305].

The luminescence of Tb^{3+} is composed of typical $^5D_4 \rightarrow {}^7F_J$ ($J = 3, 4, 5, 6$) transitions, and the main emission line is $^5D_4 \rightarrow {}^7F_5$ transition at 547 nm. Other transitions from $^5D_4$ excited level to the ground state ^7F_J ($J = 0, 1, 2$) allowed from the theory (Fig. 4.68) were not observed experimentally under UV excitation (Fig. 4.70) and are visible at VUV (Fig. 4.71).

Three bands were also observed in the VUV PLE spectra of tantalate-niobate systems activated by Eu^{3+} (Fig. 4.72a). These spectra are similar to PLE Fig. 4.71(left).

The absorption band *C* for Eu^{3+} associated with CTB of $O^{2-} \rightarrow Eu^{3+}$ [289, 290]. In $YNbO_4$ this band is peaked at 260 nm, while in Y(Ta,Nb)O_4 is shifted to 250 nm and in $YTaO_4$ host lattice is shifted to 240 nm. The nature of bands *B* and *A* is similar to previous explanation: *B* from 180 to 220 nm is associated with the absorption of the host lattices, and *A* peaked around 160 nm is related to the host charge-transfer transition. In VUV emission spectra (Fig. 4.72b) the weak emission in the vicinity of 590–600 nm is due to the magnetic dipole transition $^5D_0 \rightarrow {}^7F_1$. The strong emission around 610–630 nm is due to the hypersensitive electric dipole transition of $^5D_0 \rightarrow {}^7F_2$ induced by the lack of inversion symmetry at the Eu^{3+} site. No additional band in emission spectra in comparison with UV excitation (Fig. 4.39) was registered.

One more advantage of VUV investigation is illustrated in the emission spectra. Comparing the emission VUV (147 nm) and

UV (254 nm) spectra, one can see that the spectra are similar. Nevertheless, some additional transitions in Tb^{3+} emission from $^5D_4$ excited level to the ground state ^7F_J(J = 1, 2) are appeared under VUV excitation. Usually, these emissions are very weak and could not be seen under UV excitation. In Eu^{3+} activation the transition $^5D_0 \rightarrow ^7F_4$ under VUV excitation becomes stronger.

4.2.6.4 X-ray excitation

Under high energy X-ray excitation almost all possible allowed electron transitions from excited levels $^5D_0$ (for Eu^{3+}) and $^5D_4$ (for Tb^{3+}) to the ground state ^7F_J (J = 0–6) appear and can be observed in XL spectra. One of the examples of X-ray excited luminescence from $YTaO_4$:Tb^{3+} is shown in Fig. 4.73.

It is not surprising to distinguish weak electron transitions under X-ray excitation, because empty levels which usually exist in any materials can be readily filled under high-energy excitation and

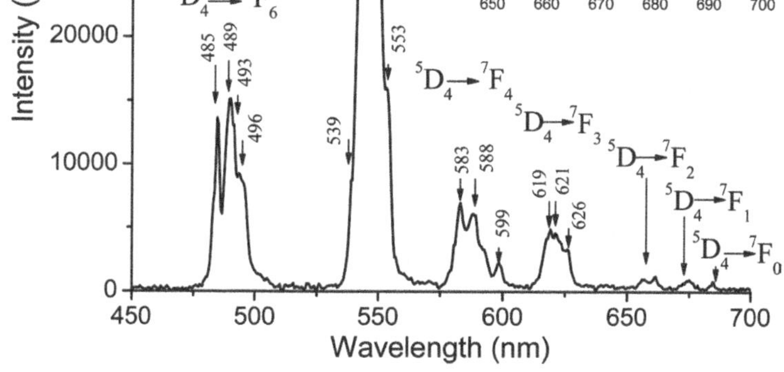

Figure 4.73 Emission spectra of $YTaO_4$:Tb^{3+} under X-ray excitation. Fragments of the transitions $^5D_4 \rightarrow ^7F_J$ (J = 2, 1, 0) illustrating the specific lines of Tb^{3+} are also shown in the figure.

radiative emissions not more trapped by the empty upper levels. This explains the increment of the luminescence intensity of Tb^{3+} emission centers under X-ray excitation. Experimental results for X-ray luminescence are in good agreement with selection rules ($4f^8$ shell) and splitting of $^{2S+1}D_J$ multiplets for Tb^{3+} in the crystal field at the C_2 symmetry site.

4.2.6.5 Low-temperature PL measurements

For low-temperature experiments the next measurement condition were used: source – 325 nm He–Cd laser, detector – PMT, grating – 1200 gr/mm, scan range – 450~720 nm, slit size – 0.03 mm, step – 0.05 nm, and temperature – 10 K.

The emission spectrum of $YTaO_4$:Tb^{3+} under 325 nm He–Cd laser excitation at 10 K is presented in Fig. 4.74. Fragments of the energy transitions $^5D_4 \rightarrow ^7F_J$ (J = 6, 5, 4, 3) and distribution the optical lines are shown in the Figs. 4.75–4.78. Other allowed transitions J = 2, 1, 0 are very weak under 325 nm excitation and will be examined in the next paragraphs under CL and X-ray excitation.

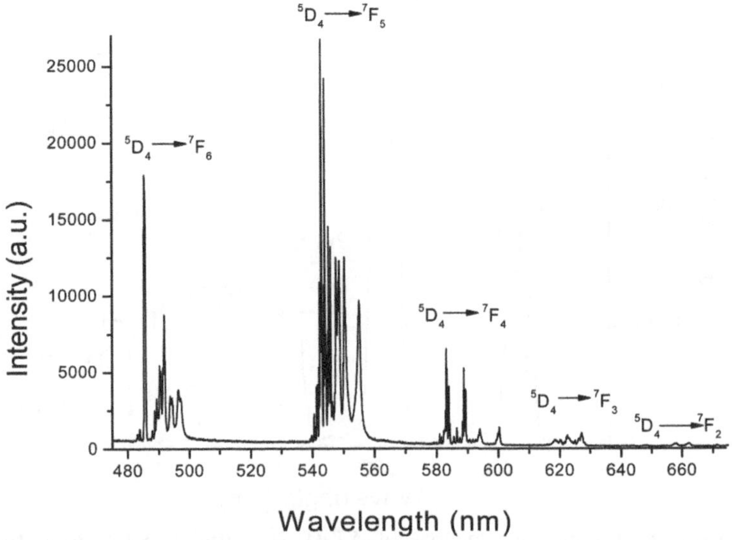

Figure 4.74 Emission spectrum of $YTaO_4$:Tb^{3+} under 325 nm He–Cd laser excitation at 10 K.

The energy levels and wave functions of the free Tb^{3+} ion have been calculated by Ofelt [235]. The main energy transitions for Tb^{3+} are from high energy excited level $^5D_4$ down to the ground levels ^7F_J ($J = 6, 5, 4, 3, 2, 1, 0$). The energy levels of a free ion are split in the crystal field produced by the host lattice. The crystal field splitting depends on the host lattice and for $YTaO_4$ and $YNbO_4$ (C_2 point group) is schematically illustrated in Figs. 4.68 and 4.69. We cannot observe in detail this splitting at room temperature because of the different temperature effects, but at low-temperature measurements this splitting can be clearly seen in Fig. 4.74. The intensity of the $^5D_4 \rightarrow {}^7F_5$ transition at about 550 nm dominates giving rise to a more or less saturated green luminescence. In general, however, the intensities of the $^5D_4 \rightarrow {}^7F_6$, $^7F_4$, $^7F_3$ we also have to take into account, but other transitions $^5D_4 \rightarrow {}^7F_2$, $^7F_1$, $^7F_0$ are negligible.

All Stark levels predicted by the theory are observed at low-temperature measurements and marked on the Figs. 4.75–4.78: $^7F_6$ level is split on 13 optical sublevels; $^7F_5$ level is split on 11 peaks, $^7F_4$ level consists from 9 separated peaks, $^7F_3$ level includes 7 peaks, etc. Moreover, fine structure can also be seen. Most of these sublevels have a complicated structure and are also split. Owing to the

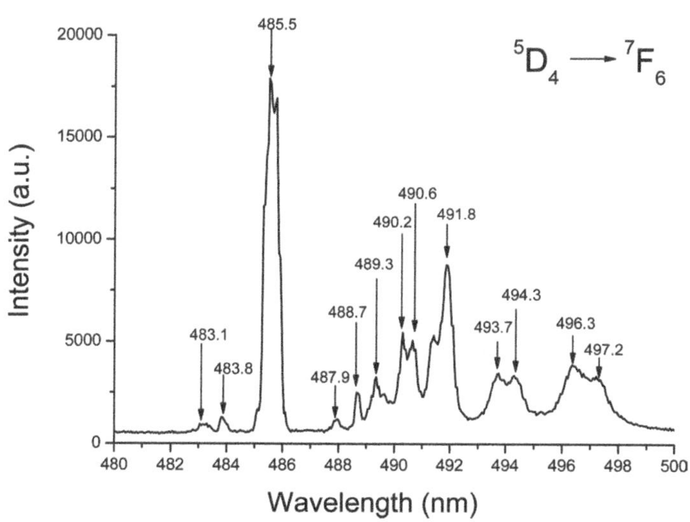

Figure 4.75 Distribution the optical lines in $^5D_4 \rightarrow {}^7F_6$ Tb^{3+} transition in $YTaO_4$:Tb^{3+} at 10 K.

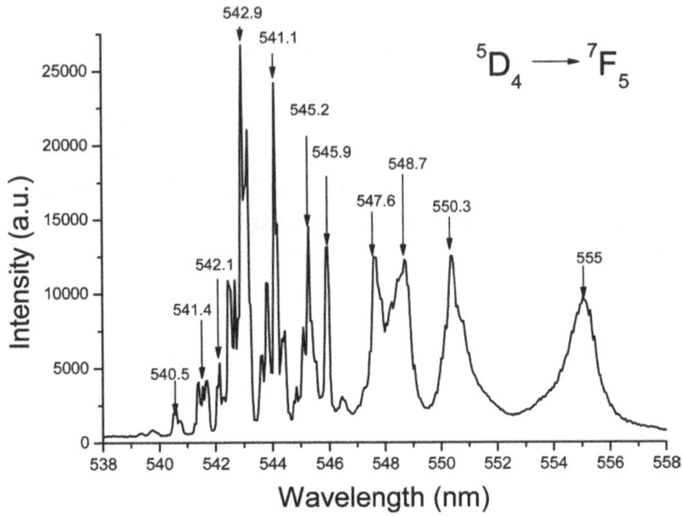

Figure 4.76 Distribution the optical lines in $^5D_4 \rightarrow {}^7F_5$ Tb^{3+} transition in $YTaO_4:Tb^{3+}$ at 10 K.

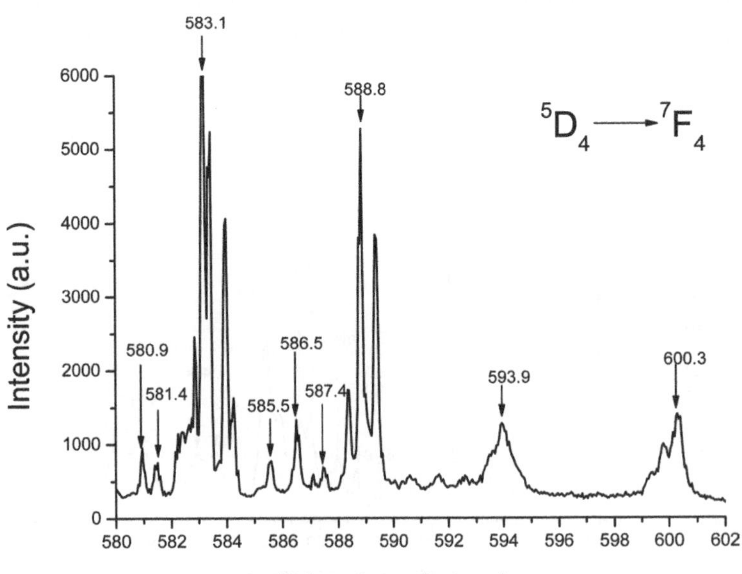

Figure 4.77 Distribution the optical lines in $^5D_4 \rightarrow {}^7F_4$ Tb^{3+} transition in $YTaO_4:Tb^{3+}$ at 10 K.

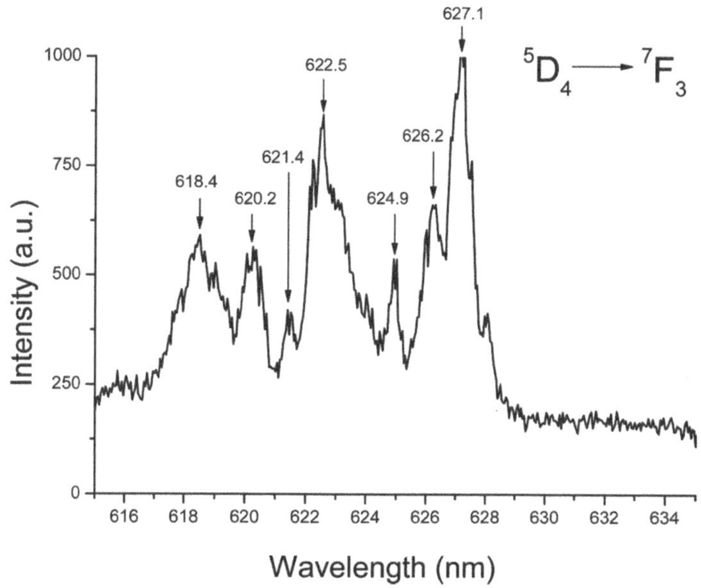

Figure 4.78 Distribution the optical lines in $^5D_4 \rightarrow {}^7F_3$ Tb^{3+} transition in $YTaO_4:Tb^{3+}$ at 10 K.

crystal-field effect, the energy levels exhibit a more or less pronounced Stark splitting, so that each line is replaced by a number of neighboring lines.

4.2.7 *First-Principles Electronic Structure Calculations*

The luminescence and structural properties of tantalate and niobate systems allowed us to calculate the crystal structure of M′-$YTaO_4$ and M-$YNbO_4$ host lattices. As is well known, performances of these phosphors and luminescence properties strongly depend on their crystal structure. However, some fundamental questions concerning host lattice emission and charge transfer transitions still remain unanswered.

In the last decade some efficient methods based on density functional theory were applied to niobate system doped by Bi [281, 306, 307]. To the best of our knowledge, no work has been reported using first-principles calculations based on the DFT to give a reasonable interpretation to host lattice. Even though tantalate and niobate phosphors are well known, there have been no theoretical approaches to interpret their luminescence. In this chapter electronic band structure, and density of states (DOS) calculations are performed via the full potential linear-augmented plane wave approach, as implemented in the WIEN2k code [244], within the framework of DFT. WIEN2k is based on the method which is among the most precise and reliable ways to calculate the electronic structure of solids. We believe that this approach provides a suitable first-principles framework for making direct comparisons with the experimental spectra. The calculating scheme is described in Chapter 3 (3.7. Computational Materials Science) and so we provide only the bare essentials here. Thus, the nature of optical excitations near the band gap of $YTaO_4$ and $YNbO_4$ was clarified using the first-principles calculations. The efficient band gap, corresponding to real data, and simple model illustrating the emission process in tantalate and niobate systems are discussed.

4.2.7.1 Samples preparation

Several series of yttrium tantalate and niobate phosphors were prepared by solid state reaction method from homogeneous mixture consisting of Y_2O_3 (99.9%) and Nb_2O_5 or Ta_2O_5 (Optipur) [176]. Na_2SO_4 was used as flux. The mixtures were homogenized with a ball mill, in acetone medium, and dried at 70°C. The phosphor samples were baked at 1200°C for 4 h and slowly cooled to room temperature. Finally, the samples were water-washed, dried, and sieved.

4.2.7.2 Densities of states calculations

The electronic structure of host lattices $YTaO_4$ and $YNbO_4$ can be calculated in the framework of density functional theory. The exchange correlation energy of electrons is described in the generalized gradient approximation within the scheme of Perdew et al. GGA96 [308] to calculate the total energy, band structure and DOS. The crystalline structures of phosphors were determined by XRD measurements on the basis of synchrotron X-ray diffraction patterns (Pohang Accelerator Laboratory, Korea) and the atomic positions of all the elements are presented in the Table 4.16.

Table 4.16 Atomic positions and lattice parameters in $YTaO_4$ and $YNbO_4$

M′-YTaO$_4$	*Atom*	*x*	*y*	*z*	*F*
$a = 5.29568$ Å $b = 5.445447$ Å $c = 5.10940$ Å $\beta = 96.39119$	Y	0.25	0.76523	0	1
	Ta	0.25	0.30632	0.5	1
	O(1)	0.49619	0.43402	0.27800	1
	O(2)	0.09912	0.09079	0.24181	1
M-YNbO$_4$	*Atom*	*x*	*y*	*z*	*F*
$a = 7.61851$ Å $b = 10.94611$ Å $c = 5.29745$ Å $\beta = 138.43617$	Y	0	0.37891	0.25	1
	Nb	0	0.85620	0.25	1
	O(1)	0.20166	0.78039	0.20515	1
	O(2)	0.25050	0.96246	0.65526	1

The total DOS for the two compounds $YTaO_4$ (a) and $YNbO_4$ (b) within a 20 eV range, with the zero of energy taken at the top of the last occupied band, are shown in Fig. 4.79.

Partial densities are presented separately on expanded scales in Fig. 4.80 for $YTaO_4$ and in Fig. 4.81 for $YNbO_4$.

Only main *p*-like state for oxygen and *d*-like states for yttrium, tantalum, and niobium are presented in Figs. 4.80 and 4.81. The valence band (VB) of $YTaO_4$ (left side in Fig. 4.80) and $YNbO_4$ (Fig. 4.81) mostly consists from O 2*p* states and they are very

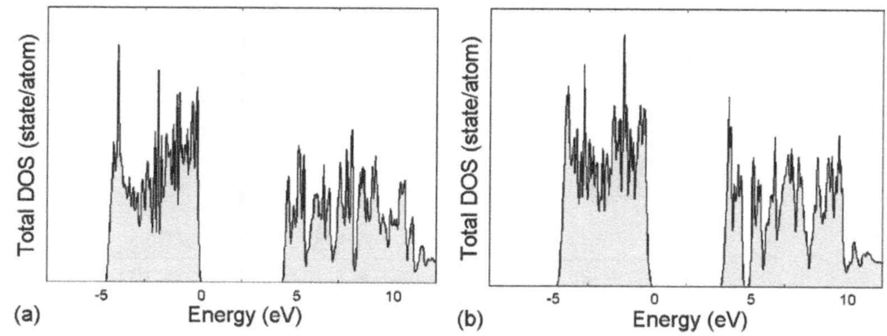

Figure 4.79 Total DOS of $YTaO_4$ (a) and $YNbO_4$ (b).

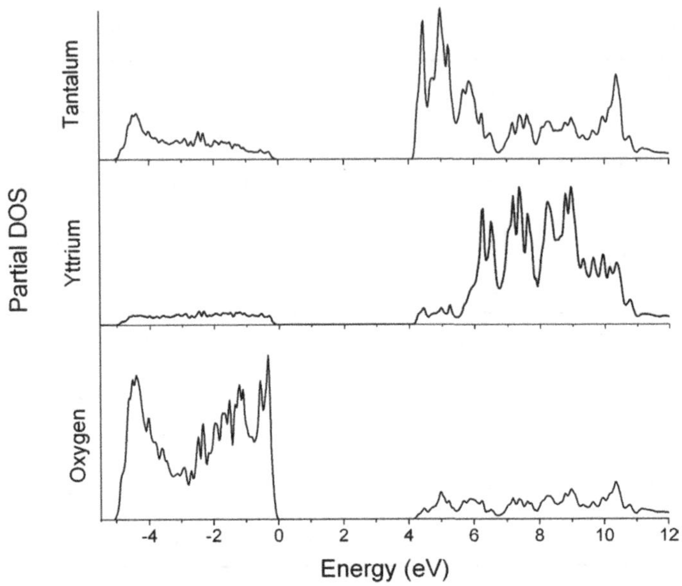

Figure 4.80 Partial densities of $YTaO_4$.

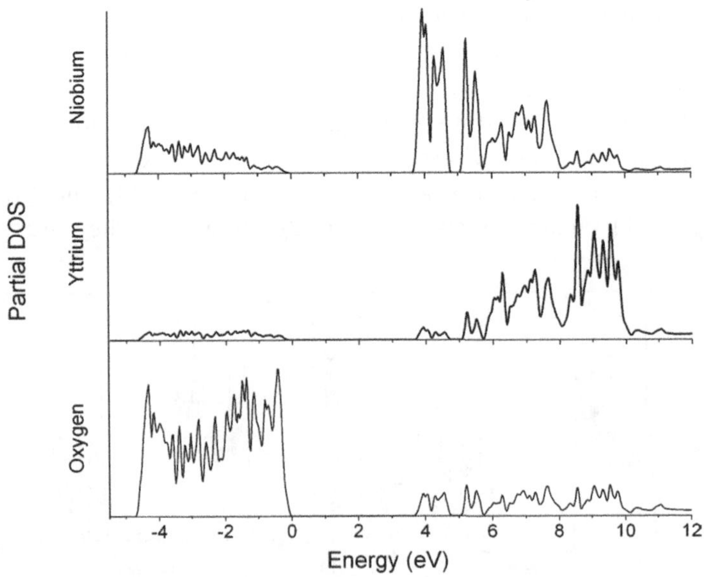

Figure 4.81 Partial densities of $YNbO_4$.

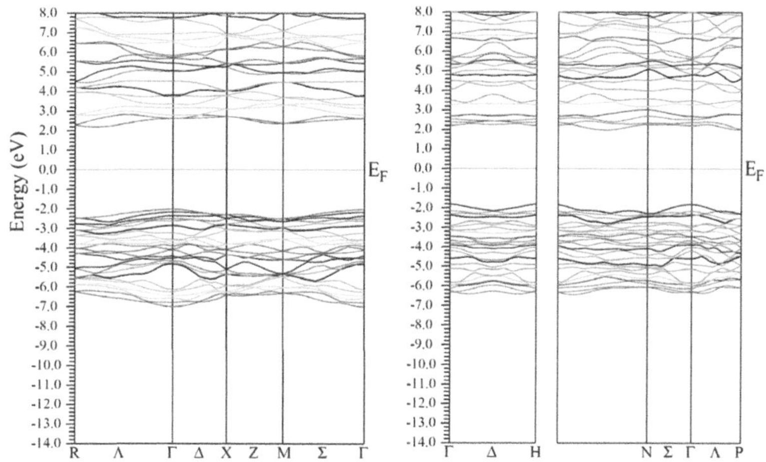

Figure 4.82 Band structure diagram for $YTaO_4$ and $YNbO_4$ plotted in the range of 14 eV below and 8 eV above the Fermi level taken as the zero of energy.

similar. Conduction band (CB) (right side in Figs. 4.80 and 4.81) is divided into two parts. The lower conduction band (from 4 to 6 eV) is mainly composed of Ta 5*d* or Nb *4d* states and the upper conduction band (6–10 eV) involves contribution mainly from Y 4*d* states. The charge transfer gap between Ta and O (Fig. 4.80) and Nb and O (Fig. 4.81) is clearly shown and it is different from DOS calculations. For tantalate it is about 4.2 eV and for niobate is approximately 3.6 eV.

Band structure diagram of $YTaO_4$ and $YNbO_4$ are shown in Figs. 4.82 (left) and (right), correspondingly.

Band dispersions for the both materials are plotted along three symmetry directions within the Brillouin zone and presented in Fig. 4.82. The shapes of the bands for $YTaO_4$ and $YNbO_4$ are very similar to each other. The valence band maxima and conduction band minima are located at the Γ point, so that these are direct gap materials.

4.2.7.3 Experimental results

Although the values of the band gaps calculated within density functional theory are known to be underestimated [309], it is

Table 4.17 Band gaps estimated from the luminescence excitation spectrum

Material	E_{gap} (eV)	Ref.
	5.0	[310]
$YTaO_4$	5.5	[311]
	5.2	This work
	4.4	[311]
	4.5	[312]
	4.4	[292]
	4.3	[307]
$YNbO_4$	4.4	[280]
	4.3	[306]
	4.3	[313]
	4.3	This work

interesting to compare the gaps taken from the luminescence and absorbance experiments for the two materials as listed in Table 4.17.

The absorbance edge can be measured directly from absorbance spectra and results are shown in Fig. 4.83. These data 5.1 eV for $YTaO_4$ and 4.1 eV for $YNbO_4$ are in good agreement with the luminescence data listed in Table 4.17.

4.2.7.4 DOS and luminescence

To give a reliable interpretation to the overall feature of the luminescence and structural properties of tantalate and niobate systems it is interesting to investigate the excitation spectra of these materials and explain them from DOS theory.

The samples were investigated using X-ray diffraction (Rigaku X-ray Diffractometer), UV and VUV excitation luminescence, and first-principles quantum-mechanical calculations were applied

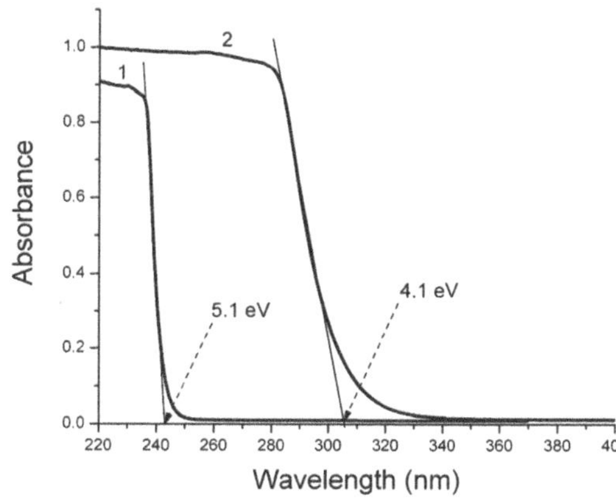

Figure 4.83 The absorbance spectra of $YTaO_4$ (1) and $YNbO_4$ (2).

in order to study their structural and luminescent properties. UV measurements were monitored using the Perkin-Elmer LS50B spectrometer with a xenon flash lamp. VUV excitation spectra of samples were measured using a VUV spectrophotometer equipped with VUV monochromator (ARC, VM 502) and a light source of 30 W deuterium lamp (ARC, DS775–100).

In Fig. 4.84 one can see the excitation spectra from $YTaO_4$ (1) and $YNbO_4$ (2). The photoluminescence excitation spectra were measured in the VUV range from 100 to 350 nm using sodium salicylate powder as a reference. The excitation spectra were obtained by observing all emission light and calibrated by that of sodium salicylate, which has constant quantum efficiency in the 115–350 nm range. Three obvious large bands A, B, and C in the VUV PLE spectra are seen for both host lattices but more clearly they are seen for $YTaO_4$ (1).

The nature of band A is not yet determined and we made here an assumption to clarify it as well as to explain the nature of bands B and C from DFT. We believe that these bands B and C observed in the VUV range in tantalate and niobate systems are related to the following processes.

The band B around 220 nm we associate with the absorption of the host lattices. The TaO_4^{3-} and NbO_4^{3-} groups can absorb excitation energy through $O^{2-}\rightarrow Ta^{5+}$ and $O^{2-}\rightarrow Nb^{5+}$ host charge transfer

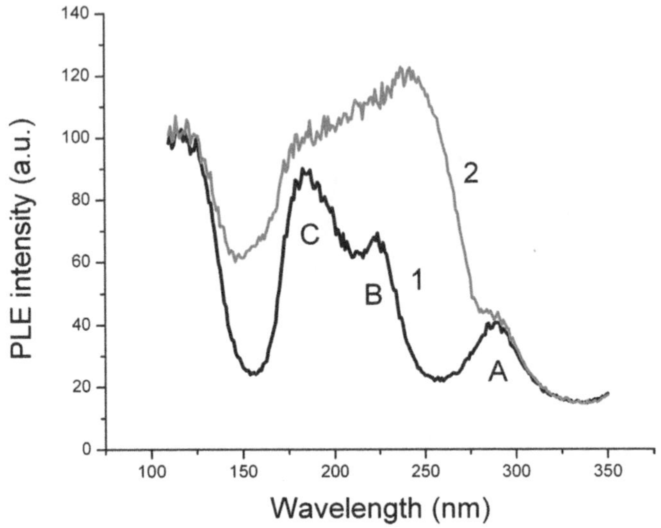

Figure 4.84 PLE spectra for $YTaO_4$ (1) and $YNbO_4$ (2) under VUV excitation.

transition correspondingly. These bands are shifted in tantalate and niobate systems, because of different structure's groups.

The band C, peaked around 170 nm, is probably related to the host absorption. The Y–O bonds are excited and the energy is transferred to host lattice. The position of this band C for Y^{3+} ions can be calculated with the help of an empirical formula given by Jorgensen [303]:

$$E_{CT} = [(X) - (M)] \times 30{,}000\ \text{cm}^{-1}$$

Here E_{CT} gives the position of the charge transfer band (CTB) in cm^{-1}, (X) the optical electronegativity of the anion, and (M) that of the central metal ion. Using Pauling scale for electronegativity [304], namely (O) = 3.44 and (Y) = 1.22, the CTB of Y–O, can be estimated near the 67,000 cm^{-1}, or around 150–160 nm. The position of the band C is the same for $YTaO_4$ and $YNbO_4$ because it is determined only by electronegativity Y and O and doesn't depend on tantalate or niobate groups.

To check this assumption about B and C bands and to explain the nature of the band A let us compare PLE spectra with DOS calculations separately for $YTaO_4$ (Fig. 4.85) and $YNbO_4$ (Fig. 4.86).

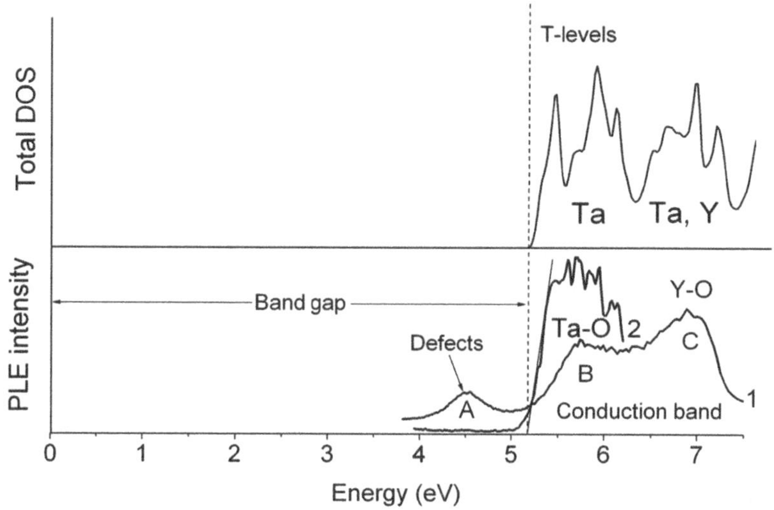

Figure 4.85 $YTaO_4$:PLE spectra under VUV (1) and UV (2) excitation (λ_{em} = 330 nm). Above is a fragment of the total density of states of $YTaO_4$ corresponding to the conduction band.

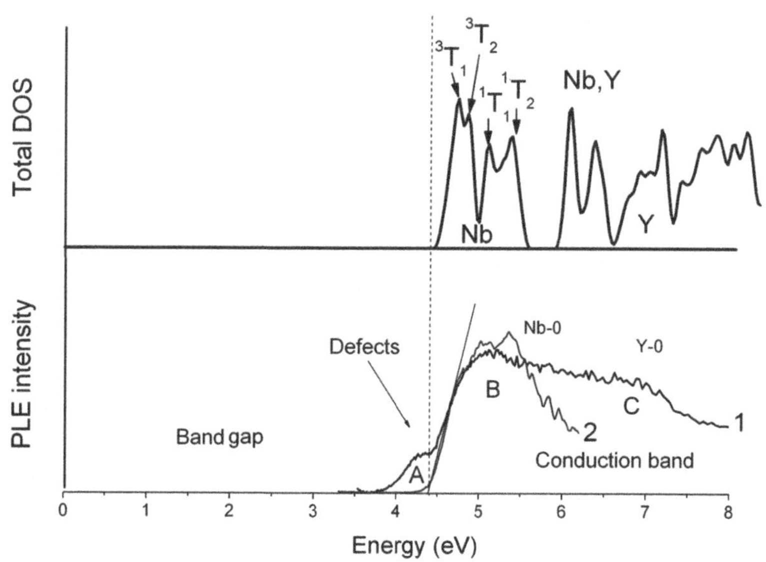

Figure 4.86 $YNbO_4$:PLE spectra under VUV (1) and UV (2) excitation (λ_{em} = 400 nm). Above is a fragment of the total density of states of $YNbO_4$ corresponding to the conduction band.

In the energy scale the position of calculated conduction band was aligned with real spectra.

In both Figs. 4.85 and 4.86 VUV (1) and UV (2) excitation spectra of the host lattices are presented and combined with total density of states. The contribution of Ta, Nb, and Y is taken from partial DOS in Figs. 4.80 and 4.81. This comparison helps us to identify and explain the nature of excitation bands and to determine correctly the band gap in $YTaO_4$ and $YNbO_4$.

As one can see, the position of band A is in the band gap. From DOS calculations for ideal structures no contribution from Ta, Nb, Y, or O is observed in this area. We can suppose that the nature of this band A is related with defects. Moreover, such defects were observed on these samples in color cathodoluminescence (Fig. 4.59). It may be a small amount of uncontrolled Tb incorporation, which has a strong *f–d* excitation band in this part of spectra (Fig. 4.25).

Tantalum and yttrium (Fig. 4.85) as well as niobium and yttrium distribution (Fig. 4.86.) in DOS explain completely the nature of bands B and C in excitation spectra. Band C in the VUV PLE spectra is associated with the absorption of the host lattice, which involves mostly the transitions between 4*d*-like states of Y and 2*p*-like states of O or Y^{3+}–O^{2-} charge transfer transition. Band C is a hybrid band in both materials and it is composed from both [Y–O] and [Ta–O] or [Y–O] and [Nb–O] charge transfer transitions.

If the high energies X-ray, electron beam or VUV excitations are applied, it would be quite reasonable to assume that the excitation energy is absorbed first by the host lattice, which involves the transition between 4*d*-like states of Y and 2*p*-like states of O. The absorbed energy may then be transferred to TaO_4 or NbO_4 groups and last transferred to the activator center if any. This process is shown in Fig. 4.87 and it explains the nature of band B.

Figure 4.87 depicts a simple model illustrating the excitation and emission processes in $YTaO_4$ and $YNbO_4$. For both phosphors the model is very similar and differs only by the excitation energy and band gap.

For $YTaO_4$ sample, under 225 nm excitation, a charge-transfer transition from the O^{2-} ion to the Ta^{5+} metal ion occurs. To excite $YNbO_4$ sample usually 254 nm excitation is enough. The possible excited states of Ta^{5+} and Nb^{5+} are $^3T_1$ (metal to ligand charge-transfer

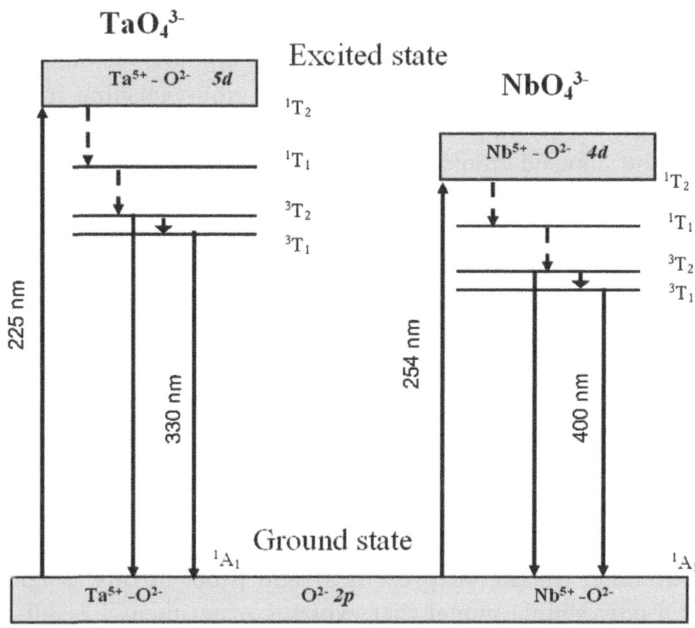

Figure 4.87 Simple model illustrating the excitation and emission processes in $YTaO_4$ and $YNbO_4$.

state), $^3T_2$, $^1T_1$, and $^1T_2$ (charge-transfer state), and the ground state is $^1A_1$ [314]. According to a simplified calculation by Ballhausen [315] the order of these levels is $^1T_2 > {}^1T_1 > {}^3T_1 > {}^3T_2$. After excitation, the transitions from the lowest $^3T_2$, $^3T_1$ excited levels to the ground state $^1A_1$ occur and give the broad UV peaked at 330 nm for $YTaO_4$ or blue emission band around 400 nm for $YNbO_4$.

For tantalate and niobate systems the filled orbitals are concentrated on the oxygen ions and the empty ones on the tantalum or niobium ions, so that the lowest optical transitions can be considered as 2p(O)→5d(Ta) or 2p(O)→4d(Nb) charge-transfer transitions. Thus, the excited state consists of an increased electron density in the vicinity of the metal ion, along the tetrahedral bonds. The ground state is a state of symmetry $^1A_1$, and the excited state produces the abovementioned four states: $^1T_2 > {}^1T_1 > {}^3T_1 > {}^3T_2$.

Electronic levels and their assignments are given schematically in Fig. 4.87. It is reasonable to assume that the maxima in DOS distribution (Figs. 4.85 and 4.86) correspond to these T-levels.

Only the transition $^1A_1 \rightarrow ^1T_2$ is allowed as an electric-dipole transition [250]. In addition to the electric-dipole transition, $^1T_1$–$^1A_1$, $^3T_1$–$^1A_1$, and $^3T_2$–$^1A_1$ also have a certain non-vanishing transition probability. The singlet-triplet transition can acquire intensity from the allowed transition by spin-orbit coupling. This will be more important for the tantalate than for the niobate, since the magnitude of the spin-orbit coupling is greater for Ta than for Nb. The $^3T_1$–$^1A_1$, and $^3T_2$–$^1A_1$ transitions will have the higher transition probability for the stronger spin-orbit coupling interaction. The $^1T_1$–$^1A_1$ transition will become more intense for the larger deviation from a cubic symmetry. In view of this point, we can assign the first strong absorption band in order of increasing energy to $^1A_1 \rightarrow ^3T_1$, $^3T_2$, transitions, and the second one to the $^1A_1 \rightarrow ^3T_1$, $^1T_1$ as shown in Figs. 4.85 and 4.86. Both excitation UV spectra for tantalate and niobate are splitted on some levels. These data are in good agreement with experimental results.

Although we cannot present a rigid proof of this assignment it is a only simple model that explains experimental results. The proposed model can be adopted for a better understanding of the transfer mechanism in the host lattices of tantalate and niobate phosphors.

4.2.8 *Red Emission Properties of (Y,M)NbO$_4$:Eu^{3+} (M:Al,Ga)*

There are few papers on the improvement of the red emission of $YNbO_4$:Eu^{3+} under nUV excitation. Some attempts have been made when other trivalent ions (Gd^{3+}, for example) partially substituted yttrium in the host lattice [257]. The effects of the substitution of trivalent ions such as Al^{3+} and Ga^{3+} for Y^{3+} was investigated by Lee et al. [316]. In this work, (Y, M)NbO_4:Eu^{3+} (M: Al, Ga) powders were synthesized by a flux method, and then the effects of the replacement of Al^{3+} and Ga^{3+} on the red emission under nUV and X-ray excitation were investigated first. Red emission (613 nm) properties of (Y,M)NbO_4:Eu^{3+} (M: Al,Ga) powders under charge transfer band (~270 nm) and near ultraviolet (397 nm) excitation were studied. In $Y_{1-x}Eu_xNbO_4$, with the increase in the Eu content (x), the red emission intensity under 397 nm excitation abruptly increased up to $x = 0.25$, and then saturated at

$x \geq 0.25$, while that under CTB excitation broadly peaked at $x = 0.1 - 0.25$. The red emissions under 397 nm excitation of $Y_{1-y}Al_yNbO_4:Eu^{3+}$ and $Y_{1-z}Ga_zNbO_4:Eu^{3+}$ were drastically enhanced, showing about two times higher intensity at y = 0.2 and z = 0.1, respectively, than that of $YNbO_4:Eu^{3+}$. This improved emission was ascribed to the local distortion and asymmetry sites in the host lattice originated from the remarkably different ionic radii of Al^{3+} and Ga^{3+} from Y^{3+}.

(Y, M)$NbO_4:Eu^{3+}$ (M: Al, Ga) powders were fabricated by firing the stoichiometric mixtures of Y_2O_3, Nb_2O_5, Al_2O_3, Ga_2O_3, and Eu_2O_3 powders. LiCl (7 wt%) was added as a flux to promote a solid state reaction and the luminescent properties by accelerating the kinetics for the compound formation due to the increased diffusion coefficients. The purities of all the powders were 99.99%. The amount of Eu_2O_3 was in the range of 0 – 45 mol%. After ballmilling the mixtures for 24 h, they were fired at 1300°C for 12 h in an electric tube furnace flowing N_2 gas.

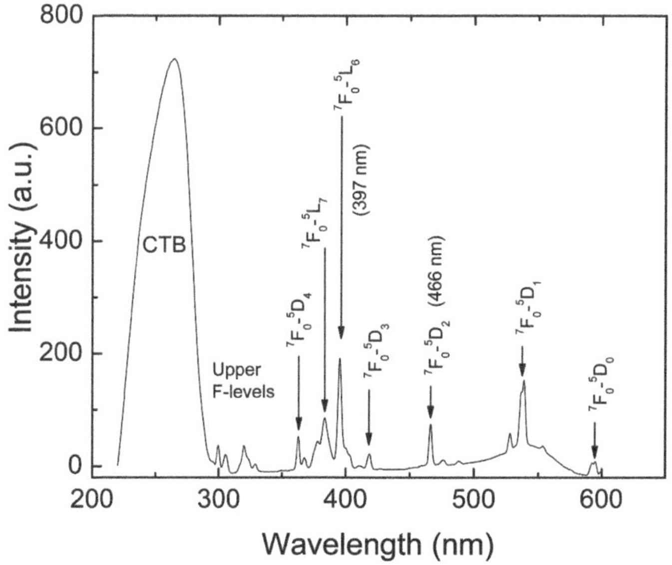

Figure 4.88 The excitation spectra of $YNbO_4:Eu^{3+}$ powder.

4.2.8.1 nUV excitation

The representative photoluminescence excitation spectrum of $YNbO_4:Eu^{3+}$ powder is shown in Fig. 4.88.

RE ion-doped $YNbO_4$ powders are known to emit visible lights under UV excitations. $[NbO_4]^{3-}$ niobate could be self-activated by UV [317], as mentioned above, and then this absorbed energy was non-radiatively transferred to Eu^{3+} ions by charge transfer mechanism, as well as CT from O^{2-} to Eu^{3+} resulting in broad CTB around 270 nm in the excitation spectra. Sharp excitation peaks due to $f \rightarrow f$ transitions of Eu^{3+} were observed as well. The strong $f \rightarrow f$ transition peaks at 386 nm, 397 nm, and 466 nm corresponded to ${}^7F_0 \rightarrow {}^5L_7$, ${}^7F_0 \rightarrow {}^5L_6$, and ${}^7F_0 \rightarrow {}^5D_2$ transitions of Eu^{3+} ions, respectively, which enable $YNbO_4:Eu^{3+}$ phosphors to be applied to white LEDs using nUV or blue chips.

The emission spectra of $Y_{1-x}Eu_xNbO_4$ powders under CTB excitation are shown in Fig. 4.89.

Weak blue emissions around 420 nm due to self-activated $[NbO_4]^{3-}$ as well as strong red emissions that originated from ${}^5D_0 \rightarrow {}^7F_J$ transitions of Eu^{3+} were attained at low x values. With increasing the

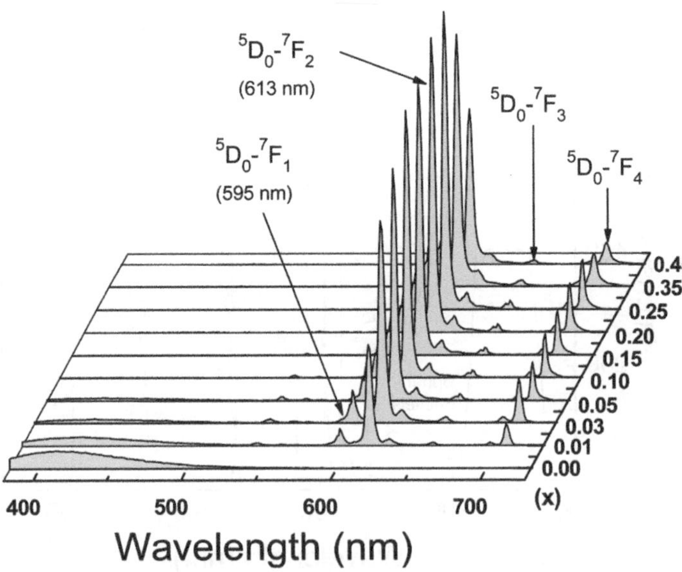

Figure 4.89 The emission spectra of $Y_{1-x}Eu_xNbO_4$ powders under CT excitation as a function of x.

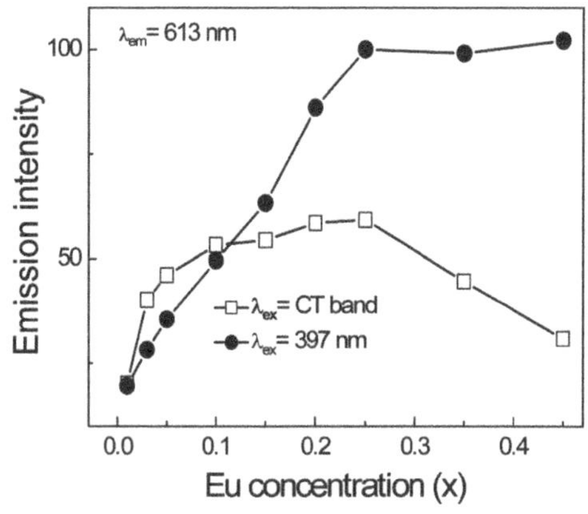

Figure 4.90 The variation in the emission intensity of $Y_{1-x}Eu_xNbO_4$ powders under CTB and nUV (397 nm) excitation.

Eu^{3+} concentration (x) blue emissions were quenched, while red peaks still appeared strong.

The variations in the respective emission intensity at 613 nm under CTB and $^7F_0 \rightarrow {}^5L_6$ excitation (λ_{ex} = 397 nm) are shown in Fig. 4.90.

In CTB excitation, the red emission intensity of $Y_{1-x}Eu_xNbO_4$ rapidly increased at x = 0.01 – 0.1, and then slightly rose up to x = 0.25. This result could be explained by the charge transfer probability of $[NbO_4]^{3-}$ to Eu^{3+}. Because the $[NbO_4]^{3-}$ group was coordinated by eight Y^{3+} ions, if uniformly substituted, each $[NbO_4]^{3-}$ could have approximately one Eu^{3+} neighbor at x = 0.1, whereas at x = 0.01 the probability that $[NbO_4]^{3-}$ could have one Eu^{3+} neighbor was about 10%. Furthermore, the charge transfer was limited to the nearest Eu^{3+} ions. Consequently, at the low Eu concentration, the probability of the charge transfer from $[NbO_4]^{3-}$ to an Eu^{3+} neighbor was very low, and so the remaining excitation energy in $[NbO_4]^{3-}$, which was not transferred to Eu^{3+}, was radiatively relaxed, showing the weak blue emission around 420 nm, as shown in Fig. 4.89. With increasing Eu concentration up to x = 0.1, each $[NbO_4]^{3-}$ could more efficiently transfer its absorbed energy to an Eu^{3+} ion due to increased charge transfer probability, leading to the stronger red emission; accordingly, the blue emission was gradually quenched and finally diminished at x = 0.1. At x = 0.25, two Eu^{3+} ions could probably occupy two Y^{3+} sites, leading to the increase in the charge transfer probability of

$[NbO_4]^{3-}$ to Eu^{3+}. Consequently, the red emission intensity under CTB excitation might be continuously enhanced at $x = 0.1 - 0.25$ in the similar way to $x = 0.01 - 0.1$. However, a slight increase in PL intensity was observed, which implied that the concentration quenching effect suppressed the emission. At $x = 0.25 - 0.45$, because Eu^{3+} ions could possibly replace more than two Y^{3+} sites, the concentration quenching effect became more effective, resulting in the rapid decrease in the emission intensity.

On the other hand, PL intensity under 397 nm excitation linearly increased up to $x = 0.25$, and then saturated at $x = 0.25 - 0.45$ due to the concentration quenching effect. Unreacted Eu_2O_3 phase and the secondary phases were not observed in XRD analysis in this experimental range of the Eu^{2+} concentration. This demonstrated that all Eu^{3+} ions were substituted and acted as activators. The phase diagram of Eu_2O_3–Y_2O_3–Nb_2O_5 is not reported, but the solubility of Eu_2O_3 in C-Y_2O_3 is about 59–70 mol% at 1250–1500°C [318]. The reason for the distinct behaviors between CTB and 397 nm excitation is unclear, but the following speculation is possible. The emission at 613 nm meant that there was a radiative transition from excited $^5D_0$ level to $^7F_2$. For CTB excitation, the energy was transferred step by step from CTB to the upper F-level, and then to ^5D_J ($J = 0, 1, 2, 3, 4$). Subsequently the $^5D_0 \rightarrow {}^7F_2$ transition was radiatively completed ($\lambda_{em} = 613$ nm) after undergoing all the traps, whereas for 397 nm excitation, the way was a little different and shorter.

The effects of the substitution of trivalent ions such as Al^{3+} and Ga^{3+} for Y^{3+} on the luminescent properties were investigated. Because the ionic radii of Al^{3+} (0.54 Å) and Ga^{3+} (0.62 Å) were much smaller than that of Y^{3+} (0.88 Å), the crystal field surrounding Eu^{3+} (0.95 Å) ions was expected to be altered considerably, resulting in the change in PL properties.

PLE and PL spectra of $Y_{1-y}Al_yNbO_4$:Eu^{3+} (10 mol%) are shown in Fig. 4.91.

The secondary phases including Al_2O_3 did not appear in the range of $y = 0$–0.4 in XRD analysis, indicating that Al^{3+} ions were completely substituted for Y^{3+} sites. The EPMA elemental analysis did not show any other impurities such as Li and Cl due to a flux, LiCl. The peak positions of CT band in PLE slightly shifted to the shorter wavelengths with the increase in y values (0.1–0.4), whereas those of $f \rightarrow f$ transitions

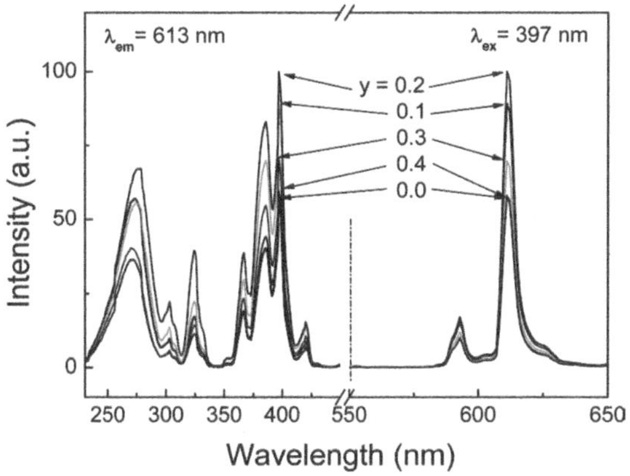

Figure 4.91 PLE and PL of $Y_{1\text{-}y}Al_yNbO_4{:}Eu^{3+}$ (10 mol%).

of Eu^{3+} ions did not move, because they are not sensitive to the host crystal structure. The PL intensity at 613 nm (λ_{ex} = 397 nm) varied as a function of the Al^{3+} concentration, peaking at y = 0.2.

XRD patterns of $Y_{1\text{-}z}Ga_zNbO_4{:}Eu^{3+}$ (10 mol%) are shown in Fig. 4.92.

Unlike $Y_{1\text{-}y}Al_yNbO_4{:}Eu^{3+}$, an unreacted Ga_2O_3 phase appeared with increasing the Ga_2O_3 content (z = 0.2–0.4), whereas these Ga_2O_3 peaks were not observed at z = 0.1, meaning the complete substitution for

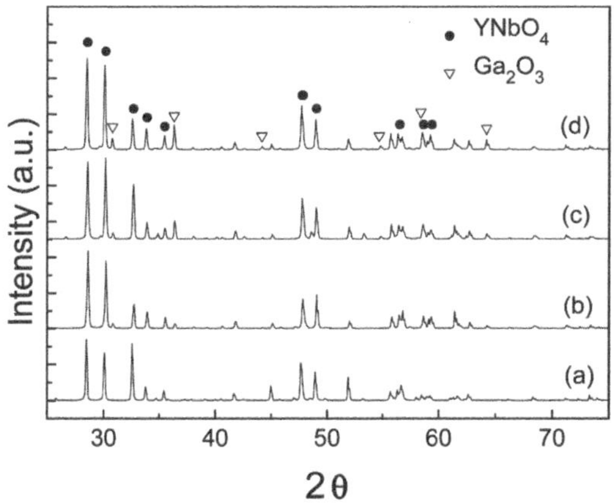

Figure 4.92 The XRD patterns of $Y_{1\text{-}z}Ga_zNbO_4{:}Eu^{3+}$ (10 mol%) as a function of z. (a) z = 0.1, (b) 0.2, (c) 0.3, (d) 0.4.

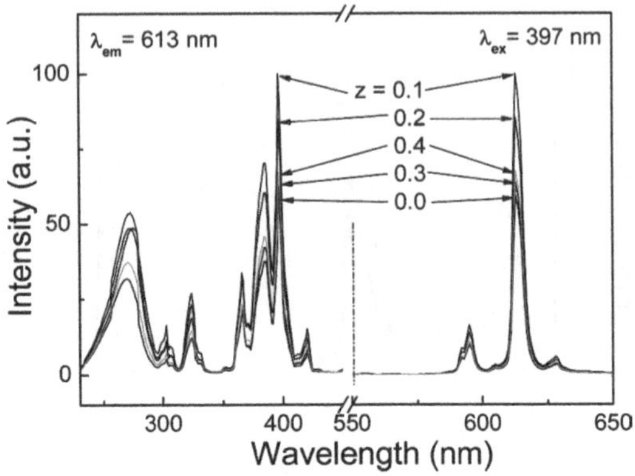

Figure 4.93 PLE and PL of $Y_{1-z}Ga_zNbO_4:Eu^{3+}$ (10 mol%).

Y^{3+} ions. It could be speculated that the distinct properties between Al^{3+} and Ga^{3+} ions such as the ionic radius, the bond strength, and the formation energy of the solid-solution caused different solid solubility. The corresponding PLE and PL spectra are shown in Fig 4.93.

The slight blue shift of CT band in PLE was exhibited with increasing z values from 0.1 to 0.4. The PL intensity under 397 nm excitation peaked at z = 0.1, and then continuously decreased with an increase in Ga content.

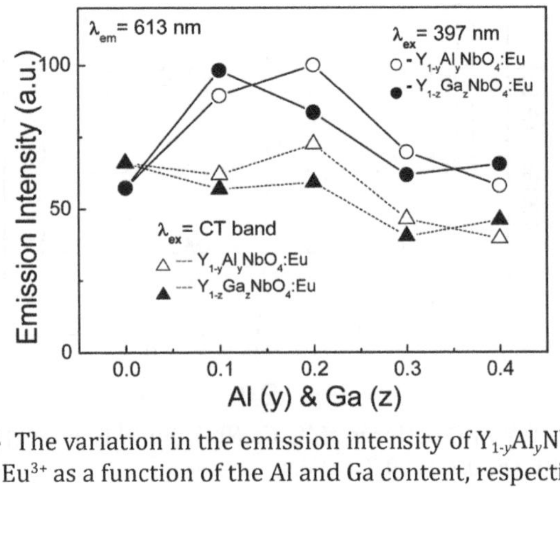

Figure 4.94 The variation in the emission intensity of $Y_{1-y}Al_yNbO_4:Eu^{3+}$ and $Y_{1-z}Ga_zNbO_4:Eu^{3+}$ as a function of the Al and Ga content, respectively.

Respective red emission intensity (613 nm) variation of $Y_{1-y}Al_yNbO_4:Eu^{3+}$ and $Y_{1-z}Ga_zNbO_4:Eu^{3+}$ under CTB and 397 nm excitations is shown in Fig. 4.94.

Under CTB excitation, the PL intensity both of $Y_{1-y}Al_yNbO_4:Eu^{3+}$ and $Y_{1-z}Ga_zNbO_4:Eu^{3+}$ decreased slightly up to y (z) = 0.4. This demonstrated that the substitution of Al^{3+} or Ga^{3+} for Y^{3+} was uneffective on CTB excitation, even though the change of the crystal field between $[NbO_4]^{3-}$ and Eu^{3+} ions due to the distinct ionic radius was expected. However, the emission intensities under 397 nm excitation exhibited the different behavior. When increasing the Al^{3+} (y) and Ga^{3+} (z) content, they increased drastically compared with pure $YNbO_4:Eu^{3+}$, peaking at $y = 0.2$ for Al^{3+} and $z = 0.1$ for Ga^{3+}, where their emission intensities were dramatically about two times as high as that of pure $YNbO_4:Eu^{3+}$. Xiao and Yan investigated the emission properties of $(Y,Gd)NbO_4:Eu^{3+},Tb^{3+}$ system under CT band excitation, demonstrating that their variations were assigned to the symmetry change of $(Y,Gd)NbO_4$ for Eu^{3+} and Tb^{3+} ions as a function of the concentration ratio (Y/Gd) [319]. This work showed that the red emission intensity (613 nm) of $Y_{0.6}Gd_{0.4}NbO_4:Eu^{3+}$ was higher than that of $YNbO_4:Eu^{3+}$, but about 10% higher at most. Compared with the $(Y,Al)NbO_4$ and $(Y, Ga)NbO_4$ systems, this small enhanced emission intensity of $(Y, Gd)NbO_4$ might be attributed to the discrepancy of ionic radii Y^{3+} (0.88 Å), Al^{3+} (0.54 Å), Ga^{3+} (0.62 Å), and Gd^{3+} (0.94 Å). Namely, the difference in the ionic radius between Y^{3+} and Gd^{3+} was small. Shi et al. [320] also investigated the effects of cation substitutions in $(Ca, Eu, M)WO_4$ (M: Mg, Zn, Li), in which the red emission intensity under nUV (395 nm) excitation increased with the incorporation of M ions.

To examine the effects of the replacement of Al^{3+} and Ga^{3+}, the symmetries of the Eu^{3+} sites in $Y_{1-y}Al_yNbO_4$ and $Y_{1-z}Ga_zNbO_4$ were evaluated by measuring the ratio of I_{595} to I_{613} (R_I), where I_{595} and I_{613} were the emission intensities of $^5D_0 \to {}^7F_1$ (595 nm) and $^5D_0 \to {}^7F_2$ (613 nm) transitions of Eu^{3+}, respectively, as shown in Fig. 4.95.

$^5D_0 \to {}^7F_1$ is a magnetic dipole transition and insensitive to a local symmetry, whereas $^5D_0 \to {}^7F_2$ is an electric dipole transition and hypersensitive to a local symmetry. The significant strong intensity

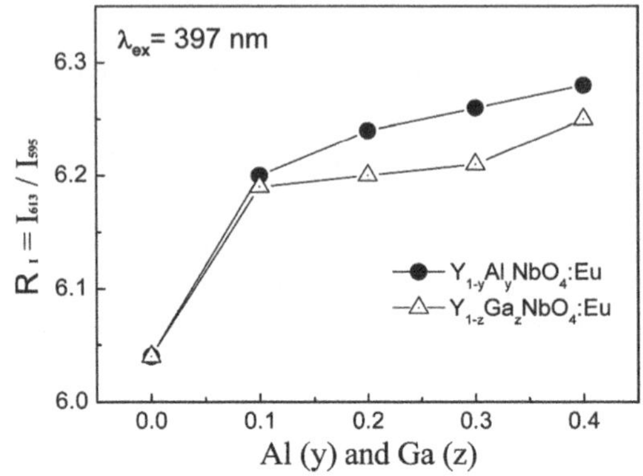

Figure 4.95 I_{613}/I_{595} (R_I) of $Y_{1-y}Al_yNbO_4$:Eu^{3+} and $Y_{1-z}Ga_zNbO_4$:Eu^{3+} as a function of the Al and Ga content, respectively.

of $^5D_0 \rightarrow {}^7F_2$ transition in (Y, M)NbO_4:Eu^{3+} system revealed that Eu^{3+} ions are located in the asymmetry sites [220]. The addition of Al^{3+} and Ga^{3+} in $YNbO_4$:Eu^{3+} led to the abrupt increase in $R_I = I_{595}/I_{613}$ at $y = 0.1$ and $z = 0.1$, respectively. This demonstrated that the crystal field surrounding Eu^{3+} ions was highly changed due to the distinct ionic radii of trivalent cations, resulting in the local distortion and lower symmetry. Even though R_I increased slightly above y = 0.1 and z = 0.1 (Fig. 4.95), the red emission intensity at 613 nm decreased above y =0.2 and z = 0.1 as shown in Fig. 4.94. It could be speculated that the addition of Al^{3+} and Ga^{3+} over the optimum content caused the excess distortion, decreasing the emission. For $Y_{1-z}Ga_zNbO_4$:Eu^{3+}, the unreacted Ga_2O_3, which was observed at z = 0.2 and more, as shown in Fig. 4.92, additionally diminished the emission.

Regarding atomic positions of Al^{3+} and Ga^{3+} in $YNbO_4$, we could speculate that both ions were substituted for Y^{3+} sites but not for Nb^{5+} ones. If they replaced Nb^{5+} ions, oxygen vacancies, which could act as luminescent killers, might be created for the charge compensation, leading to the significant decrease in PL intensity. However, PL intensity increased with doping Al^{3+} and Ga^{3+} in this experiment. Also, a previous paper for the Y(Nb,Ta)O_4:Eu,Tb system [176] verified that Eu^{3+} and Tb^{3+} ions were substituted for Y^{3+} sites using the Rietveld method.

4.2.8.2 X-ray excitation

Now we show the luminescence properties of $YNbO_4$ phosphors doped by Eu^{3+}, Ga^{3+}, Al^{3+} under X-ray excitation. The improved about 1.5 times emission of these phosphors, respectively, than that of $YNbO_4:Eu^{3+}$ was ascribed to the redistribution of energy transfer from excited levels. To better understand the luminescence mechanism and crystallographic structure of these phosphors the Raman spectroscopy was applied. The Raman spectroscopy for investigation non-activated $YNbO_4$ was first applied by Blasse [267]. Till now the vibrational spectra of doped yttrium niobate systems have not been reported.

To study phosphor luminescence under X-ray excitation, the Bremsstrahlung radiation from a Rigaku rotating anode generator operating in the range from 20 to 50 kV and 100 mA was used. Powder phosphors were illuminated into the broadband X-ray radiation. Samples were placed at a position close to exit of Be X-ray windows to minimize the loss of X-ray flux by air scattering, and rotated by 45° from the incident X-rays. Since the luminescence

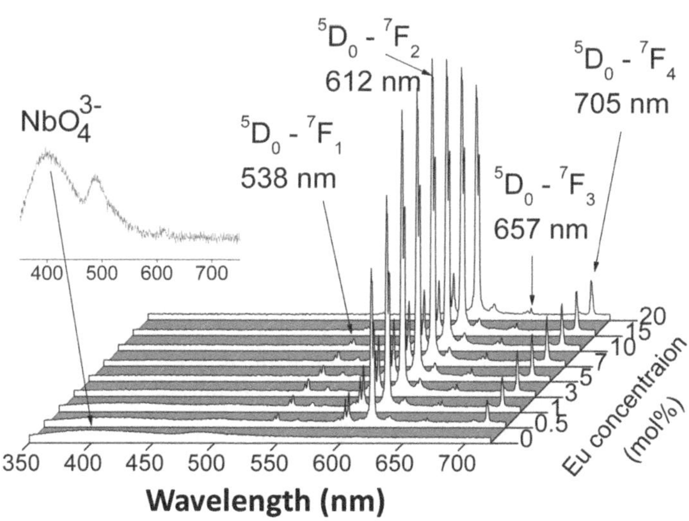

Figure 4.96 Emission spectra of $YNbO_4:Eu^{3+}$ phosphors under X-ray excitation (U = 30 kV, I = 100 mA) depending on europium concentration. The emission of NbO_4^{3-} groups in undoped material is shown in the inset.

signals were irradiated into all directions, a collimating lens was used to collect the light effectively. The emission spectrum was recorded with a spectrometer (SM240, Spectral Products) in a range of 200 ~1200 nm.

Figure 4.96 shows the luminescence spectra of M-$YNbO_4$ activated by different concentration of Eu^{3+} under X-ray excitation. The activators can be excited directly by excitation source or by energy transfer transition from the host lattice. In the lattice of rare earth doped yttrium niobate the doped luminescent centers, such as Eu^{3+}, occupy the Y sites. The NbO_4^{3-} groups can absorb excitation energy through $O^{2-} \rightarrow Nb^{5+}$ charge transfer transition respectively, and transfer the energy to rare earth luminescent centers, which give rise to the corresponding characteristic emission. Low intensity of host lattice luminescence is observed (see inset in Fig. 4.96) in non-doped phosphor and disappeared due to energy transfer in activated samples. The system $Y_{1-x}Eu_xNbO_4$ was studied more thoroughly because it illustrates clearly the importance of excitation energy transport through the lattice in order to obtain efficient phosphors. Brixner [249] mentions the intensity and color of the luminescence of a number of fergusonites. Later on this author reported on the luminescence of the Eu^{3+} ion in niobates and tantalates with fergusonite structure [294]. That study contains two remarkable conclusions, firstly that concentration quenching occurs at an extremely high Eu concentration, and secondly, that the efficiency of the tantalate phosphors relative to the niobate phosphors is very low.

We have found the shift of Eu concentration quenching from 15 mol% under UV (254 nm) excitation to 7 mol% under X-ray excitation (Fig. 4.97).

It is not astonishing because under X-ray excitation, it is quite reasonable to assume that the excitation energy is absorbed first by the host lattice, which involves the transition between 4*d*-like states of Y and 2*p*-like states of O. The absorbed energy may then be transferred to NbO_4^{-3} group, and at last transferred to the Eu^{3+} emission centers. Under UV excitation only direct Eu excitation and charge transfer band mechanism is possible.

Another advantage of X-ray excitation in comparison with UV we can see from Fig. 4.98.

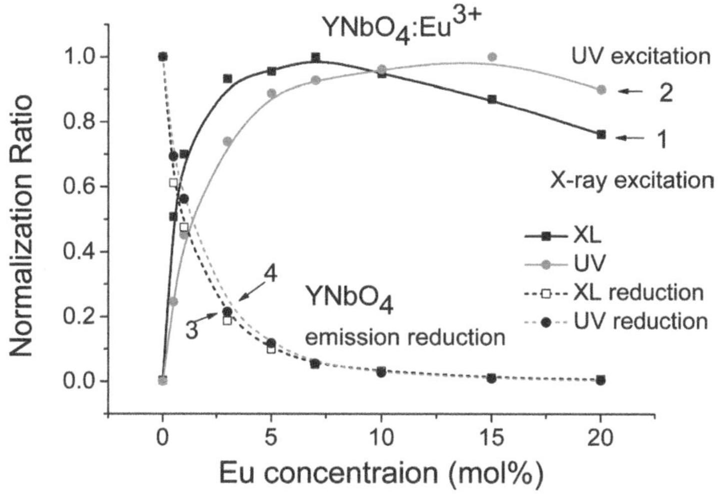

Figure 4.97 Normalized intensities of $YNbO_4:Eu^{3+}$ (612 nm) under X-ray (1) and UV (2) excitation versus europium concentration. (Dashed curves show the host lattice $YNbO_4$ emission (peaked at 400 nm) under UV (3) and X-ray (4) excitation, correspondingly).

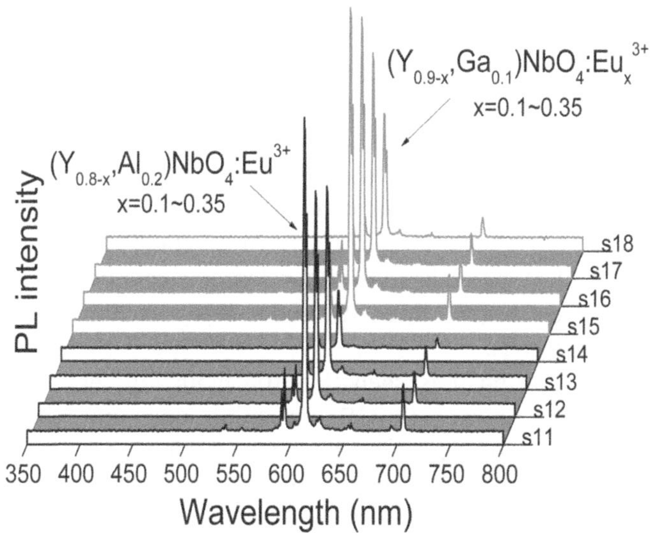

Figure 4.98 The variation of the emission intensity of $Y_{0.8-x}Al_{0.2}NbO_4:Eu_x^{3+}$ phosphors (S11–S14) and $Y_{0.9-x}Ga_{0.1}NbO_4:Eu_x^{3+}$ (S15–S18) as a function of the Eu^{3+} content, respectively: $x = 0.10, 0.15, 0.25, 0.35$.

Under X-ray excitation the relatively strong $^5D_0 \rightarrow {}^7F_4$ transition intensity is appeared. Usually this emission is very weak under UV excitation [265]. Furthermore, comparing the luminescence under UV (254 nm) and the broad X-ray excitation, it was found that the relative intensity of $^5D_0 \rightarrow {}^7F_J$ ($J = 1, 2, 4$) emission peaks of Eu^{3+}, so-called branch ratios, varies significantly with the activation excitation energy. Under UV excitation, the $^5D_0 \rightarrow {}^7F_4$ emission is mostly trapped by the empty upper levels. On the contrary, it is more likely that such empty levels can be readily filled under X-ray excitation. This explains the appearance of visible $^5D_0 \rightarrow {}^7F_4$ transition and the increment of the luminescence intensity of Eu^{3+} emission centers under X-ray excitation.

The influence of codoping $YNbO_4:Eu^{3+}$ by Al^{3+} and Ga^{3+} on the luminescence properties under X-ray excitation was studied. First, the optimum concentration of Al^{3+} = 2 mol% and Ga^{3+} = 1 mol% was established [316] and then we checked the variation of the emission intensity of $Y_{0.8-x}Al_{0.2}NbO_4:Eu_x^{3+}$ (S11–S14 of Fig. 4.98) and $Y_{0.9-x}Ga_{0.1}NbO_4:Eu_x^{3+}$ (S15–S18 of Fig. 4.98) as a function of the Eu^{3+} content. The results are shown in the Fig. 4.98.

Since the ionic radii of Al^{3+} (0.54 Å) and Ga^{3+} (0.62 Å) were much smaller than that of Y^{3+} (0.88 Å), the crystal field surrounding Eu^{3+} (0.95 Å) ions was expected to be altered considerably, resulting in the change of photoluminescence properties. The highest intensities under X-ray excitation were found for samples ($Y_{0.7}Al_{0.2}NbO_4:Eu_{0.1}^{3+}$) and ($Y_{0.8}Ga_{0.1}NbO_4:Eu_{0.1}^{3+}$) doped by Eu^{3+} 10 mol%. As an example we show in Fig. 4.99a the intensity of the main Eu^{3+} emission peak at 612 nm only for one sample ($Y_{0.8}Ga_{0.1}NbO_4:Eu_{0.1}^{3+}$) in comparison with $YNbO_4:Eu^{3+}$ phosphor. All other samples codoped by Al^{3+} and Ga^{3+} also show higher intensity under X-ray excitation relatively undoped materials.

From Fig. 4.99a it is seen that the main Eu^{3+} emission peak corresponding to the $^5D_0 \rightarrow {}^7F_2$ transition does not change his position after codoping. This indicates that all Eu^{3+} ions are situated on the non-central symmetric C_2 site of Y^{3+} in M-type monoclinic fergusonite crystal structure. According to prior analysis, the space group of the M-type structure is I_2/a [185]. This structure has no symmetric center. In the case of low symmetric lattices, Eu^{3+} $^5D_0 \rightarrow {}^7F_2$ the transition emission appears as the strongest peak.

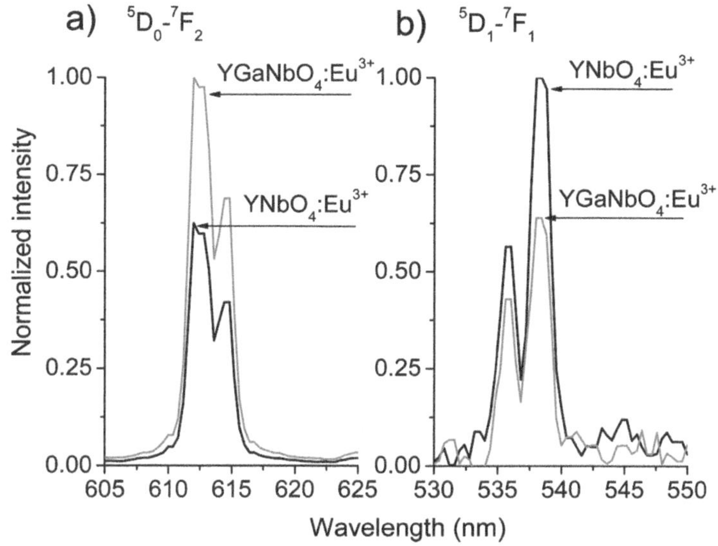

Figure 4.99 a) Luminescence intensity of the main emission peak $^5D_0$–$^7F_2$ (612 nm) in $Y_{0.8}Ga_{0.1}NbO_4:Eu_{0.1}{}^{3+}$ in comparison with $YNbO_4:Eu^{3+}$ under X-ray excitation. b) Luminescence intensity of $^5D_1$–$^7F_1$ (538 nm) transition in $Y_{0.8}Ga_{0.1}NbO_4:Eu_{0.1}{}^{3+}$ in comparison with $YNbO_4:Eu^{3+}$.

To examine the effects of the replacement of Al^{3+} and Ga^{3+}, the symmetries of the Eu^{3+} sites in $(Y,Al)NbO_4$ and $(Y,Ga)NbO_4$ in comparison with $YNbO_4$ were evaluated by the method proposed by Kirby et al. [321] and applied in previous work [322].

The magnetic-dipole transition $^5D_0 \rightarrow {}^7F_1$ is relatively insensitive to the local symmetry and it can be used as an internal standard to account for the ligand differences. The intensity ratio of the magnetic dipole transition to the electric dipole transition, which is hypersensitive to a local symmetry, measures the symmetry of

Table 4.18 The intensity ratio of $^5D_0 \rightarrow {}^7F_2$ to $^5D_0 \rightarrow {}^7F_1$ of Eu^{3+} in investigated materials

Phosphor formula	$YNbO_4:Eu^{3+}$	$Y_{0.8}Ga_{0.1}NbO_4:Eu_{0.1}{}^{3+}$	$Y_{0.7}Al_{0.2}NbO_4:Eu_{0.1}{}^{3+}$
$R = I_{612}/I_{595}$	4.99	5.02	5.05

the Eu^{3+} surrounding [321]. Subsequently we calculated the intensity ratio of the $^5D_0 \rightarrow ^7F_2$ (612 nm) emission transition to the $^5D_0 \rightarrow ^7F_1$ (595 nm) emission transition in order to study the symmetry of the Eu^{3+} site surroundings. Table 4.18 presents the result on the intensity ratio, which shows that the symmetry of activated center has not changed significantly with incorporation Al and Ga in the host lattice.

To better understand this phenomenon and to check the crystalline structure the Raman investigations were carried out. Raman spectra were obtained using the 632.8 nm excitation wavelength from a continuous-wave He–Ne laser at room temperature. Incident light is focused on a sample through optical microscope with a spatial resolution of <2 μm in the backscattering geometry. Scattered light was dispersed using triple monochromator system (TRIVISTA, ACTON, Fig. 3.30). Two of three monochromators in the triple monochromator system were used for laser line rejection to filter out Rayleigh scattering from the sample upon laser excitation. In subtractive mode, the Raman scattering signature as close as 20 cm^{-1} to

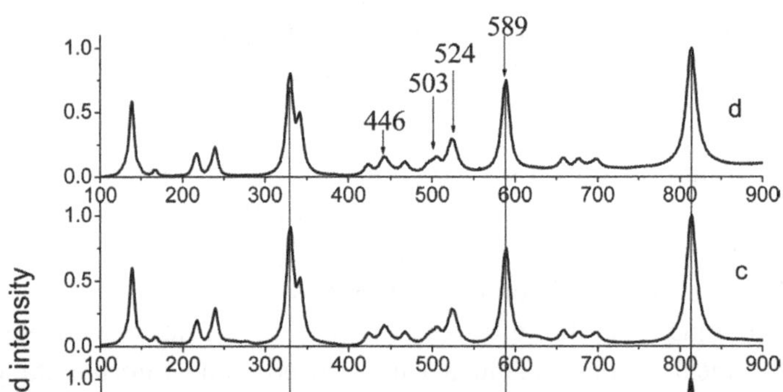

Figure 4.100 Raman spectra of undoped $YNbO_4$ (a) and doped phosphors: $Y_{0.9}Eu_{0.1}NbO_4$ (b), $Y_{0.8}Ga_{0.1}Eu_{0.1}NbO_4$ (c), and $Y_{0.7}Al_{0.2}Eu_{0.1}NbO_4$ (d).

the laser excitation line could be observable. The signal was recorded using a thermoelectrically cooled charge-coupled device detector. The Raman method is very sensitive to the lattice vibrations and any structural modifications if any can be observed.

In the Fig. 4.100 we compared Raman spectra of $YNbO_4$ (a), $Y_{0.9}Eu_{0.1}NbO_4$ (b), $Y_{0.8}Ga_{0.1}Eu_{0.1}NbO_4$ (c), and $Y_{0.7}Al_{0.2}Eu_{0.1}NbO_4$ (d).

Vibration modes ν_1, ν_2, ν_3, and ν_4 from the host lattice $YNbO_4$ (spectrum (a) in Fig. 4.100) were observed, which correspond to a regular NbO_4 tetrahedron with no interactions and distortions at 815 (ν_1), 650 (ν_3), 420 (ν_4), and 340 cm^{-1} (ν_2) predicted by Blasse [267]. Raman bands appearing at 815 and 340 cm^{-1} are due to Nb–O symmetric modes of the NbO_4 tetrahedral structure, and Raman bands appearing at 650 and 420 cm^{-1} are due to Nb–O antisymmetric modes of the NbO_4 tetrahedral structure. The phonon energies below 300 cm^{-1} are assigned to external vibrations. Four new peaks in Raman spectroscopy around 446, 503, 524, and 589 cm^{-1} in all doped samples (Fig. 4.100b,c,d) are the luminescent peaks related to europium activation and these peaks were explained in detail in section 4.2.4 in Figs. 4.46 and 4.47.

From Fig. 4.100 one can see that all doped samples have the same monoclinic M-phase as $YNbO_4$ with no interactions and distortions at 815 (ν_1), 650 (ν_3), 420 (ν_4), and 340 cm^{-1} (ν_2). It means that no considerable changes in the crystalline structure are observed.

Let us return to luminescence properties. The emission spectra of all Eu^{3+} phosphors show similar features, and have been involved in the following emission lines: $^5D_0 \rightarrow {}^7F_1$, $^5D_0 \rightarrow {}^7F_2$, $^5D_0 \rightarrow {}^7F_3$, and $^5D_0 \rightarrow {}^7F_4$. Other transitions from $^5D_0$ to $^7F_0$, $^7F_5$, and $^7F_6$ levels are also allowed from the selection rules for the transition between Stark components of Eu^{3+} in symmetry of the actual crystal field [261], but luminescence intensities are too weak. Furthermore, there appears to be an energy transfer from host to activator (inset in Fig. 4.96). Additionally, the transition from $^5D_1$ to $^7F_1$ level is also possible and we examined this transition more carefully. The results are shown in the Fig. 4.99b and they are just opposite to Fig. 4.99a.

Position of the $^5D_1 \rightarrow {}^7F_1$ transition for all the samples is the same (538 nm) and no shift was observed. It means that no phase

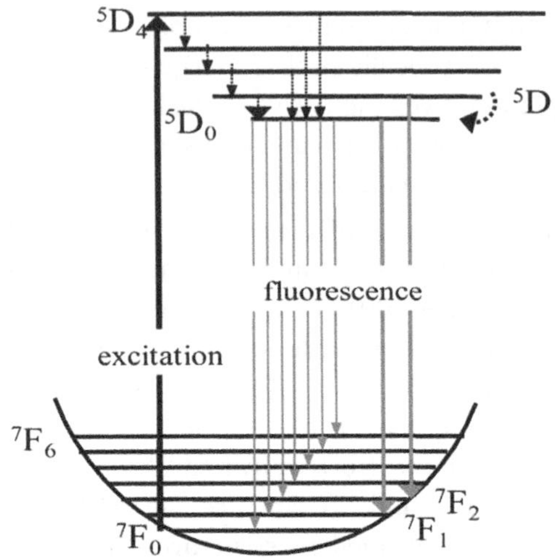

Figure 4.101 A schematic model of stimulated Eu^{3+} luminescence in (Y,Al,Ga)NbO_4 phosphors.

changes with incorporation Ga and Al is appeared. The abrupt decrease of luminescence in $Y_{0.8}Ga_{0.1}NbO_4:Eu_{0.1}^{3+}$ (absolutely the same for $Y_{0.7}Al_{0.2}NbO_4:Eu_{0.1}^{3+}$) in comparison with $YNbO_4:Eu^{3+}$ about 1.5 times can be explained by the redistribution of energy transfer from excited Eu^{3+} state $^5D_1$. One additional part of energy from excited level $^5D_1$ transfers non-radiative to the lowest level $^5D_0$ and then radiative to all the levels of ground state decreasing intensity in $^5D_1 \rightarrow {}^7F_1$ transition and increasing general luminescence in other transitions. We checked the intensities of Al and Ga doped materials in all emission lines including the most important $^5D_0 \rightarrow {}^7F_2$ (612 nm) emission transition and found that the intensity increased about 1.5 times. The proposed mechanism is schematically presented in Fig. 4.101.

The samples codoped by Al^{3+} and Ga^{3+} show higher intensity under nUV and X-ray excitation in comparison with $YNbO_4:Eu^{3+}$ phosphors. Monoclinic fergusonite crystal structure and excellent luminescent properties under nUV and X-ray excitation allow recommend these phosphors as good candidates for different applications.

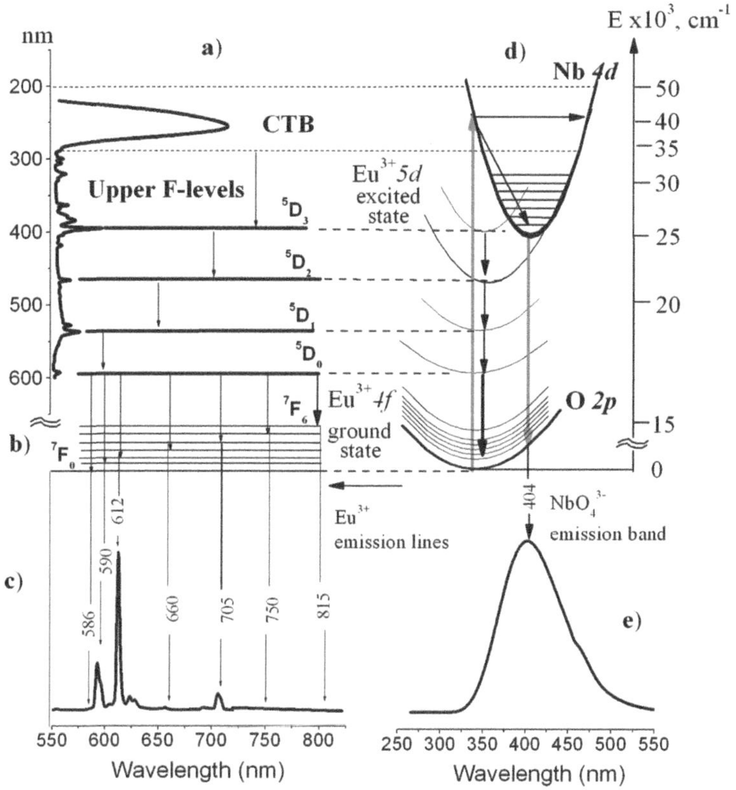

Figure 4.102 Excitation (a) and emission (c) spectra of $YNbO_4:Eu^{3+}$ (λ_{exc}= 254 nm) and configurational coordinate diagram (d), corresponding to real data and showing the energy transfer from excited to ground state in Eu^{3+}. Emission spectra of NbO_4^{3-} group is presented in Fig. (e).

4.2.9 *Configurational Coordinate Diagram and Energy Transfer Model*

Summarizing the luminescence properties from PL, CL, and X-ray excitation of tantalate and niobate phosphors, a configurational coordinate diagram (CCD) can be proposed and energy transfer model will be discussed. For CCD we need to do an analysis of the experimental data of excitation and emission spectra from non doped host lattice as well as doped by one or other activator. Let us see as an example one of studied samples $YNbO_4:Eu^{3+}$. Excitation and emission Eu^{3+} spectra for this phosphor can be taken from PL measurements, for example, from Fig. 4.39. A typical emission spectrum for non-activated host lattice or emission of NbO_4^{3-} groups was shown in Figs. 4.19 and 4.20. Combing these spectra with Eu^{3+} energy diagram (Fig. 4.69), a configurational coordinate diagram can be calculated and presented in Fig. 4.102.

Experimentally observed $f \rightarrow f$ transitions in Fig. 4.79 correspond in this model to the excited states of Eu^{3+} ($^5D_0$, $^5D_1$, $^5D_2$, $^5D_3$) and charge transfer from the host lattice toward Eu^{3+} ions in real energy scale. The NbO_4^{3-} groups can absorb excitation energy through $O^{2-} \rightarrow Nb^{5+}$ charge transfer transition and transfer the energy to Eu^{3+} luminescent centers, which give rise to the corresponding characteristic emission. When the host lattice is non-activated $[NbO_4]^{3-}$ groups absorb excitation energy and emit itself, showing a wide emission band peaked at around 400 nm.

The proposed model is in good agreement with the experimental results and can be adopted for a better understanding of the transfer mechanism in the rare earth activated phosphors as well as for improving their optical properties. Similar models can be done for $YNbO_4:Tb^{3+}$ as well as for $YTaO_4:Tb^{3+}$ and $YTaO_4:Eu^{3+}$.

4.2.10 *Terahertz Spectroscopy in Study Y(Ta,Nb)O$_4$-Based Phosphors*

THz transmission spectra were measured for the first time for these phosphors to better understand the luminescence properties.

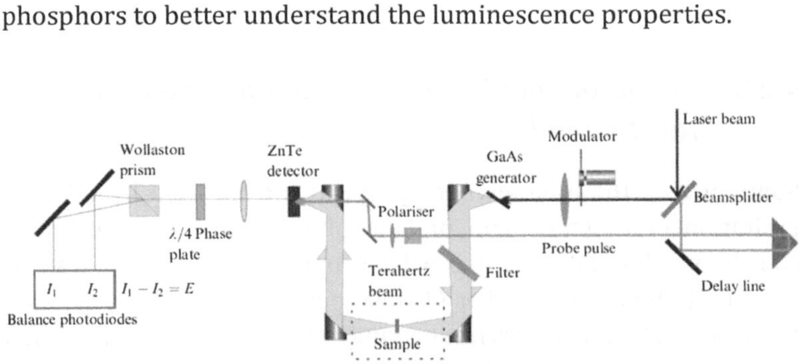

Figure 4.103 Scheme of a time-domain terahertz spectrometer.

A time-domain terahertz spectrometer scheme used in this study (Fig. 4.103) is described in detail in [323–325]. The THP detector and generator types were chosen depending on the absorption in a sample and the frequency range of interest. For example, measurements in the range from 2.5 to 3.5 THz not easily accessible for time-domain spectroscopy were performed by using 90 fs laser pulses at 790 nm. Terahertz pulses were generated by a low-temperature grown GaAs semiconductor surface; an electrooptical detector was made of a 0.3-mm-thick <110> ZnTe crystal. By using the Fourier transform of the temporal THP profile, we calculated the complex spectrum containing information on the refractive index and absorption coefficient of the medium [326].

Refraction indexes and absorption coefficients for the host lattices $YTaO_4$ (1) and $YNbO_4$ (2) (Figs. 4.104 and 4.106) as well as activated by Eu^{3+} and Tb^{3+} (Figs. 4.105 and 4.107) were calculated by using our typical THz-TDs spectrometry procedure described earlier [323].

To calculate optical properties of a powder phosphor it is necessary to reconstruct optical parameters from the measured transmission spectra $T(\omega)$ [327]. From the experiment and subsequent Fourier transform of the temporal profiles of pulses we obtain the incident terahertz pulse (THP) amplitude $[E_{ref}(\omega)]$ and the amplitude $[E_{sam}(\omega)]$ of the transmitted field and then calculate the transmission coefficient of a sample:

$$T(\omega)=\frac{E_{sam}}{E_{ref}}=T_0(\omega)FP(\omega)RL(\omega) \tag{4.26}$$

where

$$T_0(\omega)=\exp\left[-i\left(n_{sam}+n_{air}\right)\frac{d\omega}{c}\right] \tag{4.27}$$

contains the basic information on the medium through which THP has passed (the absorption coefficient and refractive index of the sample);

$$RL(\omega)=\frac{4n_{sam}n_{air}}{\left(n_{sam}+n_{air}\right)^2}=1-R^2(\omega) \tag{4.28}$$

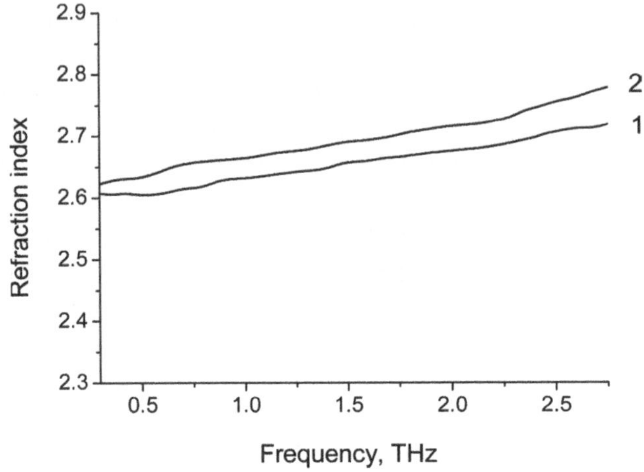

Figure 4.104 Refraction THz spectra of the $YTaO_4$ (1) and $YNbO_4$ (2).

describes reflection losses from sample boundaries;

$$FP(\omega)=\left[1-R^2(\omega)\exp\left(-i2n_{\text{sam}}\frac{d\omega}{c}\right)\right]^{-1} \tag{4.29}$$

describes multiply reflected pulses in a parallel plate–Fabry– Perot modes.

The transmission coefficient Eq. (4.26) cannot be solved analytically, but for most cases the approximate solution is valid. By assuming that RL(ω) ≈ const, and FP(ω) ≈ 1, we obtain

$$\alpha(\omega)=\frac{-\text{In}|\text{T}(\omega)|}{d}+\frac{-\text{In}\left(1-R_{\text{av}}^2\right)}{d} \tag{4.30}$$

The frequency dependence of the refractive index has the form

$$n'(\omega)=-\arg\{T(\omega)\}\frac{c}{\omega d}+1 \tag{4.31}$$

Calculated refraction indexes and absorption coefficients for tantalate and niobate phosphors are presented in Figs. 4.104–4.107.

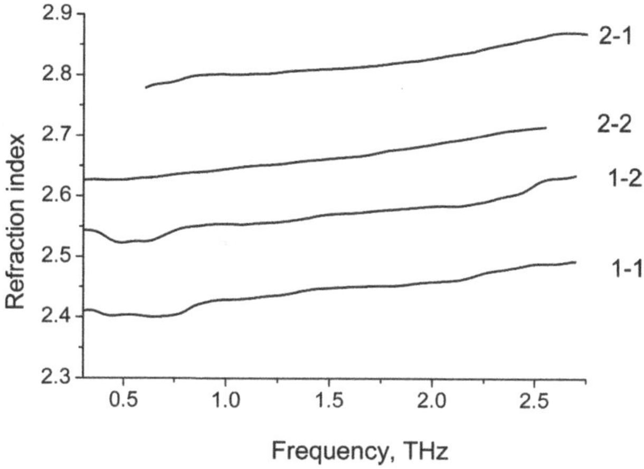

Figure 4.105 Refraction THz spectra of the $YTaO_4:Eu^{3+}$(1–1), $YNbO_4:Eu^{3+}$ (2–1), $YTaO_4:Tb^{3+}$(2–1), and $YNbO_4:Tb^{3+}$(2–2).

From THz measurements it is clear seen that the optical properties doped phosphors changed significantly with activator incorporation. Moreover, absorption coefficient increases with europium and terbium doping in both tantalate and niobate host lattices and decreases with terbium activation in $YNbO_4$-based phosphor. These results correlate with FTIR and luminescence

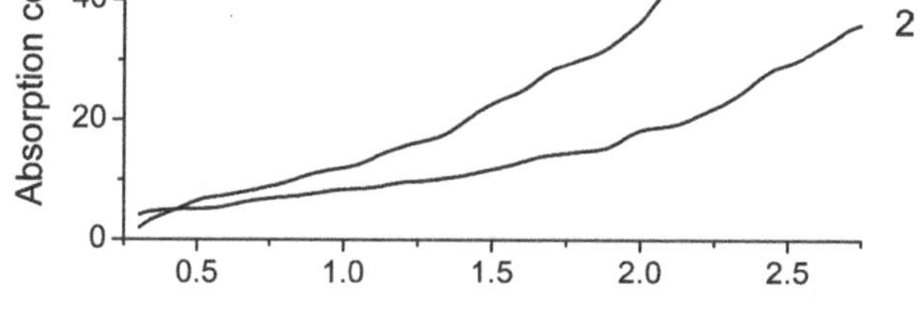

Figure 4.106 Absorption THz spectra of the $YTaO_4$ (1) and $YNbO_4$ (2).

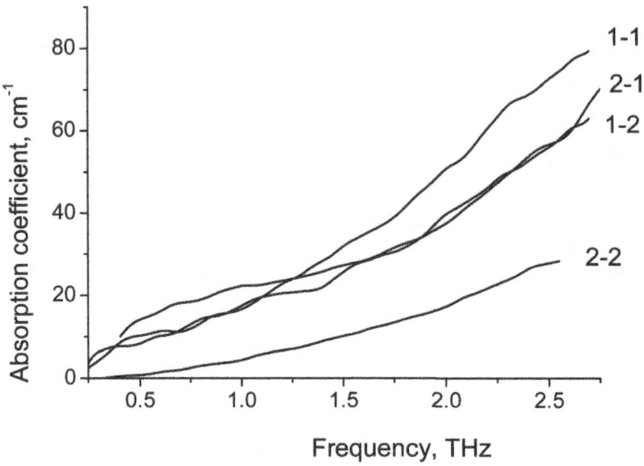

Figure 4.107 Absorption THz spectra of the $YTaO_4:Eu^{3+}$(1–1), $YNbO_4:Eu^{3+}$ (2–1), $YTaO_4:Tb^{3+}$(1–2), and $YNbO_4:Tb^{3+}$(2–2).

spectra and explain a weak Tb luminescence in niobate host lattice. The nature of this phenomenon is discussed and it may be one of the reasons behind high Eu^{3+} emission and poor Tb^{3+} luminescence properties in $YNbO_4$.

4.2.11 *Concluding Remarks*

To summarize the results of this paragraph the following is to be mentioned.

Special attention in this section has been paid to the correlation between the morpho-structural properties, luminescent characteristics and synthesis conditions of tantalate and niobate phosphors. Effect of the flux materials and terbium and europium concentration on luminescence intensity, micro- and macro-structural parameters and particle morphology was demonstrated. Structural investigations pointed up that Li_2SO_4, Na_2SO_4 and $Na_2B_4O_7$ flux materials produce homogeneous crystalline powders, whereas $Na_4P_2O_7$ and CsF form heterogeneous ones. The most effective mineralizing agents are Li_2SO_4 and $Na_2B_4O_7$ that lead to single phase powders with monoclinic $M'YTaO_4$:Tb structure and superior PL properties.

The double activation of $YTaO_4$ and $YNbO_4$ with Tb^{3+} and Eu^{3+} has permitted to synthesize phosphors with variable chromatic emissions. The addition of the rare earth ions does not change the basic structure, but increases the volume of the unit cell. Contribution of both rare earth activators Eu^{3+} and Tb^{3+} is observed in UV, VUV, CL, and XL spectra. Using the group-theoretical approach, the selection rules for the transition between Stark components of Tb^{3+} and Eu^{3+} in the symmetry of the actual crystal field and the polarization for the allowed transitions have been analyzed. Splitting of $^{2S+1}D_J$ multiplets for Tb^{3+} and Eu^{3+} in the crystal field at the C_2 symmetry site as well as the polarization rules for the dipole type optical lines are taken into consideration. The luminescence upon UV, VUV, and electron beam excitation of terbium and europium doubly activated yttrium tantalate are presented and discussed in this section. The proposed two- and three-level models are in good agreement with the experimental results and can be adopted for a better understanding of the transfer mechanism in the rare earth activated phosphors as well as for improving their optical properties.

New efficient phosphors based on $YNbO_4$ and doped by Eu^{3+}, Ga^{3+}, and Al^{3+} were synthesized with different fluxes and under different thermal conditions and investigated under X-ray excitation and Raman spectroscopy. The samples codoped by Al^{3+} and Ga^{3+} show higher intensity under nUV and X-ray excitation in comparison with $YNbO_4$:Eu^{3+} phosphors. The model of redistribution energy transfer is proposed. Monoclinic fergusonite crystal structure and excellent luminescent properties under nUV and X-ray excitation allow recommend these phosphors as good candidates for different applications. In particular, $Y_{1-y}Al_yNbO_4$:Eu^{3+} and $Y_{1-z}Ga_zNbO_4$:Eu^{3+} can be the promising red phosphors for white LEDs using nUV chips.

From THz measurements it is clear seen that the optical properties doped phosphors changed significantly with activator incorporation. Moreover, absorption coefficient increases with europium and terbium doping in both tantalate and niobate host lattices and decreases with terbium activation in $YNbO_4$ based phosphor. These results correlate with FTIR and luminescence spectra and explain a weak Tb luminescence in niobate host lattice. The nature of this phenomenon is discussed and it maybe one of the reason of high Eu^{3+} emission and poor Tb^{3+} luminescence properties in $YNbO_4$.

The tantalate and niobate phosphors presented in this chapter are found to be efficient X-ray, VUV and cathodoluminescent materials that could be used not only in X-ray intensifying screens, but also in the mercury-free fluorescent lamps, CRT and different display panels.

4.3 Phosphors Based on Europium Doped Oxides and Oxysulfides

4.3.1 *Introductory Remarks*

Phosphors based on Eu^{3+} doped RE elements such as yttrium, lanthanum, and gadolinium are widely used to provide red emission for modern display devices. The quality of phosphors is of great importance for flat panel displays performances because they directly influence their brightness and lifetime [328–333]. In this section, the main attention is paid to the performances of different red emitting phosphors based on Eu^{3+} doped solid solutions of Y_2O_3–La_2O_3–Gd_2O_3 and Y_2O_2S–La_2O_2S–Gd_2O_2S and YVO_4–$GdVO_4$ that were synthesized by solid state reaction and co-precipitation methods. Since the concentration of the activator ions is relatively low, the energy of excitation is absorbed mainly by the phosphor matrix. Therefore, the requirements for the matrix composition and crystalline structure become very important. RE oxides such as Y_2O_3, La_2O_3, and Lu_2O_3 are suitable for phosphor matrices because their cations have closed shell electronic configurations and their total orbital angular and spin momenta are equal to zero. These oxides are transparent for the visible light, have no low-energy levels and the interaction between the RE activator ions and the matrix is rather weak. The optical properties of Gd_2O_3 are close to those of the aforementioned oxides. Other promising matrices for the phosphor application are RE oxysulfides, such as Y_2O_2S, La_2O_2S, Gd_2O_2S, and Lu_2O_2S [7, 334]. They form isostructural series crystallizing in the trigonal singony which is commonly represented in the hexagonal aspect with a local symmetry of the C_{3v} lanthanide site [335]. Red-emitting YVO_4:Eu^{3+} is a well-known phosphor and nowadays this composition is widely used as the red component in cathode ray tubes. Due to its good color purity, this phosphor is also considered as a promising candidate for PDPs [336].

The aims of the present section is to describe influence of the synthesis conditions and composition of the matrix in the some complex europium activated yttrium oxide based systems on their light-emitting performances, to investigate their physical and chemical properties and optimize the synthesis conditions and parameters, to describe design of high efficient phosphor materials under electron and UV excitations. An attempt is also made to explain the photoluminescence experimental data by proposing a lattice parameter as a quantitative assessment of activator incorporation degree. The term activator incorporation degree is used to define the ability of europium to efficiently substitute yttrium from the host lattice.

4.3.2 *Eu^{3+} Activated Y_2O_3*

4.3.2.1 Synthesis of europium doped yttrium oxide phosphors

Different synthesis conditions were used to improve the light-emitting performances of $Y_2O_3:Eu^{3+}$ phosphor. Here we report on the influence of synthesis parameters on luminescence properties of $Y_2O_3:Eu^{3+}$, prepared by solid state reaction. The phosphor formation during the firing stage could be described by the equation

$$(1 - x)Y_2O_3 + xEu_2O_3 \xrightarrow{\text{flux}} Y_{2-2x}\,Eu_{2x}O_3$$

Barium chloride and sodium tetraborate (borax) were used as a flux at temperature range 1100–1400°C. More than 20 $Y_2O_3:Eu^{3+}$ samples were prepared and investigated. The PL relative intensities, lattice parameter and particle size of some phosphor samples prepared in various conditions are summarized and presented in the Tables 4.19 and 4.20.

4.3.2.2 Structural and luminescent characterization

Phosphor samples were characterized by crystalline structure and luminescent properties. Particle sizes and morphologies of

Table 4.19 General properties of some phosphor samples prepared with various activator concentrations and flux at $T = 1300°C$

Activator concentration	Sample characteristics	No flux	$BaCl_2$ (2%)	$BaCl_2$ (5%)	$Na_2B_4O_7$ (5%)
	Sample code	L1	L5	L13	L14
2% mol Eu	Lattice parameter (Å)	10.607	–	–	10.606
	PL (a.u.)	7	13	15	26
	Sample code	L2	L6	L15	L16
6% mol Eu	Lattice parameter (Å)	10.608	–	10.611	10.609
	PL (a.u.)	20	36	61	71
	Sample code	L3	L7	L17	L18
10% mol Eu	Lattice parameter (Å)	10.609	10.611	10.616	10.613
	PL (a.u.)	29	60	96	120
	Sample code	L4	L8	L19	L20
15% mol Eu	Lattice parameter (Å)	10.610	–	10.617	10.617
	PL (a.u.)	46	85	137	176

the investigated phosphors were determined by scanning electron microscope Philips-XL30SFEG. To control the particle size and to find the size distribution the Laser diffraction was carried out using HELOS particle size analysis system (the detailed analysis can be found in section 3.4). Crystalline structure of samples was determined by X-ray diffraction, using an X-ray diffractometer D/max – RC, Rigaku, Japan, with Cu Kα radiation (λ = 1.54056 Å). Photoluminescence characteristics were estimated on the basis of emission and excitation spectra registered at room temperature (Xe 500 W lamp) with DARSA PRO 5100 PL System.

Table 4.20 General properties of some phosphor samples prepared at different annealing temperatures with activator concentration 10% mol Eu^{3+}

Sample characteristics	Temperature 1100°C		Temperature 1300°C				Temperature 1400°C	
	No flux	$BaCl_2$ (2%)	No flux	$BaCl_2$ (2%)	$BaCl_2$ (5%)	$Na_2B_4O_7$ (5%)	No flux	$BaCl_2$ (2%)
Sample code	L9	L10	L3	L7	L17	L18	L11	L12
Lattice parameter (Å)	–	10.605	10.609	10.611	10.616	10.613	–	10.616
Photoluminescence (a.u.)	6	13	29	60	96	120	41	104
Particle size (μm) D_{50}%	2.5	3.0	3.0	4.0	5.5	6.8	4.0	6.0

XRD spectra of samples illustrate that all samples possess the cubic crystalline structure. (Powder Diffraction File JCPDS 65-3178 International Center for Diffraction Data, Swarthmore, 1989). The XRD curves registered for all investigated samples show that phosphors have a good crystallinity of Y_2O_3. Small amount of monoclinic yttrium borate seems to be present only in samples prepared with borax as flux (Fig. 4.108).

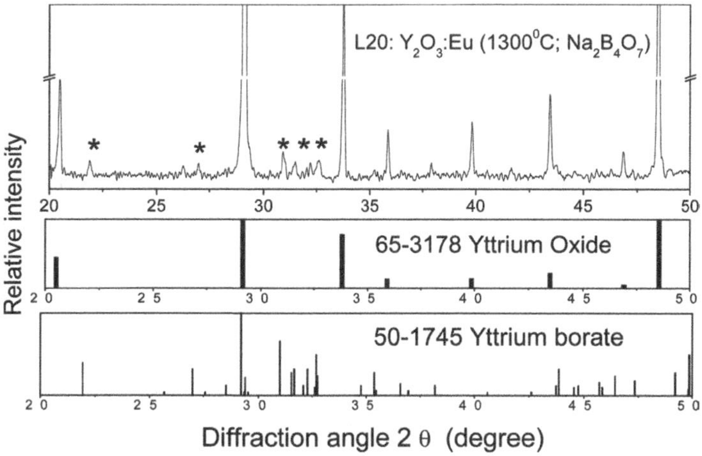

Figure 4.108 XRD spectra of Y_2O_3:Eu phosphor prepared with $Na_2B_4O_7$ as a flux.

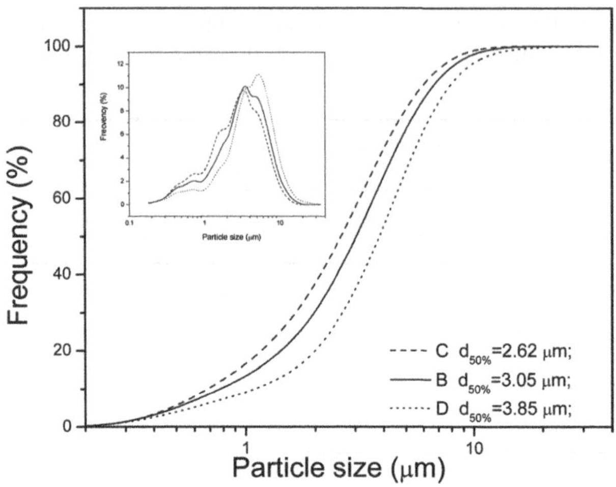

Figure 4.109 Particle size distribution of Y_2O_3:10% Eu^{3+} samples prepared at different temperatures.

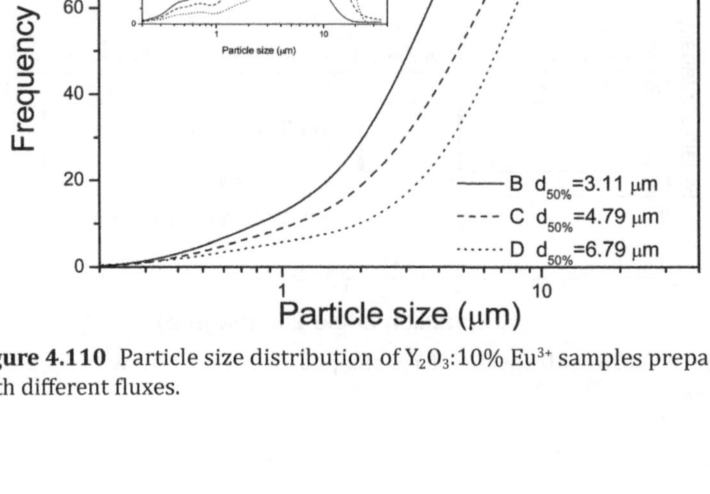

Figure 4.110 Particle size distribution of Y_2O_3:10% Eu^{3+} samples prepared with different fluxes.

Particle sizes for all samples were visualized directly from SEM images and carefully measured by using the laser diffraction method. Cumulative curves and the corresponding derivative ones for several phosphors are presented in Figs. 4.109 and 4.110.

The rising of the firing temperature is favorable to the inter-grain re-crystallization resulting in the increase of particle median diameter ($d_{50\%}$) from 2.62 at 1100°C to 3.85 μm at 1400°C. The inter-grain re-crystallization becomes important in the case of flux-assisted synthesis. The particle median diameter of samples prepared with barium chloride and borax is about 1.5 and respectively, 2.2 times larger then the samples obtained with no flux addition. The efficiency of borax mineralizing action is well illustrated by the particle size increase. As expected, $BaCl_2$ and $Na_4B_2O_7$ acts as mineralizing agent that contributes to both intra- and inter-grains re-crystallization. Intra-grain re-crystallization leads to the activator incorporation degree improvement and as a result, to superior PL characteristics.

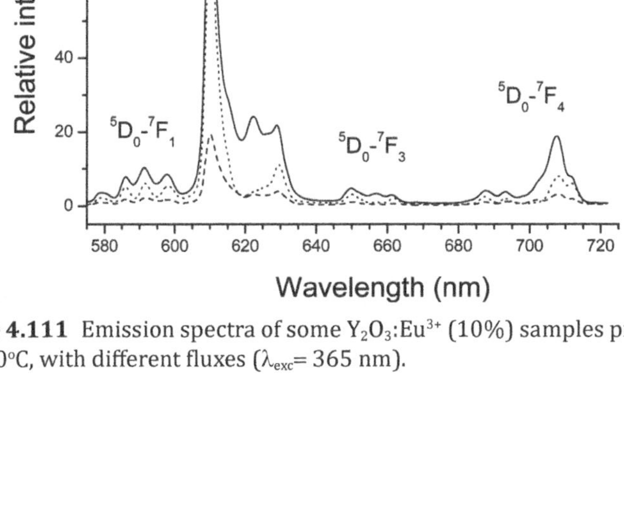

Figure 4.111 Emission spectra of some Y_2O_3:Eu^{3+} (10%) samples prepared at 1300°C, with different fluxes (λ_{exc}= 365 nm).

Average data obtained from cumulative curves for different synthesis conditions are shown in Table 4.20. It is very important that the particle size practically does not depend on the activator concentration and determined only by synthesis conditions, such as annealing temperature and flux.

Under UV and electron beam excitations, europium activated yttrium oxide phosphor shows bright characteristic red luminescence. The spectra obtained for the two modes of excitation are similar, when considered from a global point of view, showing only transitions from the $^5D_0$ level. Europium specific luminescence in the red domain of visible spectrum is the result of these electronic transitions resulting in the emission bands situated between 570 nm and 720 nm, as shown by the PL emission spectra (Fig. 4.111). The four groups of narrow lines observed can be easily assigned to the $^5D_0 \rightarrow {}^7F_{1-4}$ transitions of the Eu^{3+} ion.

Position of the bands and their relative intensity depend on the activator incorporation degree into the Y_2O_3 host lattice and this one is strongly influenced by the flux nature and firing regime used for phosphor synthesis. The emission efficiency in the group of three bands (611, 624, and 630 nm) assigned to $^5D_0 \rightarrow {}^7F_2$ electronic transitions seems to be affected by the crystalline order degree of the host lattice formed in the presence of different mineralizing agents.

PL measurements put in evidence the influence of thermal synthesis conditions on activator substitution efficiency when Eu^{3+} ion replace Y^{3+} in the host lattice. If the difference in ionic radii is sufficient, the unit cell parameter could characterize an activator incorporation degree and play the role of an indicator of activator substitution. It is shown that substitution of Eu^{3+} for Y^{3+} results in an increase in the doped phase unit cell volume as the concentration in Eu^{3+} increases. Difference in the ionic radii of yttrium and europium (0.90 Å and 0.95 Å, respectively) may account for this increase. Thus, the concentration dependence can be explained. In certain limits (for the same or similar particle sizes and before the concentration quenching), the term activator incorporation degree could be used to illustrate the phosphor performances resulting from the ratio between the concentrations of emission centers and quenching centers.

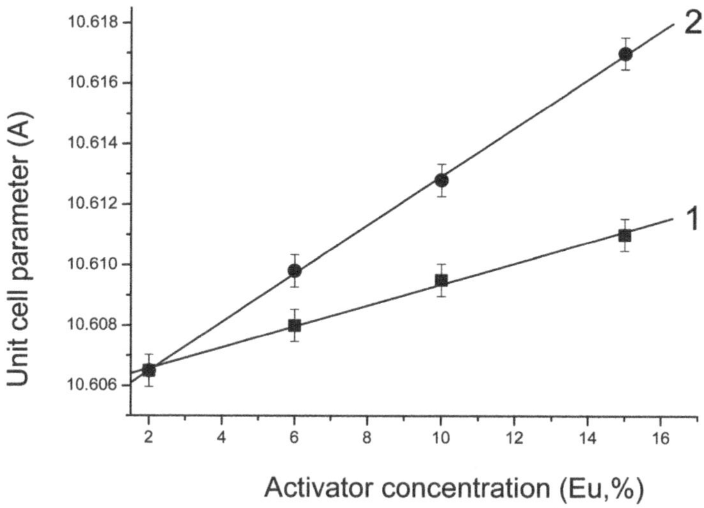

Figure 4.112 Lattice parameter dependence vs. activator concentration without flux (1) and with sodium tetraborate as flux (2). $T = 1300°C$.

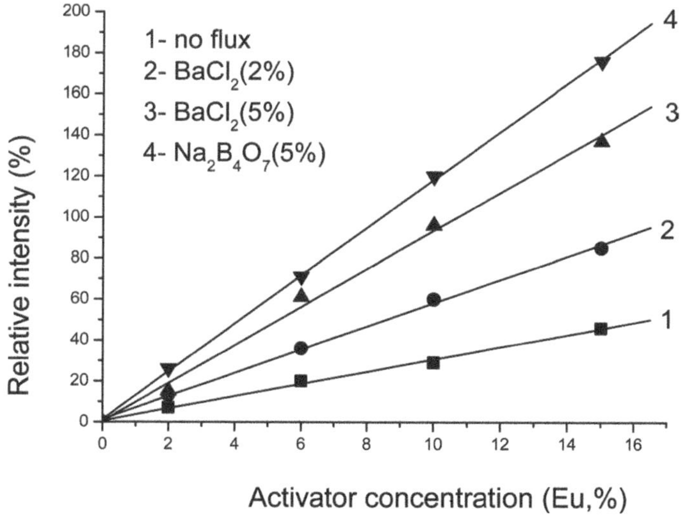

Figure 4.113 Relative PL intensities (λ_{exc} = 365 nm) vs activator concentration for different fluxes. $T = 1300°C$.

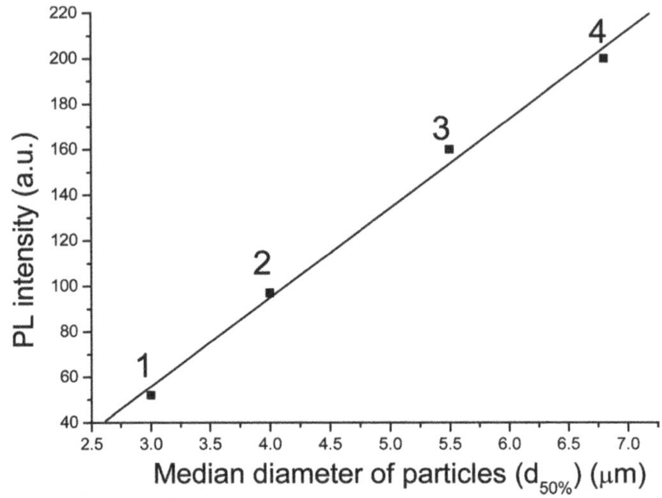

Figure 4.114 Dependence of PL intensity vs. median diameter of particles. $T = 1300°C$.

The direct correlation between the lattice parameter and luminescence properties is shown in Figs. 4.112–4.113.

The europium doping level corresponds to linear intensity vs. concentration dependence (Fig. 4.113). It should be noted that the

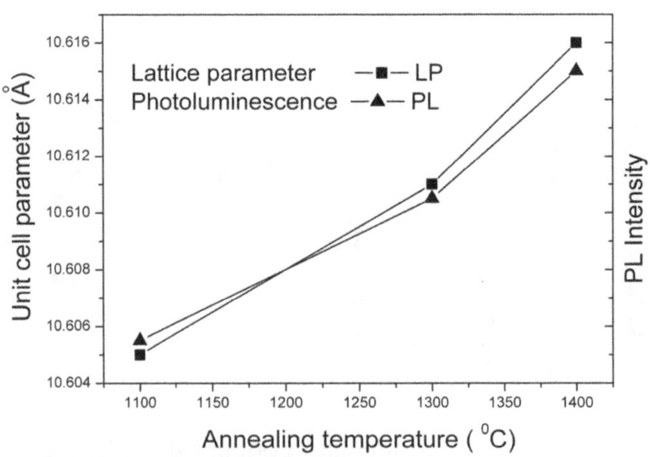

Figure 4.115 Unit cell parameter and PL intensity dependences vs. annealing temperature.

linear dependence in Figs. 4.112 and 4.113 is determined by activator incorporation degree and the inclination is defined by particle size (Fig. 4.114).

Double role of flux is evident. $BaCl_2$ and $Na_2B_4O_7$ act as mineralizing agents that contributes to both intra- and inter-grains re-crystallization. Intra-grain re-crystallization leads to the activator incorporation degree improvement and as a result, to superior PL characteristics. Inter-grains re-crystallization increases the particle size that also increases PL intensity.

Less evident is dependence of lattice parameter on the annealing temperature (Fig. 4.115).

The increased lattice constant as a function of annealing temperature was also observed (but not explained) in previous studies. This phenomenon can be explained taking into account that the activator substitution increases with the increase of the temperature and occurs more efficient. Moreover, the average particle sizes also increase with the temperature that is clearly seen from Table 4.20. The influence of annealing temperature on luminescence properties can be well observed, for example, from Table 4.20 for 10% europium and $BaCl_2$ as flux. The relative intensity increases with the annealing temperature for L10-L7-L12 series, presented in Fig. 4.115. Two important factors such as activator incorporation degree and increase of particle size intensify the luminescence enhancement. Activator substitution in general is a function of concentration, annealing temperature, flux etc, so the same replacement can be achieved by different ways: varying the concentration and flux (L15, L7) at the same temperature, or flux and temperature (L17 and 12) at the same concentration. In any case when we achieve the same lattice parameter and activator incorporation degree, the luminescence properties are very similar (it is valid, of cause, for the similar particle size).

It is found that a use of the relatively higher firing temperature and the presence of mineralizing agent contribute to both the formation of homogeneous Y_2O_3–Eu_2O_3 solid solutions and higher activator substitution efficiency for 2–15% europium into the host lattice. It has been obtained from all measurements that the samples prepared with sodium tetraborate are more efficient than the corresponding material synthesized with barium chloride as flux and significantly better than the samples prepared without any flux.

4.3.3 *Eu^{3+} Doped Y_2O_2–La_2O_2–Gd_2O_2*

4.3.3.1 Synthesis of europium doped yttrium–gadolinium–lanthanum oxide phosphors

Appropriate amounts of Y_2O_3 and/or Gd_2O_3 and/or La_2O_3 were dissolved in nitric acid ($\rho = 1.2$ g/cm^3). The activator was introduced as europium nitrate, in amounts ranging from 2.5 to 10 mol%, taken as europium oxide. The pH of the solution was brought to 2.0 by adding ammonia solution. Then a saturated solution of oxalic acid was added till complete co-precipitation of Y, Gd, and La oxalates. The precipitates were filtered, washed with distilled water and dried in air at about 80–100°C. The blends were placed into quartz crucibles and annealed in air at 1100°C during 12 h.

Alternatively, relatively large amount of sodium carbonate was added to the oxalates blend in order to decrease the firing temperature; Na_2CO_3 was introduced in fivefold concentration of the blend mass. The homogeneous mixture was annealed at 900°C for 6 h and quenched in air. All samples were grinded, washed in distilled water, and dried at 80–100°C.

4.3.3.2 Structural and luminescent characterization

Y_2O_3, Gd_2O_3, and Eu_2O_3 crystallize in cubic *C* modification and form continuous series of solid solutions. Lanthanum oxide crystallizes in hexagonal *H* modification but passes into the monoclinic *M* one after dissolution in yttrium and gadolinium oxides. The ionic radius of the La^{3+} (0.116 nm) differs from those of the Y^{3+}, Eu^{3+}, and Gd^{3+} (0.102, 0.106, and 0.105 nm, respectively). The difference being under 15%, yttrium, gadolinium, and lanthanum oxide could form, in certain conditions, solid solutions that could efficiently accommodate europium ions to form red emitting phosphors. The solid solutions Y_2O_3–La_2O_3–Gd_2O_3 can be formed over the whole composition triangle and crystallize in the cubic, monoclinic, and hexagonal modifications depending on the composition, Fig. 4.116.

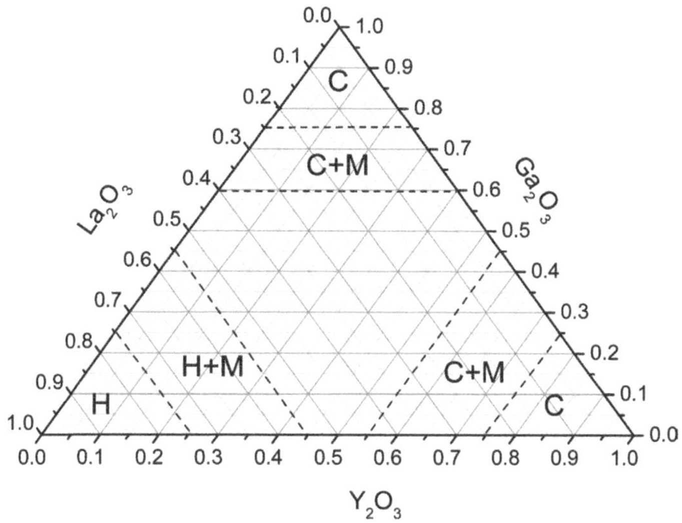

Figure 4.116 Ternary diagram in the system Y_2O_3-La_2O_3-Gd_2O_3 (C, H, M are the cubic, hexagonal, and monoclinic phases, respectively).

The ternary solid solutions in the explored areas of the composition triangle crystallize predominantly in the cubic structure. Hexagonal and monoclinic phases of solid solutions are formed only in La_2O_3-rich blends and their fractions increase

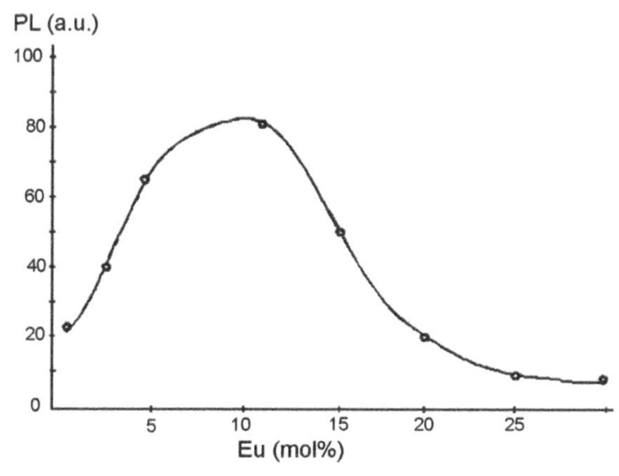

Figure 4.117 PL intensity of $(Y_X,La_{1-X})_2O_3$:Eu^{3+} phosphor vs. activator concentration.

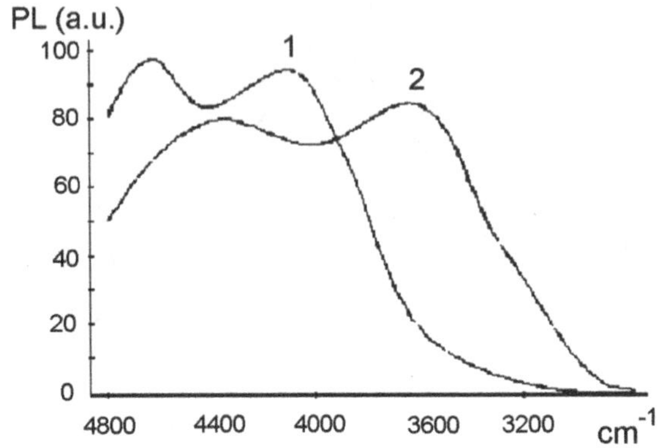

Figure 4.118 PL excitation spectra of $Y_2O_3:Eu^{3+}$ (1) and $(Y_{1-x}La_x)_2O_3:Eu^{3+}$ (2).

with the concentration of lanthanum oxide. A pure hexagonal phase without impurities was obtained only for the solid solutions containing more than 90 mol% of La_2O_3. Pure monoclinic phase was not isolated under the synthesis conditions. The introduction of Eu^{3+}activator does not influence the crystalline structure of the phosphor matrix. Maximum PL intensity (under λ = 394 nm excitation) for Y_2O_3:Eu and Gd_2O_3:Eu phosphors was achieved at 10 mol% of europium oxide (Fig. 4.117).

PL intensity of the solid solutions is increased by introducing of La_2O_3 into $Y_2O_3:Eu^{3+}$ and $Gd_2O_3:Eu^{3+}$ phosphors increases, due to a significant changes in the PL excitation spectra (Fig. 4.118).

The $(Y_{1-x}La_x)_2O_3$:Eu phosphor show enhanced PL intensity as compared to the pure $Y_2O_3:Eu^{3+}$, at lower energies of exciting photons (<4.7 eV). The use of sodium carbonate as mineralizing agent (flux) the synthesis temperature to be reduced, but with negative effects on phosphor quality. The output is diminished and the product contamination is higher in this case.

4.3.4 *Eu^{3+} Doped Y_2O_2S–La_2O_2S–Gd_2O_2S.*

4.3.4.1 Synthesis of europium doped yttrium–gadolinium–lanthanum oxysulfide phosphors

Solid solutions of yttrium, gadolinium, and lanthanum oxysulfides were synthesized by co-precipitation of oxalates with their subsequent thermal decomposition, in presence of sodium carbonate, sodium phosphate, and sulfur. For the blend synthesis, solutions of RE oxides in nitric acid (ρ = 1.2 g/cm^3) and saturated oxalic acid were used. The RE oxalate precursors were filtered and dried in air, at 80–100°C. Stoichiometric amounts of sodium carbonate, sodium phosphate ($\times 12H_2O$), and a threefold excess of sulfur were added to the oxalate precursors. The blend was wetted with ethyl alcohol to gruel consistency and dried under permanent mixing till full free-flowing. The dry blend was placed into quartz ampoules for annealing in a reducing atmosphere or vacuum ($<5 \times 10^{-3}$ Torr). The annealing was carried out within 800–1100°C during 1–8 h. After that the phosphors were cooled, powdered, washed, and filtered. The phosphors were dried in air at 80–100°C during 2 h.

Four series of two- and three-component systems of Y, Gd, and La oxysulfides were prepared according to the general formula: aY_2O_2S:bGd_2O_2S:cLa_2O_2S where $a + b + c = 1$.

Series A, system Y_2O_2S–La_2O_2S: $a = 0 \div 0.90$, $b = 0$, and $c = 0 \div 0.90$.

Series B, system Y_2O_2S–Gd_2O_2S: $a = 0 \div 0.90$, $b = 0 \div 0.90$, and $c = 0$.

Series C, Gd_2O_2S–La_2O_2S: $a = 0$, $b = 0.1 \div 0.50$, and $c = 0.50 \div 0.90$.

Series D, Y_2O_2S–Gd_2O_2S–La_2O_2S: $a = 0.34 \div 0.80$, $b = 0.10 \div 0.33$, and $c = 0.10 \div 0.33$.

4.3.4.2 Structural and luminescent characterization

Depending on the thermal treatment regime, Y_2O_2S, La_2O_2S, and Gd_2O_2S show either limited or unlimited solubility. Y_2O_2S–La_2O_2S solid solutions are formed at 900°C, when the content of Y_2O_2S does not exceed 20 mol%. When the synthesis temperature is raised up to 1000°C, the solubility is increased up to 40 mol%. The same amount of Gd_2O_2S can be dissolved in La_2O_2S at a synthesis temperature 1000 °C. Unlimited solubility is observed in Y_2O_2S–Gd_2O_2S system that can be explained by the close values of the cationic radii. As a result almost no shift of the photoluminescence bands over the whole composition range could be observed in this system. The introduction of 20 mol% of La_2O_3 into Y_2O_2S:Eu^{3+} phosphor leads to the increase PL and CL brightness with about 20–25%. This variation is accompanied by the shift of the main emission peak of the luminescent spectra towards shorter wavelength. Some

Table 4.21 Synthesis conditions and PL characteristics of phosphors based on europium doped Y_2O_2S–La_2O_2S–Gd_2O_2S

Phosphor composition			PL characteristics (λexc= 394 nm)	
Matrix composition	Eu_2O_3 mol/ 100 mol matrix	Crystalline structure	λmax	I_{rel}
$0.200Y_2O_2S$ $0.800La_2O_2S$	10	hex – $(Y_{0.2}La_{0.8})$ O_2S	611	45
$0.600Y_2O_2S$ $0.200Gd_2S$ $0.200La_2O_2S$	10	Cub–(Y, Gd, La)O_2S and hex La_2O_2S or un-reacted La_2O_3	611	59
$0.650Y_2O_2S$ $0.175Gd_2O_2S$ $0.175La_2O_2S$	10	Hex + Mcl	630	74
$0.650Y_2O_2S$ $0.175Gd_2O_2S$ $0.175La_2O_2S$	2	Hex	615	38
$0.700Y_2O_2S$ $0.150Gd_2O_2S$ $0.150La_2O_2S$	10	Cub $(Y_{0.700}Gd_{0.150}La_{0.150})O_2S$	611	45

ternary compositions Y_2O_2S–La_2O_2S–Gd_2O_2S with equal amounts of La_2O_2S and Gd_2O_2S and variable amount Y_2O_2S were synthesized at 900°C temperature. PL characteristics and crystalline structure of samples $(Y_{0.700}Gd_{0.150}La_{0.150})_2O_2S$, $(Y_{0.600}Gd_{0.200}La_{0.200})_2O_2S$, and $(Y_{0.650}Gd_{0.175}La_{0.175})_2O_2S$, doped with 10 mol% Eu_2O_3 are summarized in Table 4.21.

XRD spectra showed that the crystalline structure of $(Y_{0.700}Gd_{0.150}La_{0.150})_2O_2S:Eu^{3+}$ (10%), $(Y_{0.600}Gd_{0.200}La_{0.200})_2O_2S:Eu^{3+}$ (10%), and $(Y_{0.650}Gd_{0.175}La_{0.175})_2O_2S:Eu^{3+}$ (10%) is mainly cubic, mixed hexagonal and cubic, and mixed hexagonal and monoclinic, respectively. At 1000°C, the decrease of Y_2O_2S amount from 0.700 to 0.600 mol/(mol matrix) phosphor seems to be in the favor of the formation of the hexagonal structure. The same effect was observed for the decrease of the europium oxide concentration. Sample $(Y_{0.650}Gd_{0.175}La_{0.175})_2O_2S:Eu^{3+}$ (2%) possesses a well-formed hexagonal crystalline structure.

It seems that europium trends to stabilize the monoclinic structure of the matrix. PL spectra illustrate that for $(Y,Gd,La)_2O_2S$:Eu(10%), the strongest emission is at about 611 nm for Y-rich sample with cubic structure and at about 630 nm for sample with medium Y-content, showing mixed hexagonal-monoclinic structure.

4.3.5 *Phosphors Based on $YVO_4:Eu^{3+}$ and $(Y,Gd)VO_4:Eu^{3+}$*

4.3.5.1 Synthesis of $(Y_x, Gd_{1-x})VO_4:Eu^{3+}$ phosphor by co-precipitation and solid state reaction route

Nanopowder of $Y_{0.95}Eu_{0.05}VO_4$ phosphor was prepared by the co-precipitation method. For this aim, three aqueous solutions were prepared as follows. Appropriate amounts of Y_2O_3 and Eu_2O_3 were dissolved in hydrochloric acid (HCl 30%) to form solution *A*. V_2O_5 was dissolved in hydrochloric acid (30%) to generate solution *B* containing vanadyl VO^{2+} ions. Solution *C* is a mixture of NH_4OH (25%) and H_2O_2 (30%) taken in 3:1 ratio. Finally, a diluted solution *A* was added to a diluted solution *B* and pH of the resulting mixture was brought to 8.0 by adding the solution *C*. The final mixture was stirred and heated to about 80°C until the precipitation was over. The precipitates were washed with distilled water, dried in trays with a large exposed surface area, and fired in air during 2 h at 1100~1400°C. Some $(Y_x, Gd_{1-x})VO_4:Eu^{3+}$ (x = 0~1.0) phosphor samples were also prepared by solid state reaction (SSR)

Figure 4.119 SEM micrographs of $YVO_4:Eu^{3+}$ powder synthesized by (a) co-precipitation reaction and (b) solid state reaction.

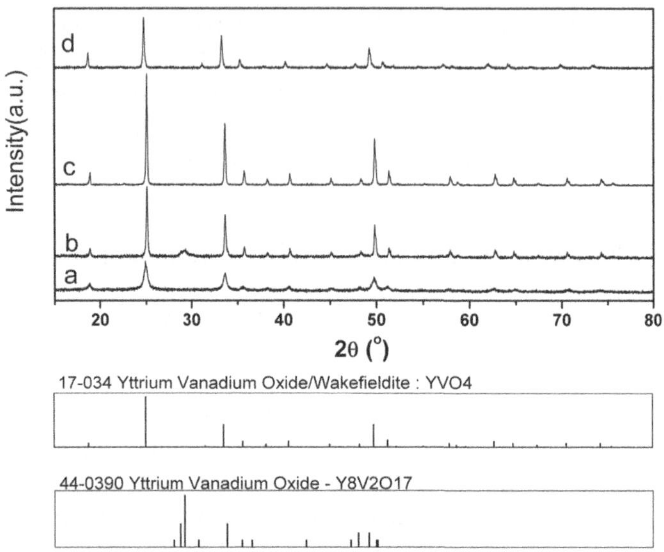

Figure 4.120 Powder XRD patterns of YVO_4:Eu^{3+} powder synthesized by co-precipitation reaction: (a) as-synthesized sample, (b) heat treated sample, (c) YVO_4:Eu^{3+}, and (d) (Y, Gd)VO_4:Eu^{3+} synthesized by solid state reaction. The 19–0341 and 44–0390 JCPDS files [338] can be also seen to identify the obtained crystalline phases.

route and compared with the samples prepared by co-precipitation method.

4.3.5.2 Structural and luminescent characterization

Figure 4.119 shows morphologies and particle sizes of YVO_4:Eu^{3+} phosphor powder prepared by the two aforementioned methods. As shown in Fig. 4.119a (inset), the particle size of YVO_4:Eu^{3+} prepared by co-precipitation reaction was below 200 nm. This picture also shows narrow particle size distribution, spherical shape and a small degree of agglomeration.

Figure 4.120a shows powder X-ray diffraction patterns of nanosize YVO_4 phosphor. Nanosize YVO_4:Eu^{3+} powder obtained after co-precipitation reaction which was done completely at 100°C showed a single and partly crystallized phase even before heat treatment. From the analysis of XRD pattern, it was understood that the introduction of activator Eu^{3+} did not influence the crystal structure of the phosphor matrix. As shown in Fig. 4.120b, after subsequent heat treatment at 1200°C for 2 h, XRD pattern of polycrystalline YVO_4:Eu powder was observed. And also small diffraction peaks are observed at 28.8° and 29.2°, which might be due to the formation of secondary phase $Y_8V_2O_{17}$. It has been reported that V_2O_5 vaporizes readily above its melting point (690°C), causing non-stoichiometry in both the vanadium and the oxygen contents [337].

As shown in Fig. 4.120c,d, $YVO_4:Eu^{3+}$ and $(Y_{0.4}, Gd_{0.6})VO_4:Eu^{3+}$ synthesized by solid state reaction showed the same crystal structure which is tetragonal with a little different lattice constant. Lattice constants of YVO_4:Eu are: $a = b = 7.112$ and $c = 6.29$. And lattice constants of $(Y_{0.4}, Gd_{0.6})VO_4:Eu^{3+}$ are $a = b = 7.178$ and $c = 6.329$, respectively. From the analysis of XRD, it was revealed that all three axes of $YVO_4:Eu^{3+}$ were elongated by an introduction of Gd because the ionic radius of Gd ion is bigger than that of Y ion [339].

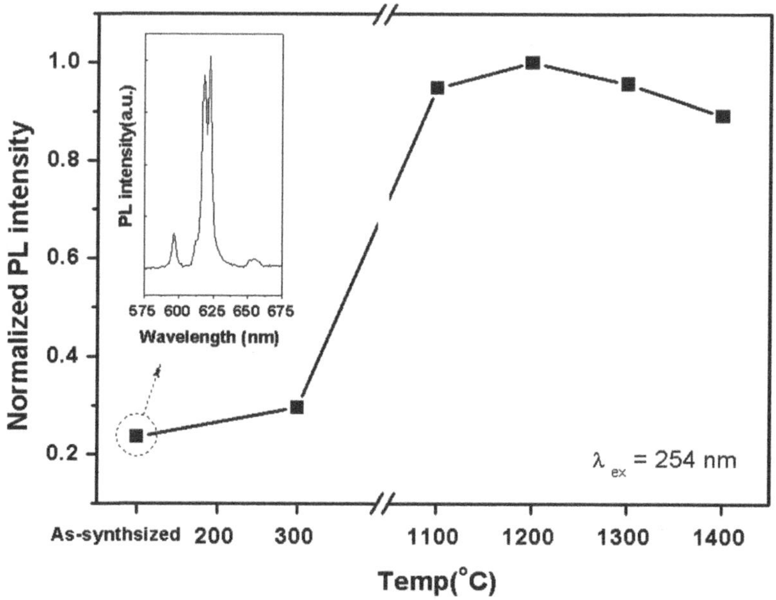

Figure 4.121 Normalized PL intensity of $YVO_4:Eu^{3+}$ synthesized by co-precipitation reaction (Inset: PL spectrum of as-synthesized sample).

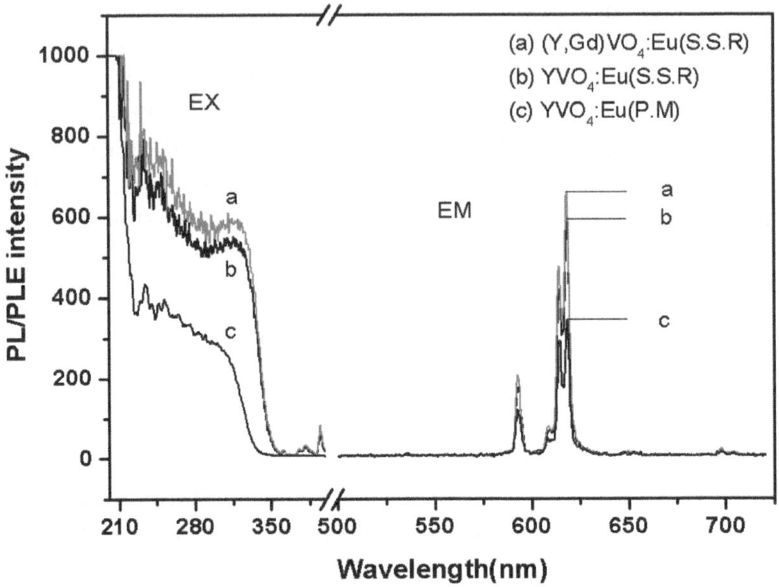

Figure 4.122 PLE and PL spectra of (a) (Y,Gd)VO_4:Eu^{3+} and (b) YVO_4:Eu^{3+} synthesized by solid state reaction, and (c) YVO_4:Eu^{3+} synthesized by co-precipitation reaction.

And as a result of that, it is believed that a small degree of distorted crystal symmetry could be obtained.

PL measurements were carried out to investigate the effect of Gd addition upon PL property of YVO_4:Eu^{3+} and to compare solid state reaction with co-precipitation reaction. As shown in inset of Fig. 4.121, as-synthesized YVO_4:Eu^{3+} phosphors by co-precipitation reaction shows specific luminescence spectrum even before heat treatment and YVO_4:Eu^{3+} fired at 1200°C shows the highest PL intensity.

Figure 4.122 shows the PL excitation and emission spectra of red-emitting YVO_4:Eu^{3+} and ($Y_{0.4}$, $Gd_{0.6}$)VO_4:Eu^{3+} phosphors excited by 254 nm.

The measured excitation spectra of YVO_4:Eu phosphors are in fair agreement with those available from the literature [340]. YVO_4 lattice is responsible for the absorption present in the spectra. This is most likely due to excitation from the filled oxygen 2*p* levels in the valence band to the empty vanadium 3*d* levels of the conduction band. In the case of emission spectrum, it has been reported that the pure YVO_4 shows blue emission, which is due to energy transition of a molecular orbital VO_4^{3-} group inside YVO_4 [52]. As shown in Fig. 4.122, however, when Eu replaces Y in the host material, very sharp peaks appear in red region instead of broad band blue emission. The emission spectrum of $YVO_4:Eu^{3+}$ presents the characteristics of Eu^{3+} emission arising mainly from the transitions from $^5D_0$ level to the ^7F_J ($J = 0, 1, 2, 3, 4$) manifolds. Among these transitions, the major emission peak of Eu^{3+} ion observed at 619 nm is assigned to the transition of the electric dipole type $^5D_0 \rightarrow {}^7F_2$ which is sensitive to the site symmetry of the activator ion.

As shown in Fig. 4.122a, when some of Y^{3+} ions in $YVO_4:Eu^{3+}$ phosphor were replaced by Gd^{3+} ions, it was found that the shape of PL spectrum was changed very little, while relative PL intensity was increased in comparison with $YVO_4:Eu^{3+}$ shown in Fig. 4.122b. It has been reported that the improvement of luminescence intensity of Y-based phosphor with Eu by adding Gd^{3+} ions is due to distortion of site symmetry around Eu^{3+} ion or the creation of Gd^{3+}–Eu^{3+} energy transfer process by introduction of Gd [341, 342].

Luminescence spectrum of nanosize $YVO_4:Eu^{3+}$ phosphor shown in Fig. 4.122c is almost the same as the one observed in Fig. 4.122b. And the corresponding excitation spectrum was also observed. This is in agreement with the mechanism known to be responsible for the luminescence of the $YVO_4:Eu^{3+}$ shown in Fig. 4.122b. In brief, the absorption of UV photons by the VO_4^{3-} groups inside the host material is followed by a non-radiative transfer to the europium [118]. Subsequently it relaxes to the ground state through a radiative transition. Under 254 nm excitation it showed lower PL intensity than $YVO_4:Eu^{3+}$ did as shown in Fig. 4.122b. Considering the existence of secondary phase $Y_8V_2O_{17}$ observed from XRD analysis, low PL intensity of $YVO_4:Eu^{3+}$ phosphor might be due to creation of defects such as vanadium or oxygen vacancy produced on surfaces of the sample [337] or reduction of VO_4^{3-} group playing a important role in luminescence of $YVO_4:Eu^{3+}$ phosphor [118].

In the case of some phosphors having Eu^{3+} ion as an activator, the $^5D_0 \rightarrow {}^7F_J$ emission is very suitable to survey the environmental effects (of surrounding ions) on the sites of Eu^{3+} ion in a host lattice lacking inversion symmetry such as YVO_4. While the $^5D_0 \rightarrow {}^7F_4$ transition is

Table 4.22 Relative PL intensity ratios (R) for the measured emission levels of Eu^{3+} doped YVO_4 phosphor

Phosphor	$\frac{^5D_0 \rightarrow {}^7F_1}{^5D_0 \rightarrow {}^7F_1}$	$\frac{^5D_0 \rightarrow {}^7F_2}{^5D_0 \rightarrow {}^7F_1}$	$\frac{^5D_0 \rightarrow {}^7F_3}{^5D_0 \rightarrow {}^7F_1}$	$\frac{^5D_0 \rightarrow {}^7F_4}{^5D_0 \rightarrow {}^7F_1}$
(Y.Gd) VO_4 : Eu (SSR)[a]	1.000	3.170	0.074	0.113
YVO_4 : Eu (SSR)	1.000	3.340	0.081	0.099
YVO_4 : Eu (CM)[b]	1.000	2.867	0.058	0.143

[a]SSR: solid state reaction.
[b]CM: Coprecipitation reaction.

sensitive to long-range environmental effects, the $^5D_0 \rightarrow {}^7F_2$ transition which originates from interactions with neighbors is hypersensitive to, especially short-range, environmental effects [343]. However, the $^5D_0 \rightarrow {}^7F_1$, the allowed magnetic-dipole transition, is not affected by the Eu^{3+} ion so that is usually utilized as an internal standard. Table 4.22 shows the relative PL intensity ratios (R) for the measured emission levels of Eu^{3+} doped YVO_4 and $(Y_{0.4},Gd_{0.6})VO_4$ phosphors. The intensity ratios of $^5D_0 \rightarrow {}^7F_2$ and $^5D_0 \rightarrow {}^7F_4$ transitions to $^5D_0 \rightarrow {}^7F_1$ transition in PL spectrum of $(Y,Gd)VO_4:Eu^{3+}$ phosphor represent that Eu^{3+} ions occupied yttrium sites of YVO_4 lattice and the short-range environments of Eu^{3+}

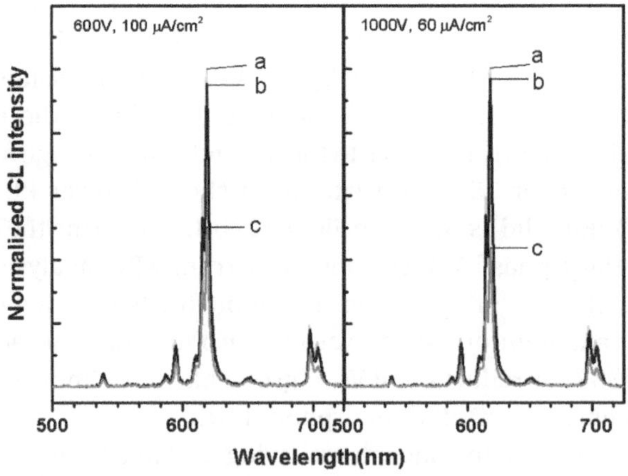

Figure 4.123 CL emission spectra of (a) $(Y,Gd)VO_4:Eu^{3+}$, (b) $YVO_4:Eu^{3+}$ synthesized by solid state reaction, and (c) $YVO_4:Eu^{3+}$ synthesized by co-precipitation reaction.

ions in ($Y_{0.4}$,$Gd_{0.6}$)VO_4 are somewhat different from that of Eu^{3+} in YVO_4, which might be due to introduction of Gd^{3+} ions.

To compare the luminescence characteristics of YVO_4:Eu^{3+} phosphors which were synthesized by solid state reaction and co-precipitation reaction, respectively, and understand the role of Gd in (Y,Gd)VO_4:Eu^{3+}, it is very useful to use low-voltage CL measurements.

Figure 4.123 shows the CL emission spectra of the samples. When we compare CL emission spectra with PL emission spectra in terms of relative intensity and spectra shape, each emission peak was shown in the same wavelength, the shape of spectra was almost identical with that of PL spectra. In addition the relative CL intensity of the samples was also in agreement with the relative PL intensity shown in Fig. 4.122.

In general the CL intensity depends not only on investigated sample but is also an unknown function of emission current and accelerating voltage.

$$I_{CL} = K f_1(J) f_2(U) \tag{4.32)}$$

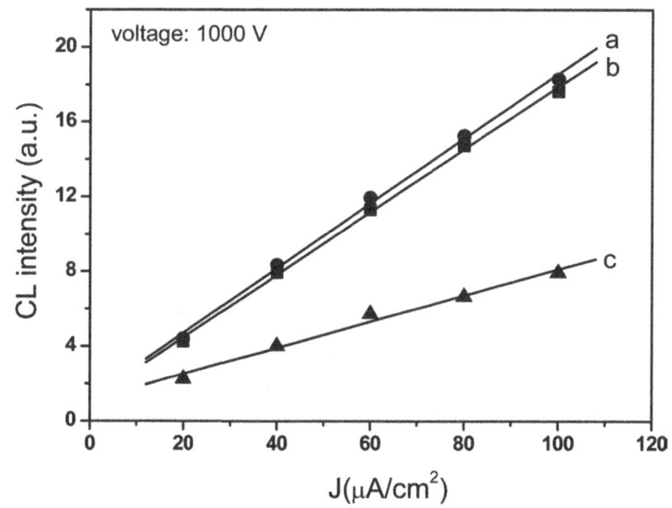

Figure 4.124 Current dependence on CL intensity of (a) (Y,Gd)VO_4:Eu^{3+}, (b) YVO_4:Eu synthesized by solid state reaction, and (c) YVO_4:Eu^{3+} synthesized by co-precipitation reaction ($U = 1$ kV).

where K is a certain constant depending on the sample, J is density of emission current, and U is accelerating voltage between cathode and anode in CL installation. Knowledge of these functions f_1 and f_2 could help us in modeling emission process and energy transfer in investigated sample. They can be received from direct experiments changing J and U. Especially, in our previous study we have investigated successfully that energy transfer from Tb to Eu occurs inside $CaWO_4:Eu^{3+}$, Tb^{3+} phosphor by analyzing relation between CL intensity and activator concentration [222]. For fixed accelerating voltage $U = 1$ kV the current dependences for the samples are shown in Fig. 4.124.

One can see the linear intensity dependence for all the phosphors

$$I_{CL} = KJ \tag{4.33}$$

The coefficient K differs for $YVO_4:Eu^{3+}$ synthesized by co-precipitation reaction. In any case if there is a linear dependence of Eu^{3+} main peak intensity (λ = 619 nm) versus emission current and no saturation one can see, we can establish that no Gd–Eu pairs are formed and all emission is attributed to the electronic transition from emitting level $^5D_0$ to ^7F_J (J = 1, 2, 3, 4, 5, 6) for trivalent ions of europium. And two-level model for CL excitation can be considered. In our previous study using two level model [222], taking into account for the stationary conditions of excitation ($dn/dt = 0$), the number of luminescent ions in excited state is given by Eq. (4.2)

$$n_1 = \frac{LJN}{LJ + \tau^{-1}}$$

where n_1 gives the number of luminescent ions in excited state, τ^{-1} is the probability for spontaneous emission excited state (1)→ ground state (0), J is an electron beam density, L is a quantum yield of CL, and N is the total number of luminescent ions ($n_1 + n_0$). The obtained expression explains the linear dependence of CL intensity on the emission current if $LJ << \tau^{--1}$. Thus, the CL intensity assumes the form

$$I_{CL} \approx n_1 = LJN\tau = KJ \tag{4.34}$$

where $K = LN\tau$.

This derived Eq. (4.34) has the same form as the experimentally derived Eq. (4.33). Therefore, from the consideration of low-voltage CL measurement of $YVO_4:Eu^{3+}$ and $(Y,Gd)VO_4:Eu^{3+}$ phosphors, it is found that the luminescence of both $YVO_4:Eu^{3+}$ and $(Y,Gd)VO_4:Eu^{3+}$ phosphor could be explained by two-level model. And for the same activator concentration and total number of luminescent ions N, and for the constant probability for spontaneous emission, the coefficient K is determined by quantum yield L. Consequently, as shown in Fig. 4.124, $YVO_4:Eu^{3+}$ phosphor synthesized by co-precipitation reaction has a low quantum yield in comparison with one synthesized by solid state reaction, which might be due to the presence of secondary phase $Y_8V_2O_{17}$.

From the literatures of Eu^{3+} ion as an activator, the analysis of XRD, PL, and low-voltage CL measurement, the effect of Gd in $(Y,Gd)VO_4:Eu^{3+}$ phosphor can be summarized as follows: The choice of host lattices for Eu^{3+} activator is very important because the luminescence properties of Eu^{3+} ion doped phosphors depend to a great extent on the site symmetry around Eu^{3+} ion. In this study, luminescence spectrum of $(Y,Gd)VO_4:Eu^{3+}$ is dominated by the peak at 619 nm due to the forced electric dipole $^5D_0 \rightarrow {}^7F_2$ transition of Eu^{3+}, which is typical behavior of Eu^{3+} ions occupying non-centrosymmetric sites. Consequently, when Gd replaces some of Y atoms in the host lattice, Gd changes the parameters of lattice and it can also influence the forced electric dipole transition of Eu^{3+} ion and improve the CL and PL intensity.

4.3.6 *Concluding Remarks*

In this section the synthesis of solid solutions Y_2O_3–La_2O_3–Gd_2O_3 and Y_2O_2S–La_2O_2S–Gd_2O_2S with different content of Y_2O_3 and Y_2O_2S was described and explained using the ternary diagram. The obtained samples possess well-formed hexagonal crystalline structure. The activator introduction does not affect the phosphor matrix structure thus confirming the substitution of RE of the phosphor matrix by europium ions.

Europium doped Y, La, and Gd oxysulfide phosphors show higher CL efficiency at low applied voltages than oxide ones. The introduction of 20 mol% of La_2O_3 into $Y_2O_2S:Eu^{3+}$ phosphor leads to the increase PL and CL brightness with about 20–25%. Optimized $(Y_{0.4}Gd_{0.6})VO_4:Eu^{3+}$ phosphors also show luminescence intensity higher in comparison with similar commercial sample.

Lattice parameter can be proposed as indicator of activator incorporation degree. In order to obtain high activator incorporation degree and consequently efficient $Y_2O_3:Eu^{3+}$ samples, flux nature and concentration, firing temperature and activator amount have to be carefully chosen.

The aforementioned results confirm that europium doped yttrium oxide prepared in the presence of barium chloride and sodium tetraborate as a flux are perspectives for PL and CL applications. Light-emitting output of samples prepared with barium chloride and sodium tetraborate as a flux exceeded the commercial ones by 20–30%. The obtained results confirmed that europium doped solid solutions of Y–Ga–La oxides, oxysulfides, and vanadates are perspectives for PL and CL application in the display manufacture.

4.4 Yttrium and Terbium Aluminate Phosphors Co-activated by Eu^{3+}

4.4.1 *Introductory Remarks*

For the purpose of development of highly energy-efficient white light sources, one needs to produce highly efficient phosphors, which can absorb excitation energy from blue LEDs and generate emissions. The development in the area of wide band gap III–V nitride semiconductors has led to the commercial production of high-efficiency LEDs [344, 345]. The recent advent of blue InGaN technology has made it possible to produce a conventional white LED in which white light is obtained by coating a $Y_3Al_5O_{12}:Ce^{3+}$ (YAG) or $Tb_3Al_5O_{12}:Ce^{3+}$ (TAG) phosphor onto a blue LED chip [346–348]. In this device known as a two-band white LED, white light is generated by additive color mixing of the blue light emitted by the blue LED and the yellow light emitted by the YAG or TAG phosphors. The development of a white LED is important because this opens new ways for LED applications such as light bulbs and fluorescent lamps with high durability and low energy consumption. However, the spectral composition of the light produced by the conventional two-band white LED differs from that of natural white light, particularly in the red region. The color properties of conventional two-band white LEDs can potentially be improved by adding another component in order to create a white LED based on three emission bands (a three-band white LED). Blue LEDs have been combined with phosphors in order to convert part of the blue light to red and green light, to produce white light. Suitable phosphors must have a high efficiency of excitation at about 420–480 nm, and a wide chromaticity zone. A luminescent material absorbs the energy in one region of the electromagnetic spectrum and emits radiation energy in another region of the spectrum. Typically the energy of the photons emitted is lower than the energy of the photons absorbed (the origin of this phenomenon will be discussed in this chapter).

YAG:Ce^{3+} is used as a phosphor for white light emitting diodes that employ the principle of luminescence conversion. However, the color impression of a such conversion on the base of a blue diode and YAG:Ce^{3+} is too "cold" while a red component is missing. The common way to improve the emission spectra of lighting devices is co-activation of the conversion phosphors, for example with Eu^{3+} which shows an intense line emission in the red region. An intense $5d \rightarrow 4f$ photoluminescence of Ce^{3+} in the Tb^{3+} site in TAG powder has been recently reported [349] as well as the possibility of its use as a luminescence converter. In comparison with YAG:Ce^{3+}, the spectrum of the TAG:Ce^{3+} is slightly shifted to the red, probably because of a reduced local symmetry. However, this shift is not sufficient to compensate the lack of the red component. A co-doping with Pr has already been applied successfully [350].

Results of study of co-doping TAG:Ce^{3+} with Eu^{3+} are presented in this section. Several series of samples with constant Ce^{3+} concentration and different Eu^{3+} concentrations are synthesized and their luminescence and structural properties are investigated. Comparison of luminescence mechanisms in YAG and TAG is also made. Although the Ce^{3+} luminescence in YAG and TAG matrices is very similar there are immense differences in the case of Eu^{3+} co-activation: while in YAG:Ce^{3+},Eu^{3+} the Eu^{3+} luminescence cannot be sensitized by Ce^{3+} [351], in TAG:Ce^{3+}, Eu^{3+} both Tb^{3+} and Ce^{3+} serve as sensitizers.

We begin this section from the short review of physico-chemical aspects related with the synthesis process.

4.4.2 *Physico-chemical Aspects Related with the Synthesis of YAG-Based Phosphors*

Generally, the formation of the phase pure $Y_3Al_5O_{12}$ (YAG) from each oxide ingredient by the conventional solid state reaction requires the high temperatures more than 1100°C for eliminating several by-products such as $Y_4Al_2O_9$ (monoclinic yttrium aluminate, YAM) and $YAlO_3$ (yttrium aluminum perovskite, YAP) [77, 352] The YAG-based materials are normally synthesized at a temperature >1500°C with prolonged heating time. On the contrary, YAG directly crystallizes from the amorphous precursor at the temperature as low as 700°C in the sol–gel method [72].

A number of investigations proved that the wet chemical methods could decrease the sintering temperature efficiently [68–73] that is very important to obtain, for example, ultrafine YAG powders at relatively low temperatures. Wet chemical processes, e.g. the hydrothermal method under the supercritical condition and the solvothermal method enable us to obtain the crystalline YAG below 400°C [86, 112]. Recently, Hukuta et al. [86, 106] have reported on a hydrothermal method for the rapid production of fine YAG:Tb^{3+} phosphors in supercritical water (374°C, 22.4 MPa). Li et al. [113] reported for the first time the preparation of YAG:Tb^{3+} nanosize phosphors via a mixed alcohol–water solvent method using inexpensive inorganic salts as precursors at relatively low temperatures (300°C) and pressures (10 MPa). The advantages of the mixed solvothermal method include the following: low-cost precursors, low-temperature preparation, low operating cost, and an environmentally friendly solvent. Inoue et al. developed the solvothermal method in glycol solvents, called as the glycothermal method. They reported the formation of the phase pure YAG from aluminum compounds and yttrium acetate in 1,4-butylene glycol (1,4-BG) [83, 353]. Kasuya and Isobe have reported that Ce^{3+} ions can be incorporated into the YAG nanocrystal by the glycothermal method [354]. When the YAG:Ce^{3+} nanocrystal phosphor was heated to 1000°C to burn out the organic species on its surface, the photoluminescence intensity decreased by a factor of 1/7. This result is entirely opposite to the PL enhancement of the conventional bulk phosphors by heat treatment. They focused on the effect of surface modification by organic species on the PL enhancement of the YAG:Ce^{3+} nanocrystal and reported the effect of the surface modifier, poly(ethylene glycol) (PEG), on the chemical interaction and the PL properties for the YAG:Ce^{3+} nanocrystal. $Y_3Al_5O_{12}$:Ce^{3+} nanocrystals were synthesized in 1,4-butylene glycol (BG) with and without poly(ethylene glycol) (PEG) by the glycothermal method. The internal quantum efficiency of the photoluminescence corresponding to the $5d \rightarrow 4f$ transition of Ce^{3+} in the YAG:Ce^{3+} nanocrystal increased from 21.3 to 37.9% by addition of PEG, while no appreciable change in the primary particle size, the crystallite size, and the lattice distortion was recognized by transmission electron microscopy and X-ray diffractometry.

An evolutionary optimization process involving combination chemistry was employed by Pereira et al. in an attempt to develop YAG [355]. The combination chemistry process utilized in this work consisted the doping of the YAG host with appropriate amounts of red (R), green (G), and blue (B) dopants in a single layer, for use in tricolor white light. The doped YAG was achieved by means of the non-hydrolytic sol–gel route. Four samples were prepared, three of which were mono-doped samples containing 1.0% of a certain lanthanide (Eu^{3+}, Tb^{3+}, or Tm^{3+}) ion, while the fourth contained the three ions. The samples were characterized by X-ray diffractometry and photoluminescence. The diffraction pattern of the mono-doped samples synthered at 800°C for 16 h displayed peaks corresponding to the YAG phase, while the sample doped with the three ions revealed the presence of a mixture of YAG and YAM phases. The emission spectra of the three mono-doped YAG samples displayed the typical bands of the blue, green, and red emission of the corresponding lanthanide ions. As for the sample doped with the three lanthanide ions; it simultaneously emitted R, G, and B lights. The green emission (546 nm) was more intense and narrow in relation to the red and blue emissions, which may be due to differences in the size of the three incorporated ions.

Rapid synthesis of Nd:YAG nanopowder by microwave flash combustion was proposed by Choudhary et al. [356] Nd:YAG nanopowders were successfully synthesized by a novel approach using rapid microwave drying of sol. Citrate nitrate sol was treated in a microwave oven for a few minutes. The as-prepared precursor was calcined at 900°C for 2 h. XRD showed the formation of phase pure polycrystalline YAG. The particle size was found to be 24 nm (estimated from Scherrer's equation). TEM indicated that the particle size was in the 24–70 nm range. Nd^{3+} doping was confirmed by SEM–EDX. Nd:YAG nanopowder synthesized by this method was of low agglomeration and the particles were spherical in shape.

Europium-ion doped YAG:Eu^{3+} garnet nanophosphors with wide ranging size tunability (40–150 nm) have been prepared by Lu et al. via a sol–gel pyrolysis method employing a fuel system that combined urea and polyvinyl alcohol (PVA) [357]. Well-dispersed nanoparticles were prepared at 1000°C. This temperature is lower than that required for synthesizing YAG:Eu^{3+} via the solid state reaction route. The particle size and morphology of the synthesized powders are found to have critical dependence on the oxidizer (metallic nitrates) to fuel ratio. The importance of using an organic polymeric dispersion matrix to obtain well-dispersed YAG:Eu^{3+} nanoparticles has been demonstrated. The photoluminescence properties of the prepared YAG:Eu^{3+} phosphors are profoundly dependent on the preparation conditions. The emission intensity of well-dispersed YAG:Eu^{3+} nanoparticles is found to be much stronger than that of the bulk sample. The excitation spectrum of well-dispersed nanoparticles shows an extension of the excitation peak in the high-energy region. These unique properties of YAG:Eu^{3+} nanoparticles can be rationalized by considering numerous surface states due to the large surface area to volume ratio of the nanoparticles. In addition, using the hypersensitive $^5D_0 \rightarrow {}^7F_2$ transition of Eu^{3+} as a local probe, the role of surface states that modify the optical properties of YAG:Eu^{3+} nanophosphors has also been illustrated.

The synthesis of cerium activated YAG:Ce^{3+} by the wet-chemical synthesis route is reported by Popovici et al.[274]. Y–Ce–Al precursors were prepared using the reagent simultaneous addition technique – SimAdd from Y–Al–Ce nitrate mixture and urea – and subsequently transformed into phosphor samples. The influence of the thermal synthesis regime and flux on the phosphor quality was investigated in order to obtain $Y_3Al_5O_{12}$:Ce^{3+} fine powders with pure cubic structure and quite good photoluminescent properties. Cerium activated yttrium aluminate phosphors were synthesized by homogeneous precipitation, using the SimAdd technique [358, 359]. For this purpose, Y–Al–Ce precursors were prepared from $Y(NO_3)_3{\cdot}5H_2O$ (99.9%Aldrich), $Al(NO_3)_3{\cdot}9H_2O$ (Merck), $Ce(NO_3)_3{\cdot}5H_2O$ (extrapure, Merck) as metallic ion sources and urea H_4CN_2O (99.3% AlphaAesar) as anion source. The nitrate mixture contains Y^{3+}, Ce^{3+}, and Al^{3+} in stoichiometric amounts, corresponding to the molar ratio Y: Ce: Al = 2.91: 0.09: 5. Equal volumes of solutions of nitrate mixture (0.5 M) and urea (2.0 M) were simultaneously added into the urea diluted solution. The precipitation was performed under constant stirring, at 80°C and under pH control. After maturation, the Y–Al–Ce precursor was well washed, centrifuged, and dried. The phosphor thermal synthesis was performed at 900–1600°C, in N_2 atmosphere, for 1–4 h. YAG:Ce^{3+} powders were water-washed, dried, and sieved; no ball milling was used during the phosphor processing. In order to study the flux influence, prior the thermal treatment, precursors were mixed with NaCl, NH_4Cl, Li_2CO_3, NaF, MgF_2, CaF_2, or H_3BO_3 as flux. During the thermal synthesis stage, the flux interacts with the Y–Al–Ce precursor and promotes both the incorporation of the Ce activator and the crystallization of the yttrium aluminate host lattice. The photoluminescence and morpho-structural properties of YAG:Ce^{3+} samples were investigated in dependence on the thermal synthesis regime i.e. the firing temperature and duration and also the flux addition. All phosphors obtained at temperature higher than 900°C are well-crystallized powders with cubic structure. This illustrates that the composition of precursors is appropriate for the formation of the pure garnet structure. No secondary products such as YAP and YAM are noticed in YAG:Ce^{3+} powders. The reagent simultaneous addition technique SimAdd is an alternative wet-chemical synthesis route for phosphor synthesis, which has the advantage of facilitating the control of precursor composition and morpho-structural characteristics. This study revealed that the thermal synthesis regime (e.g. flux and firing regime) used for the conversion of precursor into phosphors influences not only the PL properties but also the microstructure and morphology of YAG:Ce^{3+} particles. Some of the optimal thermal synthesis parameters for this synthesis route could be defined, i.e. the mineralizing agent (ammonium chloride) and the firing regime

(2 h × 1200–1400°C). Moreover, some peculiar aspects related with the flux role in the Y-Al-Ce precursors conversion into YAG:Ce^{3+} phosphors could be put in evidenced. Microcrystalline $Y_3Al_5O_{12}$:Ce^{3+} powders with bright yellow luminescence and good sensitivity to blue radiation could be prepared by SimAdd route.

4.4.3 *Luminescence Properties of $Tb_3Al_5O_{12}$:Ce^{3+} Eu^{3+}*

For the solid state lighting, LED, different types of display we need microparticles with high brightness and the solid state reaction remains the main synthesis route. In this chapter we continue investigation YAG and TAG phosphors by the classical way.

4.4.3.1 Preparation of the phosphors

The powder samples have been prepared by a conventional solid state reaction. According to the nominal composition of $Tb_3Al_5O_{12}$:Ce^{3+}, Eu^{3+}, the starting materials Tb_4O_7, Al_2O_3, CeO_2, and Eu_2O_3 (99% for Al_2O_3, 99.9% for all other oxides) are weighed and milled. Synthesis took place in a furnace heated by induction using a sealed alumina tube in a reducing atmosphere of H_2/N_2 to stabilize the Tb^{3+} state. The crucible was made out of Al_2O_3. BaF_2, LiF, and H_3BO_3 as flux materials were used. The synthesis lasted approximately 2 h at temperature of about 1500°C. The samples were prepared with different Ce (1, 3, 5 mol%) and Eu (1, 3, 5, 7, 9, 11, 13 mol%) concentrations.

4.4.3.2 Measurements

Phosphor samples were characterized by crystalline structure and luminescence properties. Particle sizes and morphologies of the investigated phosphors were determined by scanning electron microscope Hitachi-S-3000N. To control the particle size and to find the size distribution the Laser diffraction was carried out using HELOS particle size analysis system. The crystal structure and the composition of powder samples were characterized by X-ray diffraction using a X-ray diffractometer D/max – RC, Rigaku, Japan, with Cu Kα radiation (λ = 1.54056 Å).

Optical spectroscopy and PL characteristics are studied on the basis of emission and excitation spectra registered at room temperature (Xe 500 W lamp) with DARSA (Diode Array Rapid Scan Analysis, Professional Scientific Instrument Co, Korea) PRO 5100 PL System. Excitation spectra were corrected for the energy distribution of the Xe-lamp. The excitation was performed with a 460 nm radiation which usually used in blue LED as well as with selectively excited emission to establish the luminescence mechanism in cerium-europium double activated terbium aluminum garnet phosphor. The measurement system and mathematical procedure were developed for determining the quantum efficiency of luminescent materials.

4.4.3.3 Luminescence properties

For the application in the blue LED a yellow phosphor with improved luminescence properties it is interesting to obtain Ce^{3+} and Eu^{3+} luminescence simultaneously in garnets, especially for the low energy Ce^{3+} excitation around 460 nm. The energy transfer between Ce^{3+} and Eu^{3+} ions in garnets has not yet been investigated in detail. However, this pair is well known for YOCl:Ce^{3+},Eu^{3+} [360, 361] and $YAlO_3$:Ce^{3+},Eu^{3+} [362]. In the latter case, the energy transfer should be highly effective due to the overlap of the luminescence of Ce^{3+} and the strong absorption of the $^7F_0{\rightarrow}^5L_6$ transition of Eu^{3+} at about 390 nm. In garnet, the photon energy in the Ce^{3+} luminescence is different from the $^7F_0{\rightarrow}^5L_6$ and also from the $^7F_0(^7F_1){\rightarrow}^5D_2$ (464 nm) transition energies.

The emission spectra of double-doped TAG are presented in Fig. 4.125: (a) TAG:(1 mol% Ce^{3+}), Eu^{3+}; (b) TAG:(3 mol% Ce^{3+}), Eu^{3+}, and (c) TAG:(5 mol% Ce^{3+}), Eu^{3+} at room temperature. The luminescence was excited with the 460 nm radiation of Xe lamp.

Doping with 1–3-5 mol% Ce yields a broad-band of $5d{\rightarrow}4f$ luminescence Ce^{3+}. In addition to the direct Ce^{3+} excitation, the Eu^{3+} contribution to the luminescence is also expected. Doping with 1mol% Eu leads to a quenching of the Ce^{3+} luminescence and to appearance of the Eu^{3+} luminescence due to the $^5D_0{\rightarrow}^7F_J$ transitions of the Eu^{3+} ion at 591, 597, 610, 631, and 708 nm. The most intensive peak corresponds to the allowed magnetic dipole $^5D_0{\rightarrow}^7F_1$ transition.

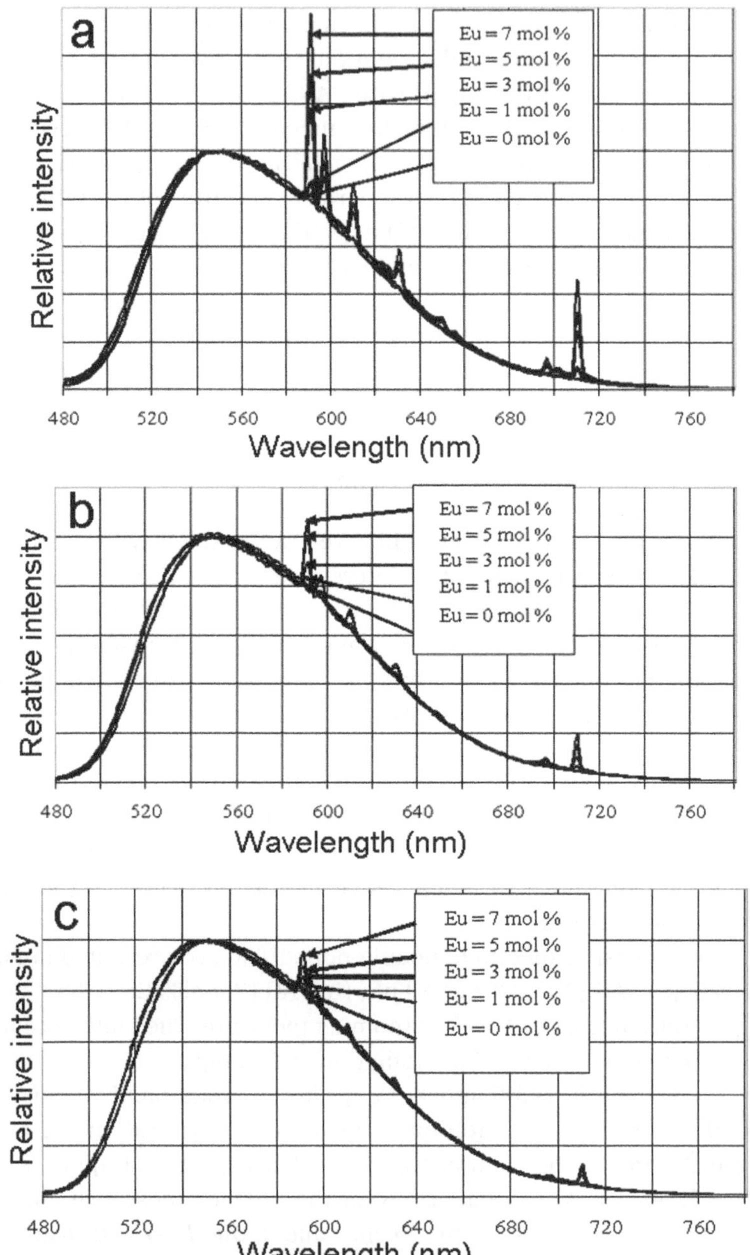

Figure 4.125 The emission normalized spectra of double-doped TAG: (a) TAG:$Ce_{0.01}$, Eu; (b) TAG:$Ce_{0.03}$, Eu; and (c) TAG:$Ce_{0.05}$, Eu.

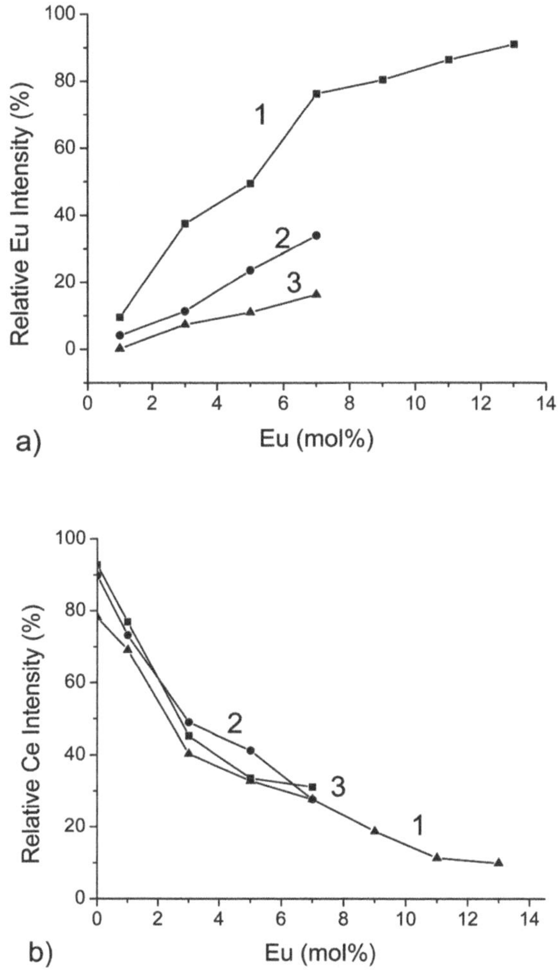

Figure 4.126 (a) The relative contribution of Eu^{3+} in intensity at 591 nm in dependence on Eu^{3+} and Ce^{3+} concentration and (b) quenching of the Ce^{3+} luminescence: (1)-TAG:Ce$_{0.01}$, Eu; (2)-TAG:Ce$_{0.03}$, Eu; (3)-TAG:Ce$_{0.05}$, Eu.

The relative intensity of Eu^{3+} peak (591 nm) in dependence on Eu^{3+} and Ce^{3+} concentration and quenching of the Ce^{3+} luminescence are presented in Fig. 4.126.

Excitation and emission spectra of YAG and TAG phosphors were measured at room temperature (Fig. 4.127). From

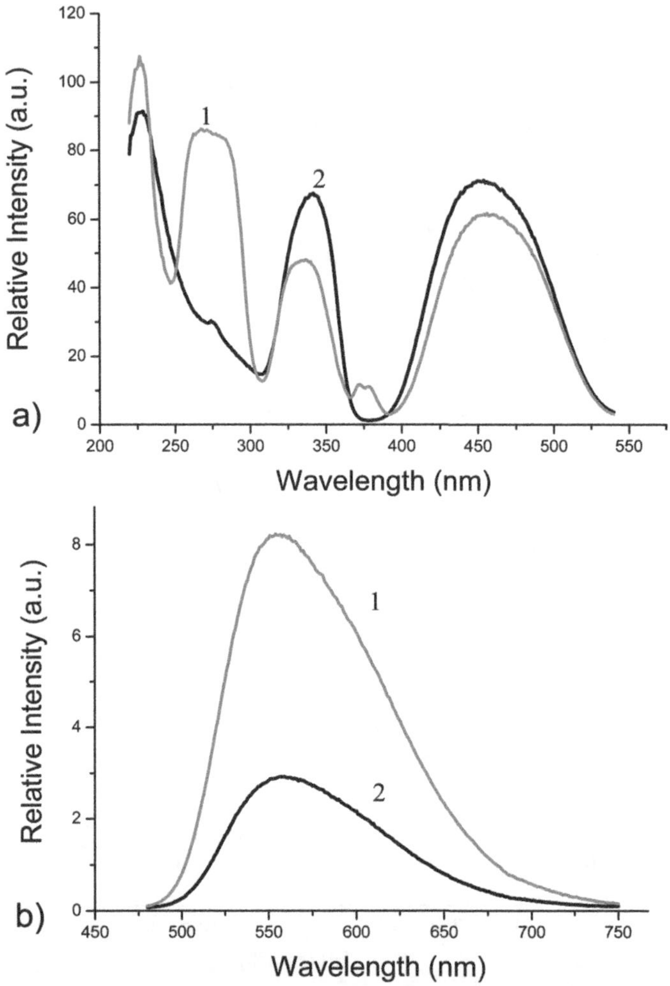

Figure 4.127 Excitation (a) (λ_{em} = 550 nm) and emission (b) (λ_{ex} = 277 nm) spectra for TAG:Ce^{3+} (1) and YAG:Ce^{3+} (2) phosphors.

comparison excitation spectra of YAG and TAG (Fig. 4.127a) one can see the principal distinction in one additional Tb broad band (1) at approximately 277 nm that corresponds to the $4f \rightarrow 5d$ transitions in Tb^{3+} ion. It is not surprising that the emission under this excitation must be much stronger in TAG phosphor than that in YAG (Fig. 4.127b). Since this excitation is absorbed in Tb^{3+} but

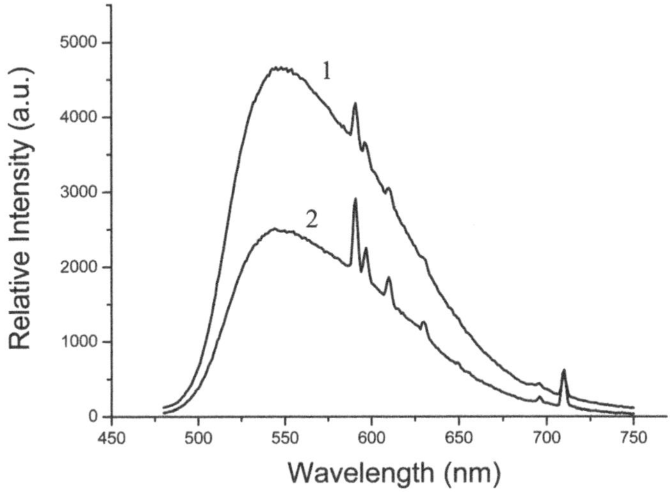

Figure 4.128 Selectively excited emission of TAG:Ce^{3+}(5 mol%), Eu^{3+}(5 mol%) with 460 nm (1) and 277 nm (2).

is emitted by Ce^{3+} the energy transfer $Tb^{3+} \rightarrow Ce^{3+}$ is involved. Two broad bands from 310 to 380 nm in the long wavelength UV region and from 400 to 540 nm in the blue spectral region (Fig. 4.127a) correspond to the $4f \rightarrow 5d$ transitions of Ce^{3+} in YAG and TAG [360, 363] and the broad band with maximum at 227 nm corresponds to the host lattice absorption and leads to the host-excited luminescence.

4.4.4 *Luminescence Mechanism and Energy-Level Diagram*

To understand the luminescence mechanism in cerium–europium double activated terbium aluminum garnet phosphor the selectively excited emission was measured (Fig. 4.128).

The emission spectra consist of a broad band corresponding to the Ce^{3+} $5d \rightarrow {}^2F_J$ transition and of the series of sharp lines arising from Eu^{3+} ${}^5D_0 \rightarrow {}^7F_J$ transitions. It is clearly seen from Fig. 4.128 that the intensity ratio $I(Ce^{3+})_{554}/I(Eu^{3+})_{591}$ strongly depends on the excitation energy and varies from 1.1 at 460 nm up to 0.84 at 277 nm, i.e. in 1.3 times. This can be explained by the presence of

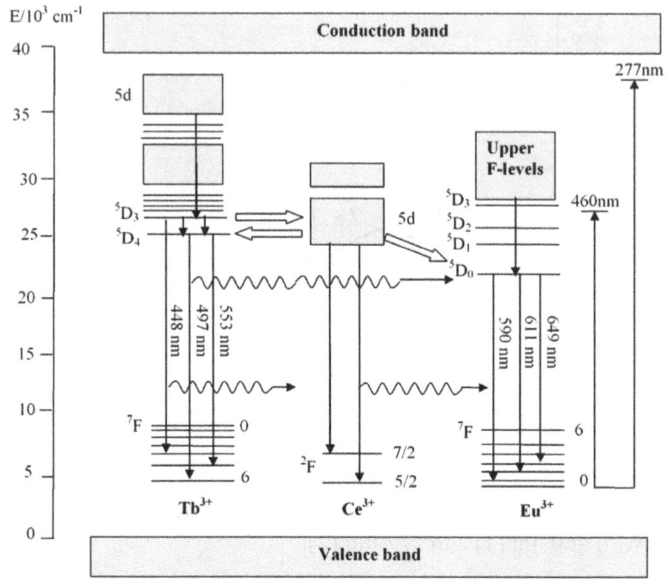

Figure 4.129 Energy-level diagram and transfer pathways for Tb^{3+} and Ce^{3+} in TAG:Eu^{3+},Ce^{3+} garnet.

the energy transfer from Ce^{3+} to Eu^{3+} at the high energy excitation. Therefore, the Ce^{3+} emission decreases while Eu^{3+} emission increases.

From the analysis the experimental results one can conclude that the excitation of Eu^{3+} occurs via the Ce^{3+} quenching luminescence. This means that the Ce^{3+} ion acts as a sensitizer for the Eu^{3+} luminescence. The Eu^{3+} luminescence can also be activated by a Tb^{3+} sensitizing which is not too surprising as an energy transfer from Tb^{3+} to Eu^{3+} is well known [222, 226, 298, 364]. Therefore the Tb^{3+} host lattice acts as a sensitizer for the activator Eu^{3+} but also as an activator itself showing host lattice emission. The fluorescence of Tb^{3+} in the TAG is due to the transitions from the Stark components of the $^5D_4$ level to the Stark components of the ^7F_J levels but the low energy excitation used in blue LED (460 nm) is not enough to excite these levels. 227 nm excitation and more is needed to excite the host lattice. Nevertheless one can expect the transfer of energy from the Ce^{3+} to Tb^{3+} lattice ion due to a slight overlap of the Ce^{3+} luminescence in TAG with the $^7F_6 \rightarrow ^5D_4$ absorption band of Tb^{3+} (483 nm).

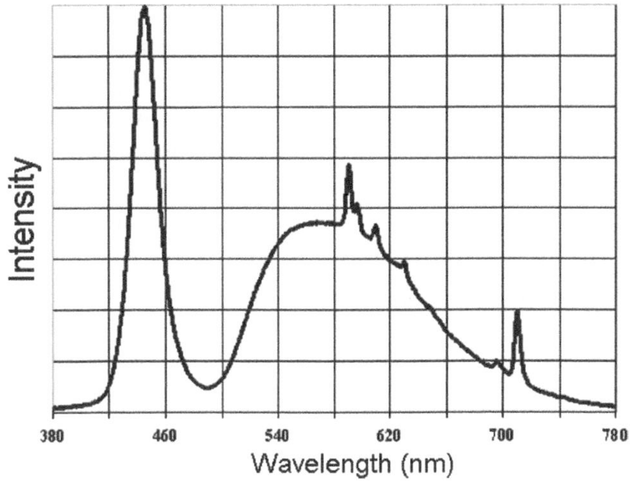

Figure 4.130 Emission spectra of the blue LED including TAG:Ce^{3+}, Eu^{3+}.

The luminescence mechanism could be understood from the energy-level diagram (Fig. 4.129).

In the TAG co-doped with Ce^{3+} and Eu^{3+} the excitation into the Ce^{3+} band at 460 nm leads to the Eu^{3+} emission while in the case of the YAG lattice it does not [351]. This means that in TAG:Ce^{3+} can be used as sensitizer which transfers its energy via the Tb sublattice to the activator Eu^{3+}. If this Tb sublattice is not available such a sensitizing does not work. The first step of the path through which the activation energy is transferred from the sensitizer Ce^{3+} to the activator Eu^{3+} involves transfer between the $5d \to 4f\ ^2F_{5/2;7/2}(Ce^{3+})$ and $^7F_6 \to {}^5D_4$ (Tb^{3+}) transitions. After migration in the Tb sublattice the activation energy is transferred from Tb^{3+} ($^5D_4 \to {}^7F_J$) to a Eu^{3+} activator ($^7F_J \to {}^5D_J$). Moreover, an additional way of transfer energy is possible, namely, the radiation transitions in Tb may excite directly the neighboring Eu and Ce atoms. In such a case Tb^{3+} in TAG does not serve as a doping element but becomes a part of the host structure. After migration in the Tb sublattice the activation energy is transferred from Tb^{3+} ($^5D_4 \to {}^7F_J$) to a Eu^{3+} activator ($^7F_J \to {}^5D_J$). This part of the process was demonstrated for the single-doped case of TAG:Eu^{3+} in Ref. [20]. In YAG, only a direct simultaneous excitation of

Ce^{3+} and Eu^{3+} at about 370 nm leads to a weak emission from both of the ions [351]. Emission spectra of the blue LED including of cerium–europium double activated terbium aluminum garnet phosphor with improved luminescence properties in the light emitting device is demonstrated in Fig. 4.130. Some additional Eu^{3+} peaks and broadening of the spectra in the red region significantly improves the color rendering index (CRI) and luminescent properties of LED.

4.4.5 *Quantum Efficiency of YAG and TAG Phosphors*

There is a need for a simple and accurate method of determination of the quantum efficiency of luminescent materials. Ideally, such a technique for express analysis should require minimal sample preparation and should be simple in application. In this section we report a technique that satisfies these criteria. The quantum efficiency of a luminescent material is defined as the ratio of the number of photons emitted to the number of photons absorbed. This ratio is also referred to as quantum yield. In contrast, the radiant efficiency is the ratio of the output power to the input power. Quantum efficiency has long been used as a criterion for the selection of luminescent materials for application in fluorescent lighting [251, 365, 366], at panel displays [367, 368] and, more recently, solid state lighting. A problem with some quantum efficiency measurement techniques described in the literature is the lack of an acceptable calibration standard. For example, quantum efficiency measurements of powder samples are often made in a system that cannot accommodate liquid fluorescence standards. There are three predominant techniques in the literature for measuring quantum efficiency: absolute optical, relative optical and photocalorimetric. In optical approaches, a primary challenge is to determine precisely the amount of incident light absorbed by the phosphor. The amount of incident light absorbed can be determined directly or calculated from the sample reflectance. Most of the quantum efficiency measurements in powder phosphors use the latter method and the reflectivity of standards such as MgO or $BaSO_4$. Photocalorimetry and photoacoustic spectroscopy have been applied as alternatives to optical methods for quantum efficiency measurements [369–371]. In photoacoustic spectroscopy a black absorbing standard is used. The method is based on the optoacoustic effect where modulated light incident on a sample will produce a temperature rise in the sample that manifests itself by a periodic pressure wave that can be measured with a microphone transducer. The magnitude of the temperature wave depends on the sample absorption and the efficiency of non-radiative transitions. In this technique, the thermal energy of the Stokes shift contributes to the photoacoustic spectroscopy signal, as do any other non-radiative processes, but the luminescence energy does not. Photoacoustic spectroscopy and photocalorimetry are difficult because good thermal contact is critical, small temperature fluctuations (mK) are difficult to detect, and thermal isolation from the surroundings is a challenge.

The technique described herein is simple and can be applied effectively to characterize the powder phosphors. The luminescence spectrum is obtained by plotting the wavelength vs. intensity of the emitted light from a sample excited by an appropriate excitation source of constant energy. The excitation source can be light, an electron beam, heat, X-rays, or radiation from radioactive materials. The spectrum is obtained using a monochromator equipped with an appropriate light detector. The typical apparatus for measuring the spectral characteristics of phosphors is shown in Fig. 3.22.

Emission, excitation and reflection spectra have to be measured for calculation of the quantum efficiency of phosphors. $(YGd)_3Al_5O_{12}:Ce^{3+}$ phosphor with known quantum efficiency could be proposed as standard test sample. Spectral photon irradiance is expressed as:

$$E_p(\lambda) = \frac{dE_p}{d\lambda} \tag{4.35}$$

$$E_p = \int_0^{\infty} E_p(\lambda)\,d\lambda \tag{4.36}$$

For quantum efficiency of investigated sample and standard one we have:

$$\frac{\alpha_x q_x}{\alpha_{st} q_{st}} = \frac{E_{P_x}}{E_{P_{st}}} \tag{4.37}$$

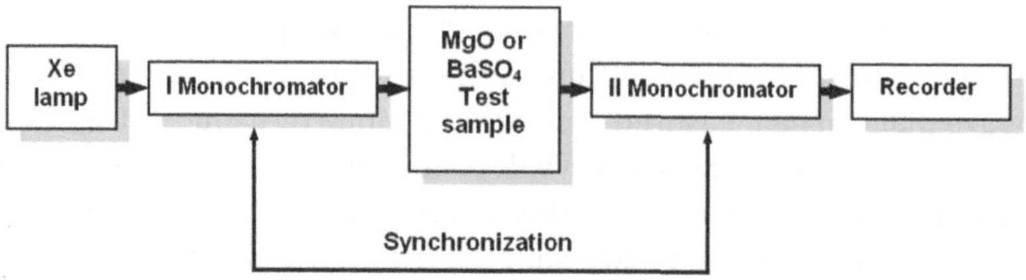

Figure 4.131 The scheme of quantum efficiency measurement.

where q_x and q_{st} are the quantum efficiencies of the specimen investigated and the standard (commercial YAG material, for example), and α is the absorption coefficient.

$$q_x = \frac{(1-R_{st})\int_0^{\infty} E_{P_x}(\lambda)d\lambda}{(1-R_x)\int_0^{\infty} E_{P_{st}}(\lambda)d\lambda} q_{st} \tag{4.38}$$

The following formula is applied in the calculations:

$$q_x = \frac{(1-R_{ST})\Phi_X}{(1-R_x)\Phi_{ST}} q_{ST} \tag{4.39}$$

Here R_X and R_{ST} are the reflectance of the specimen and the standard, and Φ_X and Φ_{ST} are the total measured luminescence efficiency (the integrated area under the emission spectrum) of the specimen and the standard, respectively. The reflectance R_X and R_{ST} of the investigated and standard samples respectively are calculated from the formula

$$R_N = \frac{A_{RN}}{A_{RB}} 0.97 \tag{4.40}$$

where R_N stands for the sample reflectance (N = X and ST for investigated powders and the YAG standard, respectively), and A_{RN} and A_{RB} are total areas under the measured reflection spectra for the sample investigated and the MgO, respectively. In this formula the number 0.97 represents the reflectance of the MgO powder. (Instead of MgO, $BaSO_4$ with index 0.91 can be used).

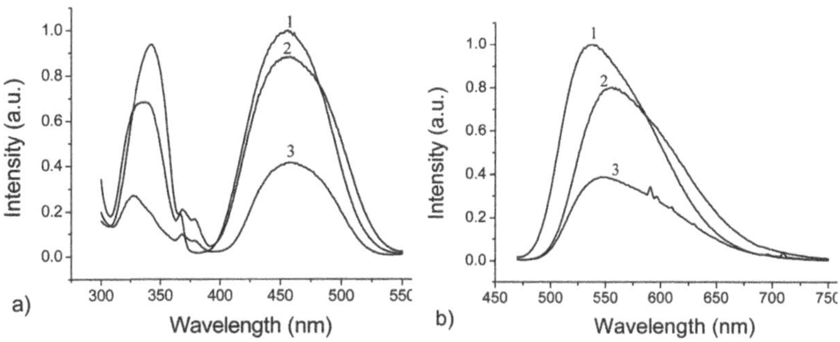

Figure 4.132 Excitation (a) and emission (b) spectra of investigated samples $Tb_3Al_5O_{12}:Ce^{3+}$(2) and double activated $Tb_3Al_5O_{12}:Ce^{3+},Eu^{3+}$ (3) with commercial $(YGd)_3Al_5O_{12}:Ce^{3+}$ (1).

In Fig. 4.131 the scheme of experiment which is realized with photoluminescent DARSA system for reflectance measurements is shown:

If the spectroscopic system does equipped with a special integrated sphere (e.g. ISO40-SF/IG) and does not allow to measure the quantum efficiency directly, the reflectance spectra measurements have to be carried out. Two monochromators are to be synchronized at the excitation wavelength. At the fixed excitation wavelength of the first monochromator the emission spectra around this wavelength are measured by the second monochromator for the reference sample (MgO) and examined phosphors. Because of the Stock's shift (phosphors emit the light at the other wavelength that excite) the measured spectra are the reflectance spectra. The emission and excitation spectra of examined and standard samples are also registered.

Figure 4.132 compares excitation and emission spectra of investigated samples $Tb_3Al_5O_{12}:Ce^{3+}$ and a double activated $Tb_3Al_5O_{12}:Ce^{3+}$, Eu^{3+} with commercial $(YGd)_3Al_5O_{12}:Ce^{3+}$

It is seen from Fig. 4.132 that the luminescence spectra of specimens are very similar in the shape and varies in intensities. Two broad bands from 310 to 380 nm in the long-wavelength UV region and from 400 to 520 nm in the blue spectral region correspond to the $4f \rightarrow 5d$ transitions of Ce^{3+} in YAG. The comparison of the luminescence spectra of the TAG-based phosphors with the emission of the commercial YAG convinces us that the latter can serve as a very

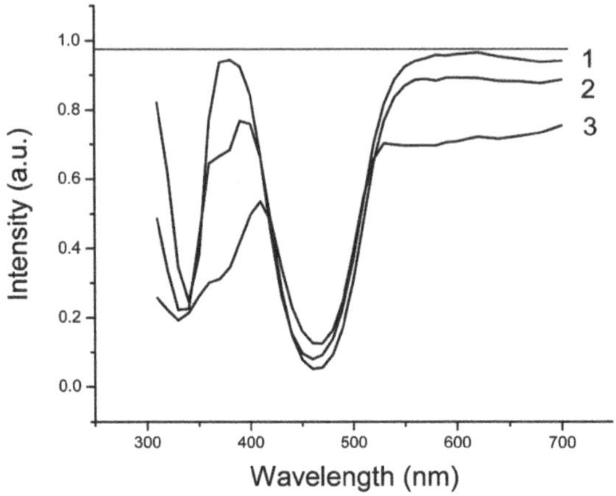

Figure 4.133 Diffuse reflection spectra of the $Tb_3Al_5O_{12}:Ce^{3+}$(2), double activated $Tb_3Al_5O_{12}:Ce^{3+}$, Eu^{3+} (3) and commercial $(YGd)_3Al_5O_{12}$:Ce^{3+} (1) in comparison with MgO.

convenient reference material against which the emission efficiency of our phosphors can be judged. Since the emissions of all materials investigated fall within the same spectral region there is no need to correct the emission spectra for the photomultiplier response to make a reasonable calculation of the quantum efficiencies of the emissions of the various TAG powders. Reliable judgment about the real efficiencies of the emissions from the various specimens requires an accurate determination of the amount of light absorbed by each of the phosphors. This is especially important in case under investigation since the various specimens are characterized by different microstructures and the activator concentration. To compare the amount of light absorbed by an actual specimen, we measured reflection spectra for each of them and related them to the reflectivity of our standard commercial YAG phosphor and to the reflectivity of MgO. Equation (4.40) shows the relationship between the reflectance of the specimen investigated and the reflectance of the standard powder MgO.

In Fig. 4.133 we show reflection spectra for the commercial YAG and all the TAG samples (one of them (2) is doped with 6 mol% of Ce and other (3) with 3 mol% of Ce and 3 mol% of Eu), together with the spectrum of MgO. The results leave no doubt that for reliable

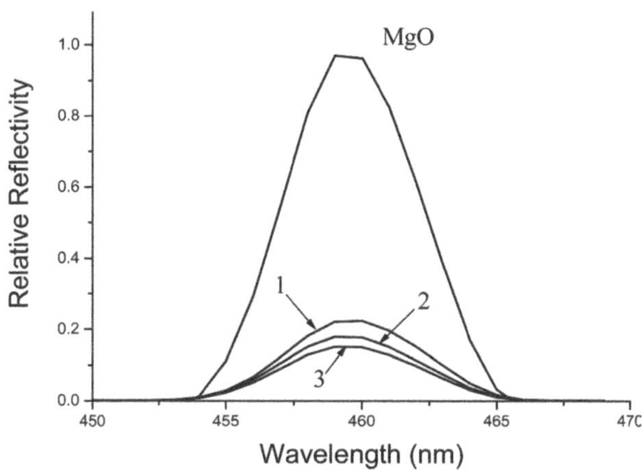

Figure 4.134 Diffuse reflection spectra of the $Tb_3Al_5O_{12}:Ce^{3+}$(2), double activated $Tb_3Al_5O_{12}:Ce^{3+},Eu^{3+}$(3) and commercial $(YGd)_3Al_5O_{12}:Ce^{3+}$ (1) in comparison with MgO at 460 nm excitation.

estimation of the quantum efficiency of the emission it is definitely not enough to compare just the luminescence efficiency. In fact, it is also necessary to take into account the variations in the amounts of light absorbed by each of the samples. Despite the fact that all TAG specimens presented in Fig. 4.133 have the same host lattice their capabilities for reflecting the incident light vary strongly. It can be explained by the different Ce^{3+} activator concentration in the samples (2) and (3), as well as influence of Eu^{3+} on the luminescence properties of the sample (3). Thus the amount of light accessible for absorption by the Ce^{3+} ions (able to excite the phosphor) must also alter. Generally, as expected, with rising Ce content (6 mol% in the sample 2 and 3 mol% in the sample 3) less and less light is being reflected, which means that a still increasing amount of the light is being absorbed by the phosphor. On the basis of the observation of the excitation (Fig. 4.132) and reflection (Fig. 4.133) spectra one can conclude that for the measurements of the quantum efficiencies of the mentioned phosphors the most convenient excitation wavelength would be 460 nm. The diffuse reflection spectra at this wavelength are presented in the Fig. 4.134.

The quantum efficiency of phosphors calculated according to Eqs. 4.39 and 4.40 and by using the experimental data from Fig.

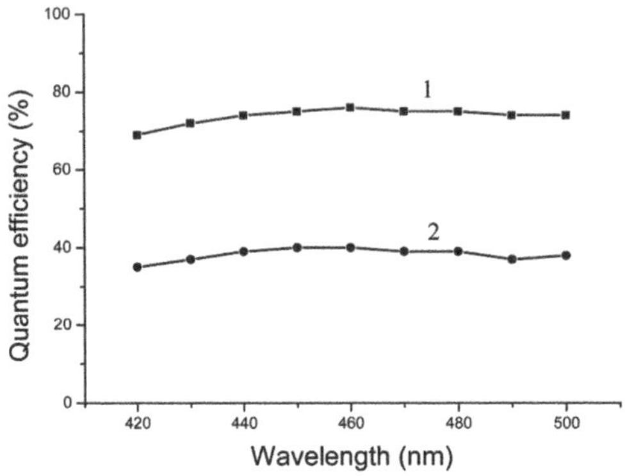

Figure 4.135 The quantum efficiency of (1) $Tb_3Al_5O_{12}:Ce^{3+}$ (6 mol% Ce) and (2) $Tb_3Al_5O_{12}:Ce^{3+},Eu^{3+}$(3 mol% Ce and 3 mol% Eu) in dependence of excitation wavelength.

4.134, gives us the 76% for $Tb_3Al_5O_{12}:Ce^{3+}$ (6 mol% Ce) and 40% for $Tb_3Al_5O_{12}:Ce^{3+}$, Eu^{3+} (3 mol% Ce and 3 mol% Eu) relatively 90% $(YGd)_3Al_5O_{12}:Ce^{3+}$. It is not astonished, because the main reason of a low intensity of the last sample is a lack of Ce^{3+} activator and weak emission intensity in comparison with standard sample. The quantum efficiency practically does not depend on excitation wavelength in the large excitation area. It is good seen from the Fig. 4.135 for our samples and could be explained by the same processes of energy absorption, energy transfer and emission at different excitations. The quantum efficiency strongly depends on the sample structure and activator concentration.

Energy absorption event does not necessarily take place in the activator ion but can occur at a random place in the lattice. This implies that the energy transfer of the absorbed energy to the luminescent center takes place before emission can occur. The migration of the energy absorbed by the lattice can take place through one of the following processes:

- migration of electric charge (electrons, holes)
- migration of the excitons
- resonance between atoms with sufficient overlap integrals
- reabsorption of photons emitted by another activator ion or sensitizer

The occurrence of energy transfer within a luminescent material has far-reaching consequences for its properties as a phosphor. On the one hand the absorbed energy can migrate to the crystal surface or to lattice defects where it is dispersed through the radiationless deactivation. As a consequence the phosphor energy yield that is the quotient of emitted light energy to initially absorbed energy, drops. In the same phosphor structure with the same defect concentration the energy loss are also the same and emission is determined by the activator concentration.

4.5 Thiogallate Luminescent Materials

4.5.1 *Introductory Remarks*

Thiogallate compounds represented by general formula MGa_2S_4 (where M = Mg^{2+}, Ca^{2+}, Sr^{2+}, Ba^{2+}, Pb^{2+}, Zn^{2+}, Eu^{2+}, Cd^{2+}, or some other bivalent ions) have been well known for a long time as phosphor materials for CRT, EL devices, or LED applications. Usually, thiogallates based on Ca^{2+}, Sr^{2+}, and Ba^{2+} are used more often due to their high efficiency.

Luminescent properties of barium, strontium and calcium thiogallates ($BaGa_2S_4$, $SrGa_2S_4$, and $CaGa_2S_4$) doped with europium activators have been investigated since 1970 [372–378]. Thiogallates are much less sensitive to the action of moisture than binary chalcogenides. Doping these materials with rare earth elements, allows to enhance the quantum yield of their luminescence due to the formation of the emitting centers in which the effective energy transfer from the excited 5*d* levels to 4*f* levels takes place. Many efforts have focused on improving the performance of thiogallate luminescent materials.

Recently, a new wave of intensive studies of these materials has begun because they have shown very attractive photoluminescence, electroluminescence and cathodoluminescence in the visible spectral range [379–386]. Because the absorption and emission of the $4f \rightarrow 5d$ transitions in the doping ions extend to the visible spectrum, these phosphors are appropriate for solid state lighting applications upon the excitation by near UV or blue emitting diodes [387]. For the purpose of developing energy-efficient white light sources, we need to produce highly efficient yellow, green and red emission phosphors which can absorb excitation energy from blue LEDs. $CaGa_2S_4:Eu^{2+}$ for example, is a yellow-emitting material, whereas $SrGa_2S_4:Eu^{2+}$ shows very high green emission and they can be efficiently excited by blue emitting diodes. They allow acceleration of charge carriers up to optical energies under high electric field and are used also in electroluminescent displays. The good linearity of the cathodoluminescence at high current densities makes them suitable for field emission displays or TV projection. While these practical applications are successfully being achieved, there are still several aspects regarding the general ways to provide the advanced light-emitting materials with improved properties. Among them, searching for novel chemical composition and improving the existing phosphors including the chemical synthesis are of great importance.

Peters and Baglio synthesized thiogallate phosphors under H_2S stream by $CaCO_3$, Ga_2O_3 and Eu_2O_3 [372]. This gas was also used in other works of preparing thiogallates [381, 382]. Because H_2S is very toxic to humans, an interesting method was attempted recently using 5% of H_2 and 95% of N_2 as reduction agents avoiding the use of such toxic gas [383]. Other attempts were launched to prepare thiogallates without H_2S. For example, Aidaev made the synthesis at low temperature from NH_4CNS, but this process is relatively complicated and has long reaction times [384]. Nevertheless, the development of synthesis methods to improve the luminescent properties of thiogallate materials is still challenging.

Some attempts to combine the luminescent data of transition of Eu^{2+} in inorganic compounds (including thiogallate phosphors) were made by Dorenbos [388], but unfortunately, in the used references the row materials and chemical methods of preparation phosphor, as well as activator concentration and temperature were different. The luminescent data also were various, because many properties like luminescence intensity and wave length peak strongly depend on activator concentration and preparation conditions.

In this book, the luminescence of the thiogallate phosphors is revisited, and the results of producing a new green multiphase thiogallate phosphor $[SrGa_2S_4+MgGa_2O_4]:Eu^{2+}$ with improved luminescence properties are presented. Photoluminescence excitation and emission spectra along with Raman spectroscopy and X-ray diffraction are analyzed to estimate the nephelauxetic effect, crystal field splitting and Stokes shift depending on activators surroundings. These results will be useful to evaluate the quality of the powders prepared for different electronic devices.

Up do date, to the best of our knowledge, only the transition metal ion doped magnesium gallate spinels were investigated, and Eu^{2+} doped magnesium gallate spinels have not been yet reported. Furthermore, understanding the fundamental of the radiative mechanisms of thiogallates needs to be enlarged. A model of $4f \rightarrow 5d$ optical transitions taking into account the electron-vibrational interaction is discussed. The influence of the temperature on the luminescence process was investigated. The photoluminescence spectra of the Sr, Ca and Ba thiogallates were analyzed and fitted using the proposal equations. The Stokes shift at room temperature was calculated and compared with experimental data. No such investigation taking into account an influence of electron-vibrational interaction on $4f \rightarrow 5d$ optical transitions at real temperature have been done for these materials hitherto. These data will be useful to evaluate the quality of the powders or thin films prepared for devices.

Polycrystalline $BaGa_2S_4:Eu^{2+}$, $SrGa_2S_4:Eu^{2+}$, and $CaGa_2S_4:Eu^{2+}$ samples were synthesized by solid state reaction correspondingly from BaS, SrS, CaS, and Ga_2S_3 sulfide powders mixed in stoichiometric composition and annealed at 900–1000°C with a carbon reduction atmosphere for 4 h. The doping ions were introduced in the form of EuS. This preparation method is a little different from that reported by Peters and Baglio [372] and by others [381–384, 389] because the toxic H_2S stream was not used in synthesis process. The method presented here provides powder samples with good crystalline properties, as shown by XRD measurements. The powder samples with 6 mol% of Eu^{2+} concentration exhibit a blue-green ($BaGa_2S_4:Eu^{2+}$), deep green ($SrGa_2S_4:Eu^{2+}$) and yellow ($CaGa_2S_4:Eu^{2+}$) colors, respectively.

Phosphor samples were characterized by crystalline structure and luminescence properties.

X-ray diffraction measurement – The crystal structures of the prepared samples were determined by XRD measurement using a goniometer (**PANalytical, X'Pert pro MPD** with Cu- Kα (λ = 1.5418 Å) at 40 kV and 30 mA. The scan speed was 3 seconds per step (0.02° step – 2θ) covering the range between 10° and 90°.

Energy dispersive spectroscopy and electron probe micro-analysis – Qualitative and quantitative non-destructive elemental analysis were performed with EPMA machine (SX-100, the electron acceleration was 20 kV, beam current was 10 nA, and the diameter of the electronic beam was 50 μm, DT:0.3 S). It is one of the most precise and accurate micro-analysis techniques available and all elements from Beryllium to Uranium can be analyzed.

Photoluminescence measurement – Optical spectroscopy and PL characteristics were performed on the basis of emission and excitation spectra registered at room temperature with a 500 W Xe lamp and DARSA PRO 5100 PL System (Professional Scientific Instrument Co, Korea). Excitation spectra were corrected for the energy distribution of the Xe lamp. The PL was measured with a 460 nm excitation, which is usually used in blue LED.

Raman spectroscopy – Raman scattering spectra of $SrGa_2S_4$, $CaGa_2S_4$, and $BaGa_2S_4$ were measured by a Renishaw 3000 spectrometer with a He–Ne laser (excitation wavelength of λ = 633 nm and λ = 785 nm) and a photomultiplier counter at room temperature in

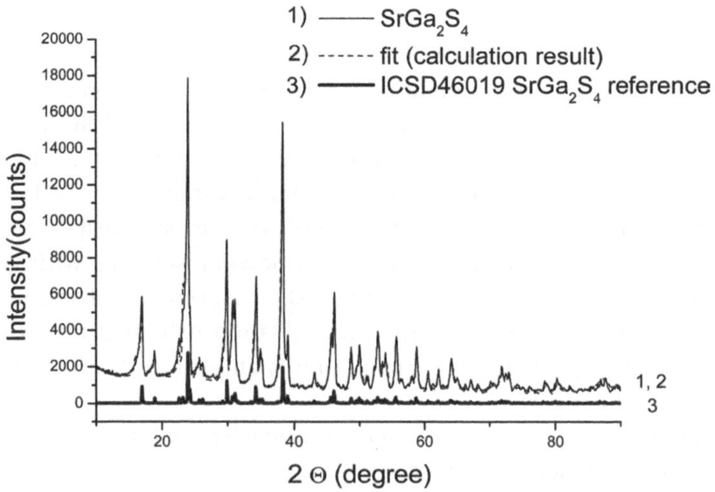

Figure 4.136 The X-ray diffraction patterns of $SrGa_2S_4:Eu^{2+}$ and it fitting in comparison with ICSD-46019.

back-scattering configuration. The spectral resolutions of spectrometer are about 4 cm^{-1} at 633 nm and 1 cm^{-1} at 785 nm.

Morphology and size measurement – Particle sizes and morphologies of the investigated phosphors were determined by scanning electron microscope Hitachi-S-3000N. To control the particle size and to find the size distribution the Laser diffraction was carried out using HELOS particle size analysis system.

The details of phosphor characterization and experimental measurements are given in Chapter 3.

4.5.2 *Structural and Luminescent Properties of Calcium, Strontium, and Barium Thiogallates*

Figure 4.136 shows the X-ray diffraction patterns of the $SrGa_2S_4:Eu^{2+}$ powder baked at 1000°C (1). The peak positions of XRD patterns are in good agreement with fitting (2) and inorganic crystallographic structure database (ICSD-46019) (3).

The solid curve (1) in Fig. 4.136 is the measured data and the dash curve (2) means the calculation result of Rietveld analysis developed by Rietveld in 1967 for structure profile refinement of X-ray powder diffraction data [285]. The Rietveld calculation gets all these crystallographic information by comparing model profile and X-ray or neutron diffraction curves using the least-squares method. The peak positions of the XRD curve are related to the unit cell lattice constants of the crystal structure, and the peak intensities are affected by atomic position, atom species, atomic occupancy, and thermal parameters. In addition, crystalline size and strain induce peak broadening can be extracted from the XRD pattern. The absence of peaks explains the crystallographic symmetry. In the case of multiphase structures, quantitative phase analysis is also available by using the Rietveld method. The detailed crystallographic information from Rietveld analysis is listed in the Tables 4.23–4.26. The calculated parameters in Table 4.23 are based on the original XRD patterns, and the corresponding ICSD data (indicated as reference) were used for Rietveld refinement.

First of all Rietveld refinement was made to check the crystallographic structure of the samples. Figure 4.137 shows that the structure of synthesized samples is very similar to the structure

Table 4.23a Crystallographic data for different ion positions in synthesized $SrGa_2S_4$

wRp = 0.12	orthorhombic		x	y	z	F	Multiple factor
Rp = 0.095	A = 20.8558(19)	Sr_1	0.125	0.125	0.125	1	8
R(F^2) = 0.174	B = 20.5115(22)	Sr_2	0.125	0.125	0.625	1	8
ICSD–46019	C = 12.2138(12)	Sr_3	0.8726	0.125	0.125	1	16
		Ga_1	−0.0033(8)	0.8040(4)	0.1633(7)	1.05(2)	32
	$\alpha = 90$	Ga_2	−0.0052(8)	0.2370(4)	0.3752(8)	1.16(2)	32
$SrGa_2S_4$ Single phase	$\beta = 90$	S_1	0.1672(10)	0.2472(13)	0.997(2)	1.15(4)	32
	$\gamma = 90$	S_2	0.1725(14)	0.5078(17)	0.262(3)	0.83(4)	32
	F d d d	S_3	0.999(2)	0.3438(8)	0.0089(15)	1.00(4)	32
		S_4	−0.006(3)	0.4202(10)	0.2701(16)	1.00(3)	32

and wRp = 0.12, Rp = 0.095, R(F^2) = 0.174.

Table 4.23b Crystallographic data for $SrGa_2S_4$ from database ICSD-46019

Orthorhombic		x	y	z	F	Multiple factor
$a = 20.932$	Sr1	0.125	0.125	0.125	1	8
$b = 20.549$	Sr2	0.125	0.125	0.625	1	8
$c = 12.227$	Sr3	0.8726	0.125	0.125	1	16
	Ga1	0.0005	0.8038	0.164	1	32
$\alpha = 90$	Ga2	0.0005	0.2374	0.373	1	32
$\beta = 90$	S1	0.168	0.2483	0.998	1	32
$\gamma = 90$	S2	0.1677	0.502	0.2539	1	32
F d d d	S3	0.9995	0.3449	0.0062	1	32
	S4	0.0021	0.4222	0.2651	1	32

of $SrGa_2S_4$ orthorhombic phase ICSD-46019 pattern (3). Thereby, we could confirm that in realized synthesis conditions a single phase of $SrGa_2S_4$ was received. Crystallographic data for synthesized $SrGa_2S_4$ and reference sample (original unit cell and atomic positions) are given in Tables 4.23a and 4.23b.

From the occupation factor F, one can see that when F is equal or close to 1, the site is occupied by an atom. When $F < 1$ (S2 position in the synthesized phosphor) the site could be substituted by a lighter atom. In this case, it might be an oxygen, for example.

The thiogallate compounds of the type $M^{II}Ga_2S_4$ (where M^{II} = Ca, Sr) belong to the orthorhombic crystal class with the space group D^{24}_{2h} *(Fddd)* [390] and $BaGa_2S_4$ to cubic structure with the space group *Pa3*(*Th6*) [376]. There are 32 formula-mass units per unit cell (z = 32) and therefore 56 atoms in a primitive cell: 8 M^{II}, 16 Ga, 32 S. According to Eisenmann et al., [390] the M^{II} atoms occupy square anti-prismatic sites formed by eight sulfur (symmetry group D_{4d}) forming $M^{II}(S)_8$ units with C_2 and D_2 symmetry. In the $SrGa_2S_4$:Eu^{2+} and $CaGa_2S_4$:Eu^{2+} compound, the cation ions are substituted by the

Table 4.24 Crystal structure of strontium surrounding in $SrGa_2S_4$ and short distance Sr–S in comparison with reference data

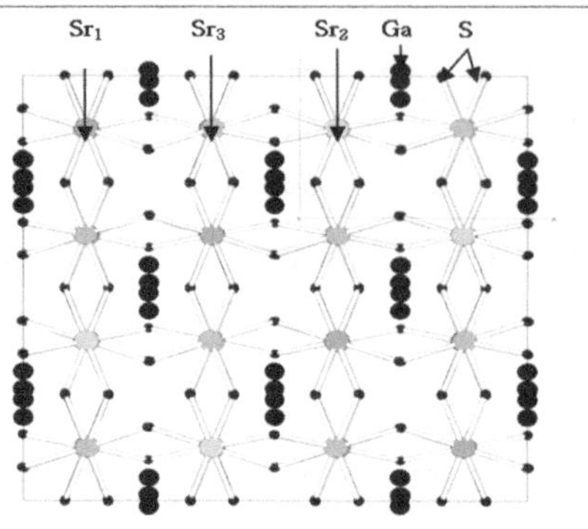

Crystal structure of $SrGa_2S_4$

Short Distance	$SrGa_2S_4$ (Å)	ICSD ref (Å)
Sr(1)–S(6)	4,778	4,778
Sr(1)–S(9)	3,131	3,134
Sr(2)–S(6)	3,451	3,452
Sr(2)–S(7)	3,108	3,110
Sr(2)–S(8)	3,119	3,122
Sr(3)–S(6)	2,878	2,881
Sr(3)–S(7)	3,115	3,117
Sr(3)–S(8)	3,095	3,098
Sr(3)–S(9)	3,099	3,102

Eu^{2+} ions. From XRD data (Table 4.23), they are situated in three slightly different sites: 8a, 8b and 16c, whose multiplicity is not equal and exhibits a 1:1:2 relative ratio dictated by symmetry. Gallium atoms are tetrahedrally coordinated to four sulfur atoms forming $Ga(S)_4$ units (symmetry group Td) and the sulfur atoms are at the center of deformed $M^{II}{}_2Ga_2$ tetrahedrons forming $(S)M^{II}{}_2Ga_2$ units. The assembly of the $M^{II}(S)_8$ anti-prismatic units with common edges forms chains parallel to the *a* axis of the unit cell. Each chain is linked to four chains by corner sharing. Gallium atoms are located between two consecutive chains. The unit cell consists of four layers along the *c* axis. The inset figure in Table 4.24 illustrates the symmetry of the S and M^{II} environments. The site symmetry of each atom can be obtained from the correlation tables [391].

According to presented data in Table 4.24, the shortest distance between Sr and S is estimated as 3.30 Å, which is in good agreement with reference ICSD data. (The measurements were made for all interatomic distances less than 5 Å). For $CaGa_2S_4$ this distance Ca–S is shorter and it was found as 3.0 Å. (Table 4.25).

For cubic $BaGa_2S_4$ the distance Ba–S is longer and it is about 3.39 Å. The main data and crystallographic structure of $BaGa_2S_4$ are given in the Table 4.26.

The compositional analysis by energy dispersive X-ray spectrometry on a JEOL 2010 CX electron microscope has been

Table 4.25 Crystal structure of calcium surrounding in $CaGa_2S_4$ and a short distance Ca–S in a synthesized sample

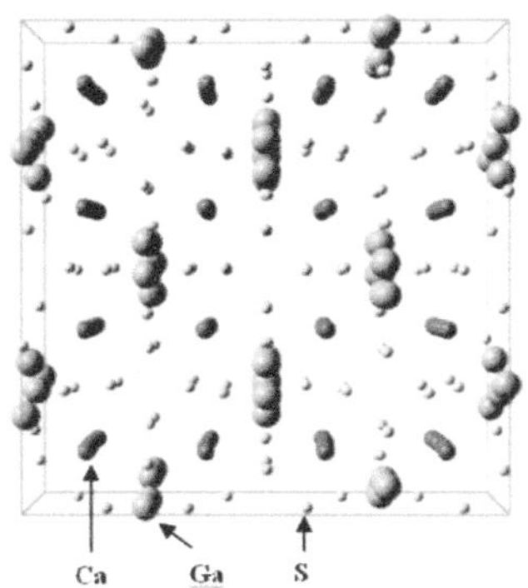

Crystal structure of $CaGa_2S_4$

	x	y	z	Short distance	(Å)
Ca(1)	0.125	0.125	0.125	Ca(1)–S(6)	2.967
Ca(2)	0.125	0.125	0.625	Ca(I)-S(9)	2.985
Ca(3)	0.8724	0.125	0.125	Ca(2)–S(6)	5.289
Ga(4)	0.0003	0.8042	0.1638	Ca(2)–S(7)	2.985
Ga(5)	0.0009	0.2383	0.3743	Ca(2)–S(8)	2.966
S(6)	0.1649	0.2474	0.0052	Ca(3)–S(6)	3.130
S(7)	0.1643	0.5023	0.2474	Ca(3)–S(7)	3.079
S(8)	0.9975	0.3438	0.0008	Ca(3)–S(8)	3.006
S(9)	0.0008	0.4184	0.2663	Ca(3)–S(9)	3.002

Table 4.26 Crystal structure of $BaGa_2S_4$ and a short distance Ba–S in a synthesized sample

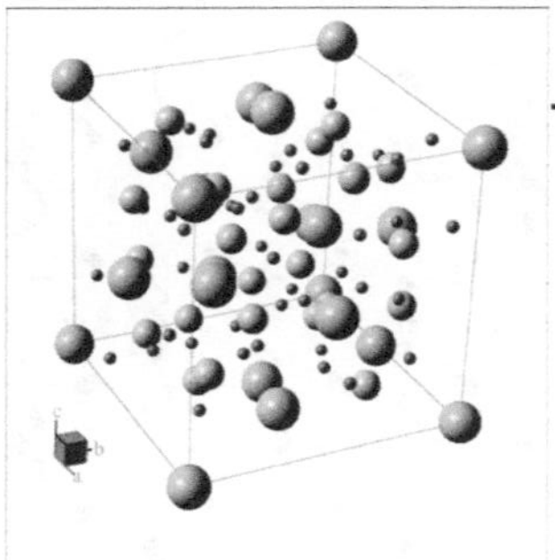

Crystal structure of $BaGa_2S_4$

$BaGa_2S_4$–ICSD–35135	Atomic distance (Å)				
	X	Y	Z		
Ba(1)	0	0	0	Ba(1)–S(4)	3.533
Ba(2)	0.3722	0.3722	0.3722	Ba(1)–S(5)	3.590
Ga(3)	0.1959	0.3568	0.1263	Ba(2)–S(4)	3.271
S(4)	0.3627	0.2935	0.1268	Ba(2)–S(5)	3.172
S(5)	0.2291	0.5267	0.0773		
	a = 12.685	b = 12.685	c = 12.685	Pa–3 symmetry	
Cubic	90	90	90		

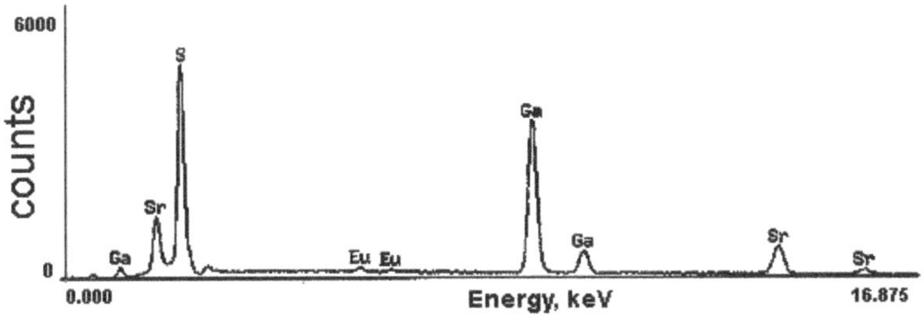

Figure 4.137 EDS patterns corresponding to $SrGa_2S_4$.

carried out. Energy dispersive spectra from single phase $SrGa_2S_4$ phosphor is shown in Fig. 4.137.

The distribution of elements is analyzed in Table 4.27.

From Fig. 4.137 and Table 4.27 it is clearly seen that at the initial 6 mol% Eu^{2+} concentration in synthesized sample the activator is fully incorporated. The degree of incorporation is an important parameter that is responsible for luminescence properties [226] rather than initial activator concentration. Similar experiments were carried out on $CaGa_2S_4$ and $BaGa_2S_4$. Unfortunately, EDS analysis could not detect the light elements such as oxygen. Thus, the compositions of the samples should also be examined by electron probe microanalysis.

The Stokes Raman spectra of the polycrystalline $SrGa_2S_4$, are given from 50 cm^{-1} to 500 cm^{-1} in Fig. 4.138. The spectra of dominant lines of $SrGa_2S_4$, $CaGa_2S_4$, and $BaGa_2S_4$ are shown in Fig. 4.139. Stokes spectra of $SrGa_2S_4$ and $CaGa_2S_4$ consist of 19 vibration lines positioned between 68 cm^{-1} and 408 cm^{-1} (the more intense lines

Table 4.27 Compositional EDS analysis

Element	Atomic%	Weight%	Wt% Error
Sr–La	6.79	9.89	0.16
S–K	29.85	15.90	0.11
Eu La	0.59	1.50	0.17
Ga–K	62.76	72.71	0.59
Total	100.00	100.00	

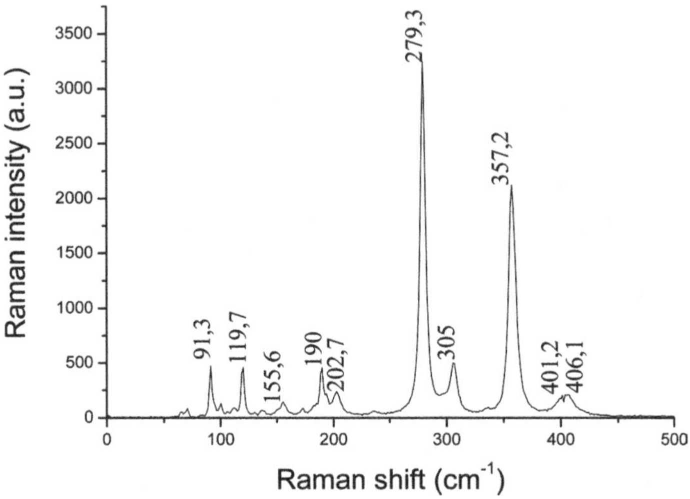

Figure 4.138 Raman spectrum of $SrGa_2S_4$, excited with the 785 nm laser line.

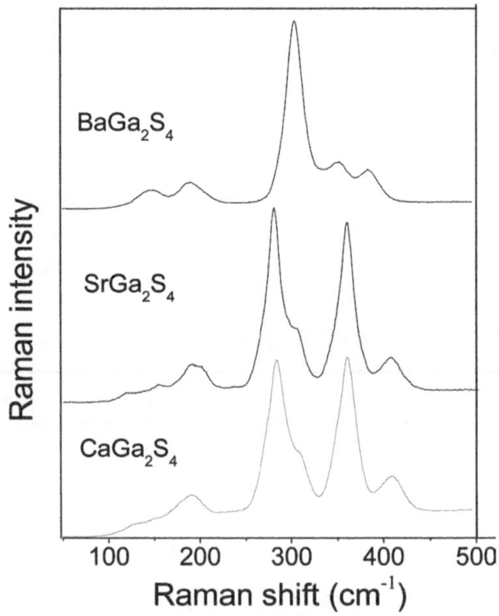

Figure 4.139 Raman spectra of dominant lines of $CaGa_2S_4$, $SrGa_2S_4$, and $BaGa_2S_4$, excited with the 633 nm laser line.

are marked in Fig. 4.139). The factor-group method predicts that the total number of Raman-active phonons is 84 for this structure [392]. The reduced modes in the Raman spectra are in favor of a molecular model to describe the vibrations of these compounds [393, 394]. Two dominant peaks at 279 cm^{-1} (34.5 meV) and at 357 cm^{-1} (44.2 meV) for $SrGa_2S_4$ and similar peaks at 284 cm^{-1} (35.2 meV) and 361 cm^{-1} (44.8 meV) for $CaGa_2S_4$ are registered. Only one intense vibration line at 302 cm^{-1} (37.5 meV) is found for $BaGa_2S_4$. (Fig. 4.139). These Raman spectra are in good agreement with those reported in the literature [385, 395] and could be used for the fitting of emission bands.

As shown for the orthorhombic chalcogallates [396], the Raman vibrations can be described by a molecular model. The orthorhombic $M^{II}Ga_2S_4$ spectra were interpreted by considering the isolated GaS_4 and $M^{II}S_8$ vibrating units. The most intense vibration line was assigned to the totally symmetrical bond-stretching modes, where the S anions move together in the cation–anion direction in GaS_4 or in $M^{II}S_8$ units. The change of Ca, or Sr cation leads to a slight translation in frequency of the overall spectrum. This shift is not caused by mass effects but is ascribed to the differences in size of the cations. The Raman frequencies increase with decreasing ionic radius of the M^{II} cation, i.e. with decreasing interatomic lengths and size of the unit cell (Table 4.24). Thus the phonon frequencies vary in the following order: $\nu_{Sr} < \nu_{Ca}$. If the vibrations of S around M^{II} are described with $M^{II}S_8$ molecular unit there is no contradiction with the fact that no motion of M^{II} cation is observed: in the $M^{II}S_8$ symmetry group (D_{4d}) the frequencies of the Raman-active modes are independent of the M^{II} mass [397]. At this point of the discussion considering the vibrating structure as isolated GaS_4 and $M^{II}S_8$ units is in agreement with the Raman data.

The spectrum of $BaGa_2S_4$ is very similar to the Raman spectra of orthorhombic $M^{II}Ga_2S_4$ compounds with M^{II} = Sr and Ca. The vibration lines are located in the same range of frequencies (between 50 and 410 cm^{-1}), indicating that the mass of Ba affects very slightly the spectrum and that no vibration involving this cation directly is present in the Raman spectrum. The most intense line of the $BaGa_2S_4$ spectrum (302 cm^{-1}) is located at higher frequency than the most intense mode of the orthorhombic $M^{II}Ga_2S_4$ spectra (between 278 and 285 cm^{-1}). Considering that the Ga–S lengths (2.27 Å) are shorter

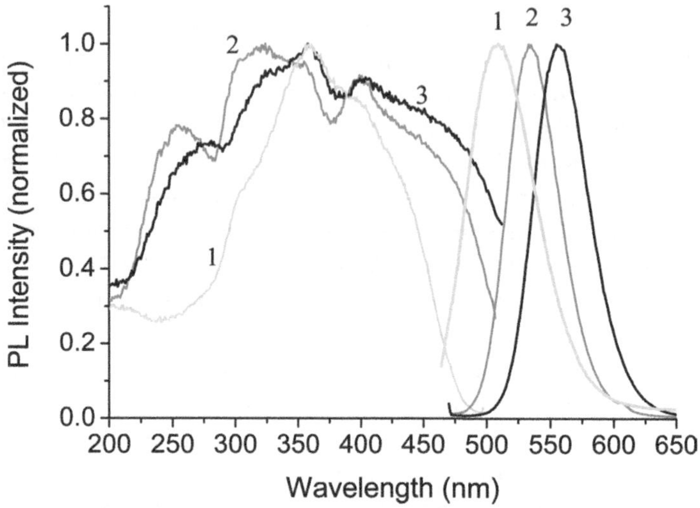

Figure 4.140 The emission (right) and excitation spectra (left) of the thiogallate compounds of the type $M^{II}Ga_2S_4$ (where M^{II} = Ba (1), Sr (2), Ca (3)) doped with 6 mol% of Eu^{2+} at 300 K.

and the Ba-S value lengths (3.39 Å) are longer in cubic $BaGa_2S_4$ than in orthorhombic $M^{II}Ga_2S_4$ (Ga–S between 2.27 and 2.29 Å, M^{II}–S between 3.00 and 3.30 Å) (Tables 4.24 and 4.26), this suggests that the unit responsible for this mode is the GaS_4 tetrahedrons and not the BaS_6 or BaS_8 units whose vibrations might lead to weaker frequency modes. So, the most intense vibration mode in $BaGa_2S_4$ can be ascribed to the totally symmetrical bond-stretching modes in GaS_4 tetrahedrons.

The emission and excitation spectra of the thiogallate compounds of the type $M^{II}Ga_2S_4$ (where M^{II} = Ca, Sr, Ba) doped with 6 mol% of Eu^{2+} at 300 K are very similar (Fig. 4.140). For all samples the emission spectra under excitation at 460 nm consist of a broad band in the visible range. The maximum wavelengths of the band is shifted from 504 nm for Ba, to 535 nm for Sr and to 558 nm for Ca. Full widths at half maximum (FWHM) are equal correspondingly to 62, 49, 50 nm. No additional band due to the emission of Eu^{2+} in investigated thiogallates is observed, which indicates that no secondary phase is present in investigated samples.

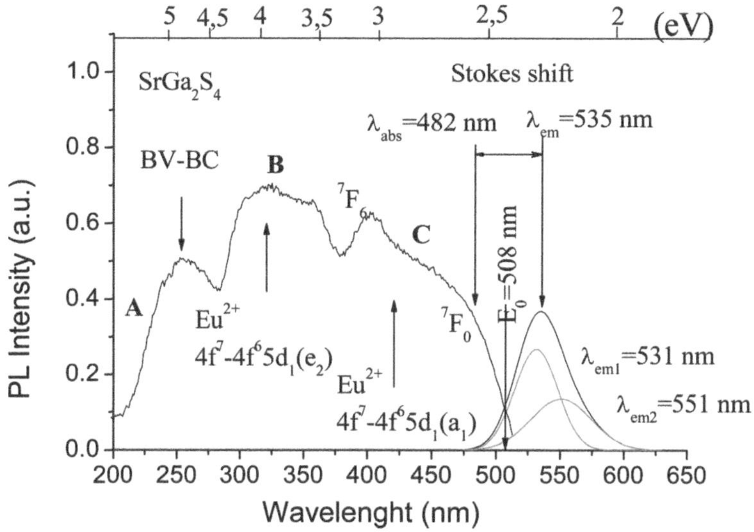

Figure 4.141 Excitation and emission spectra of the $SrGa_2S_4:Eu^{2+}$ (6 mol%) powder at 300 K.

The excitation spectrum of $SrGa_2S_4:Eu^{2+}$ (6 mol%) and the emission spectrum at 300 K, as well as its detailed analysis for 535 nm wavelength, are presented in Fig. 4.141.

$SrGa_2S_4:Eu^{2+}$ excitation spectrum is composed of three large bands in the UV–visible range:

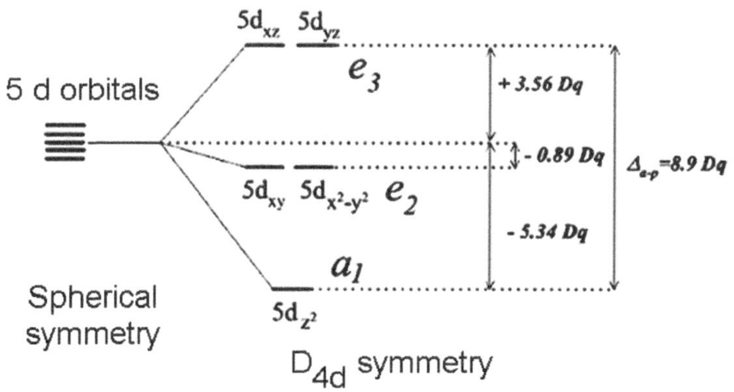

Figure 4.142 Splitting of the five 5*d* orbitals in a square antiprismatic (D_{4d}) symmetry.

The excitation band *A* (maximum at 254 nm) is ascribed to the transitions between the valence band and the conduction band of the $SrGa_2S_4$ host matrix.

The two excitation bands *B* and *C* are ascribed to the $4f^7(^8S_{7/2})\rightarrow 4f^6(^7F)5d$ transitions in Eu^{2+} electronic levels.

In the $SrGa_2S_4:Eu^{2+}$ compound, the Sr^{2+} ions are substituted by the Eu^{2+} ions. They occupy three different sites with C_2 and D_2 symmetry [376]. Because their shapes have minor differences, the emission bands of Eu^{2+} in the three available sites are expected to lie near each other. The observed emission band (Fig. 4.142, on the right) can be the superposition of these bands because they are inhomogeneously broadened by the size difference between Eu^{2+} and Sr^{2+}. As the two positions Sr_1 and Sr_2 (Table 4.23) are very close the emission band could be decompose in two spectra (Fig. 4.141). In first approximation, they are located at the center of square anti-prisms formed by the 8 S^{2-} anions (Tables 4.23 and 4.24). In this D_{4d} symmetry, the 5*d* orbitals are split in three levels [382] as illustrated in Fig. 4.142.

Band *C* is ascribed to the $4f^7(^8S_{7/2})\rightarrow 4f^6(^7F)5d\ (a1)$ transitions and the band (*B*) to the $4f^7(^8S_{7/2})\rightarrow 4f^6(^7F)5d\ (e2)$ transitions. By evaluating the energy gap between the *B* and *C* bands to 8000 cm^{-1} (1 eV), the crystal field strength, $\Delta_{a\text{-}p} = 8.9Dq$, can be estimated at about 16000 cm^{-1} (2 eV).

An almost mirror image relationship seems to hold between the emission and the excitation spectra in the energy region between the two maxima (Fig. 4.141). This relation is characteristic of a phonon-broadened emission and confirms the hypothesis made previously; it suggests that the mean phonon energy hν is the same for the $4f^7(^8S_{7/2})$ ground state as for the $4f^6(^7F)5d\ (a1)$ excited state. We can determine the energy of the zero-phonon line E_0 at the intersection of the emission and excitation spectra. From Fig. 4.141, we found $E_0 = 2.44$ eV, which corresponds to a wavelength of 508 nm (comparable with that given by Eichenauer et al. [398] and Chartier et al. [382]). Unfortunately, we can not distinguish at room temperature the fine structure due to the spin-orbit coupling in the $4f^6$ configuration leading to the splitting of the ^7F levels in seven levels: ^7F_J with $J = 0\rightarrow 6$. The energy E_{abs} of the $f\rightarrow d$ transition from the ground state to the lowest level of the $4f^65d$

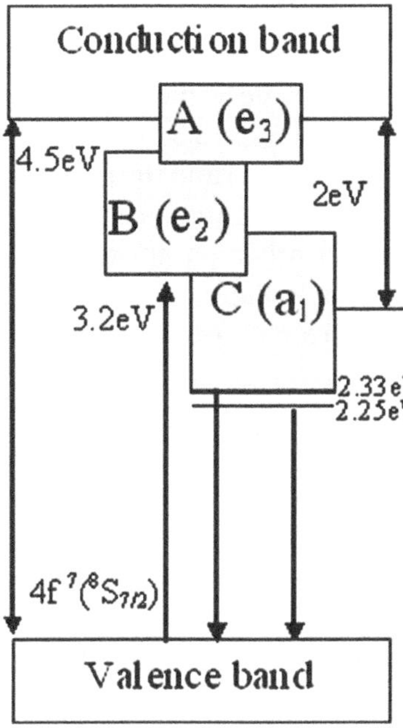

Figure 4.143 Energetic scheme and electronic levels in $SrGa_2S_4$:Eu^2.

(*a1*), corresponds to the $^7F_0$ level. Since this level is not resolved, we can only give an estimation of E_{abs}. By using the mirror image relationship between the emission and the excitation spectra, we have found E_{abs} = 2.57 eV (482 nm). The energy of the transition from the $4f^7$ ($^8S_{7/2}$) ground state to the $4f^6$ ($^7F_0$)5*d* excited state is lowered from the free ion value when the Eu^{2+} ion is brought into a crystal environment. The energy of the $f \to d$ absorption and of the $d \to f$ emission can be written according to the formalism of Dorenbos [388]

$$E_{abs} = E_{free} - D \text{ and } E_{em} = E_{free} - D - \Delta S$$

where E_{free} is the energy difference between the lowest $4f^7$ level and the $4f^6$ ($^7F_0$)5*d* level for the free or gaseous ion, *D* is the energy lowering also called redshift, and ΔS the Stokes shift. The redshift energy depends on the crystal environment and can be represented by a term, called the centroid shift or nephelauxetic (covalence) effect, concerning the shift of the average of the five 5*d* levels relative to the free ion value, and a second term associated to the crystal field splitting. For Eu^{2+}, E_{free} is equal to 4.19 eV [399]. The emission and excitation spectra provide the values of absorption and emission energies, $E_{abs} = 2.57$ eV and $E_{em} = 2.31$ eV. The Stokes shift is then $\Delta S = E_{abs} - E_{em} = 0.26$ eV (2000 cm^{-1}) and the redshift $D = E_{free} - E_{abs} = 1.62$ eV (13000 cm^{-1}). Thanks to the determination of the crystal field strength and calculation other parameters, the energetic scheme was proposed and shown in Fig. 4.143. The bottom of the first $4f^6 5d$ levels (*a1*) is consisted from two near situated lines, corresponding to different Eu sites.

For $CaGa_2S_4:Eu^{2+}$ the emission and excitation spectra (Fig. 4.140, curves 3) are very similar. It is not strange, because the crystallographic structure for calcium thiogallate is the same as for $SrGa_2S_4:Eu^{2+}$. The position of the $CaGa_2S_4:Eu^{2+}$ emission peak depends on the activator concentration in the compound. The peak position shifts from 553 nm (0 mol% of Eu) to the longwave region 565 nm (at 10 mol% Eu concentration) and approaches a constant [400]. In investigated series $CaGa_2S_4$ samples were doped by 6 mol% of Eu, as well as $SrGa_2S_4$ and $BaGa_2S_4$ samples. Eu^{2+} concentration can be increased over a wide range without the appearance of the concentration-quenching effect. This fact is verified by numerous experiments with $Ca_xEu_{1-x}Ga_2S_4$ ($0 \leq x \leq 1$) solutions and is explained by the closeness of the Eu^{2+} and Ca^{2+} ion radii. From the detailed analysis of Fig. 4.140 (curves 3), the crystal field splitting is estimated at about 16800 cm^{-1} (2.09 eV). For $CaGa_2S_4:Eu^{2+}$ above-mentioned experimental and calculation way was repeated and the emission and excitation spectra provided the values of absorption and emission energies, $E_{abs} = 2.48$ eV and $E_{em} = 2.22$ eV. The Stokes shift is then $\Delta S = E_{abs} - E_{em} = 0.26$ eV (2079 cm^{-1}) and the redshift $D = E_{free} - E_{abs} = 1.71$ eV (13700 cm^{-1}).

For $BaGa_2S_4:Eu^{2+}$ there is a little difference in spectra and crystallographic structure in comparison with strontium and calcium thiogallates. The excitation spectrum (Fig. 4.140, curve 1) consists of 3 not very good resolved at room temperature broad bands in visible range. The excitation spectrum reported by Peters et al. [372] for $BaGa_2S_4:Eu^{2+}$ (2 mol% of Eu) is similar. The spectra given at 77 K by Jabbarov et al. [381] for 1 mol% of Eu are better resolved but principally the main peaks correspond to presented here results.

In the $BaGa_2S_4$ crystal, the Ba^{2+} ions occupy two lattice sites with different symmetry and coordination number: 8 Ba^{2+} ions are 6-fold coordinated in the special position *c* with point symmetry $C_3(3)$, whereas 4 Ba^{2+} ions are in the special position *a* or *b* with point symmetry $S_6(3)$ and coordination number 12 (Table 4.26). These data correspond to Eisenmann et. al. [376]. Since Eu^{2+} ions replace Ba^{2+} ions in $BaGa_2S_4:Eu^{2+}$, they are also distributed into these two different sites. In first approximation following Ref. [381] we can consider that the 6-fold coordinated Eu^{2+} ions have an octahedral (O_h) environment and that the 12-fold coordinated Eu^{2+} ions lie in icosahedral (I_h) sites. In O_h symmetry the 5*d* orbitals are split in two levels t_{2g} and e_g, whereas they are still degenerated in I_h symmetry (h_g level). Thus, three $f \to d$ transitions are expected: the $4f^7(^8S_{7/2}) \to 4f^6(^7F)5d\ (t_{2g})$ and the more energetic $4f^7(^8S_{7/2}) \to 4f^6(^7F)5d\ (e_g)$ transitions for the 6-fold coordinated Eu^{2+} ions and the $4f^7(^8S_{7/2}) \to 4f^6(^7F)5d\ (h_g)$ transition for the 12-fold coordinated ions. Three bands are observed in the excitation spectrum. Fig. 4.140, spectrum 1, provides a crystal-field splitting equal to about 12100 cm^{-1} (1.5 eV).

The energy of the zero-phonon line, E_0, at the intersection of the emission and excitation spectra for $BaGa_2S_4$ was estimated to 2.67 eV (464 nm). By using the mirror image relationship between the emission and the excitation spectra, we have found $E_{abs} = 2.9$ eV (424 nm). The energy of the Stokes shift ΔS is evaluated from excitation and emission spectra: $\Delta S = E_{abs} - E_{emi} = 2.9 - 2.4 = 0.5$ eV (4000 cm^{-1}). The knowledge of E_{abs} allows us to determine the redshift $D = E_{free} - E_{abs} = 4.19 - 2.9 = 1.3$ eV (10,000 cm^{-1}).

We observe the increase of Stokes shift with increasing size of the divalent cation M^{II} in thiogallate $M^{II}Ga_2S_4:Eu^{2+}$ phosphors, i.e. in the following order: Ca, Sr, Ba.

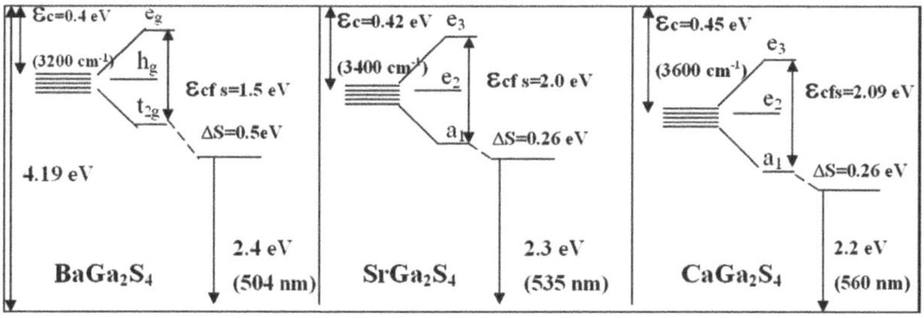

Figure 4.144 Nephelauxetic effect in Ca, Sr, and Ba thiogallates.

Table 4.28 Luminescence data for calcium, strontium, and barium thiogallates, doped with 6 mol% Eu^{2+} concentration

Formula	λ(abs) nm/ (ev)	*Eo* nm/ (ev)	λ(em) nm/ (eV)	FWHM (nm)	*S*	hv meV	Δ*S* cm^{-4} / (eV)	*D* cm^{-4} / (eV)	ε(csf) cm^{-1}/ (eV)	ε(centr) cm^{-1}/ (eV)
$BaGa_2S_4$	424 2.9	464 2.67	504 2.4	62	7	37.5	4000 0.5	10500 1.3	12100 1.5	3200 0.40
$SrGa_2S_4$	482 2.57	508 2.44	535 2.31	49	4	34.5	2000 0.26	13000 1.62	16000 2.0	3400 0.42
$CaGa_2S_4$	500 2.48	529 2.3	558 2.22	50	4	35.2	2079 0.26	13700 1.71	16800 2.09	3600 0.45

The crystal field splitting is expected to increase with decreasing size of the divalent cation M^{II}. The rule is confirmed for M^{II} = Ba, Sr and Ca: the strength of the crystal-field was estimated at 12100 cm^{-1} for $BaGa_2S_4$: Eu^{2+}, at 16000 cm^{-1} for $SrGa_2S_4$:Eu^{2+} and at 16700 cm^{-1} for $CaGa_2S_4$: Eu^{2+}.

The lower redshift in $BaGa_2S_4$:Eu^{2+} (D = 10,000 cm^{-1}) compared with $SrGa_2S_4$:Eu^{2+} (13,000 cm^{-1}) and $CaGa_2S_4$: Eu^{2+} (13,700 cm^{-1}) is mainly explained by the lower crystal field splitting but also by the lower nephelauxetic effect or centroid shift. This shift ε_c was first calculated for all tree considerate thiogallates and schematically shown in Fig. 4.144. The Ba element is less electronegative than the Sr element and Ca. The M^{II}–S lengths are larger in $BaGa_2S_4$:Eu^{2+} (3.39 Å) than in $SrGa_2S_4$:Eu^{2+} (3.30 Å) and $CaGa_2S_4$:Eu^{2+} (3.0 Å) (Tables 4.24–4.26). The Ba–S bonds are then less covalent than the Sr–S and Ca–S ones and the nephelauxetic effect is weaker in $BaGa_2S_4$ than in $SrGa_2S_4$ and $CaGa_2S_4$.

Table 4.29 Crystallographic data for calcium, strontium, and barium thiogallates.

Formula	Stricture	Space group	a (Å)	b (Å)	c (Å)	z	Shortest M–S (A)	Coord number	Symmetry
$BaGa_2S_4$	Cubic $\alpha = \beta = \gamma = 90$	Tetrahedral Th^6–Pa3	12.685	12.685	12.685	12	3.39	6.12	Pa-3
SrG_2S_4	Orthorhombic	D_{2h}^{24}–Fddd	20.855	20.511	12.213	32	3.30	8	D_2,C_2
$CaGa_2S_4$	Orthorhombic	D_{2H}^{24}–Fddd	20.122	20.09	12.133	32	3.0	8	D_2,C_2

Summarized luminescence and crystallographic data for calcium, strontium and barium thiogallates, doped with 6 mol% of Eu^{2+} concentration are given in Tables 4.28 and 4.29.

One of the important questions for practical application of thiogallates in LED, displays and other devices is radiant efficiency and stability. Usually these materials are very bright and have very high efficiency, but sometimes their stability toward hydrolysis and temperature is not sufficient. Ternary compounds are more stable than binary ones. Thiogallates are much less sensitive to the action of moisture than binary chalcogenides. The radiant efficiency of the Eu^{2+} activated thiogallates decreases in the order Ca > Sr > Ba and the stability decreases in the opposite order Ba > Sr > Ca [372, 378]. This tendency is in good correlation with the crystal field splitting (Table 4.28). Less splitter powders show more stable properties.

4.5.3 *New Efficient Multiphase Green Phosphor [$SrGa_2S_4$ + $MgGa_2O_4$]:Eu^{2+}*

Polycrystalline single phase $SrGa_2S_4$:Eu^{2+} and multiphase [$SrGa_2S_4$+$MgGa_2O_4$]:Eu^{2+} phosphor samples were synthesized by

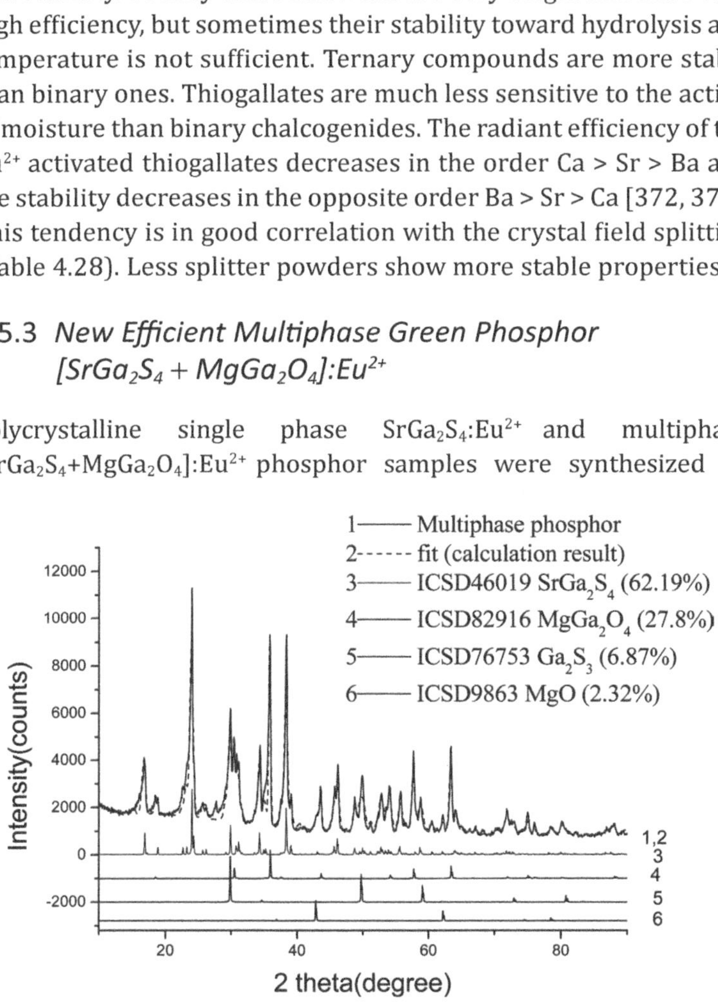

Figure 4.145 The X-ray diffraction patterns of multiphase phosphor made of strontium thiogallate and the fitting in comparison with the reference ICSD data.

Table 4.30 Crystallographic data for single phase $SrGa_2S_4$ and multiphase phosphor

					Calculated structure		
Sample	Phase	Wt%	Structure	Reference	a (Å)	b(Å)	c (Å)
$SrGa_2S_4$	$SrGa_2S_4$	100	Orthorhombic	ICSD46019	20.855	20.511	12.231
	$SrGa_2S_4$	62.2	Orthorhombic	ICSD46019	20.860	20.488	12.219
Multiphase phosphor	$MgGa_2S_4$	27.8	Cubic Fd -3m	ICSD82916	8.301		
	Ga_2S_3	6.9	Cubic F4 -3m	ICSD76753	5.195		
	MgO	2.3	Cubic Fm -3m	ICSD9863	4.211		

solid state reaction. For $SrGa_2S_4$, sulfide powders SrS and Ga_2S_3 were initially mixed in stoichiometric composition and annealed at 900–1000°C with a carbon reduction atmosphere for 4 h. For multiphase samples, we used the same raw materials with addition of MgS and $Mg(NO_3)_2$ at the same synthesis conditions. The doping ions were introduced in both cases in the form of EuS. The powder samples were prepared with 6 mol% of Eu^{2+} concentration and these thiogallate single and multiphase samples exhibit a deep green colors. The phosphor samples were characterized for crystalline structure and luminescence properties as was described previously in Introductory remarks.

Figure 4.145 shows the XRD patterns of the multiphase phosphor made of strontium thiogallate powder baked at 1000°C (1). The peak positions of XRD patterns are in good agreement with fitting (2) and ICSD (ICSD: Inorganic Crystallographic Structure Database) data (3–6).

The solid curve (1) in Fig. 4.145 is the measured data and the dash curve (2) means the calculation result of Rietveld analysis [285]. In case of multiphase structures, quantitative phase analysis is also available by using Rietveld method. The detailed crystallographic information from Rietveld analysis with "GSAS" program is listed in Tables 4.30, 4.23a, and 4.24. The calculated parameters in Table 4.27 are based on our original XRD patterns, and corresponding ICSD data (indicated as reference) were used for Rietveld refinement.

Table 4.31 Crystallographic data for different ion positions in synthesized multiphase phosphor ($MgGa_2O_4$ phase) and short distance Mg–O in comparison with reference data

Atoms	Atomic position $MgGa_2O_4$			Short distance	$MgGa_2O_4$ (Å)	ICSD ref (Å)
	x	*y*	*z*			
Mg1	0.125	0.25	0.125	Mg(1)–O	1.832	1.829
Ga1	0.125	0.125	0.125			
Ga2	0.5	0.5	0.5			
Mg2	0.5	0.5	0.5	Mg(2)–O	2.055	2.052
O	0.2573	0.2573	0.2573			

Figure 4.145 shows that in realized synthesis conditions we acquired not only one single phase of $SrGa_2S_4$, but also some other different phases. The most important of them is $MgGa_2O_4$ because this phase affects the luminescence. The other phases could be neglected or excluded during the synthesis.

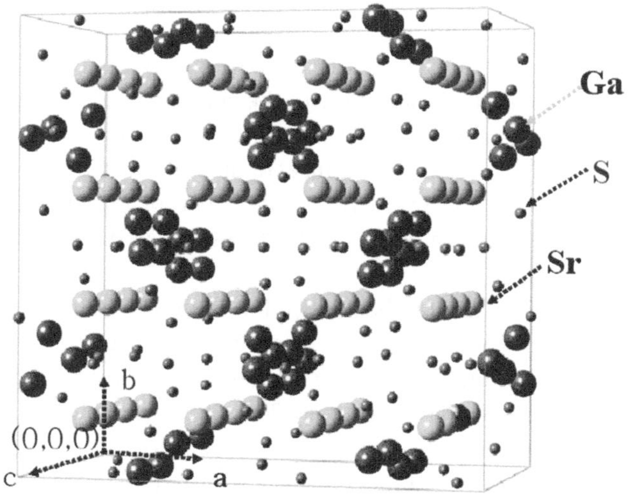

Figure 4.146 Crystal structure of $SrGa_2S_4$.

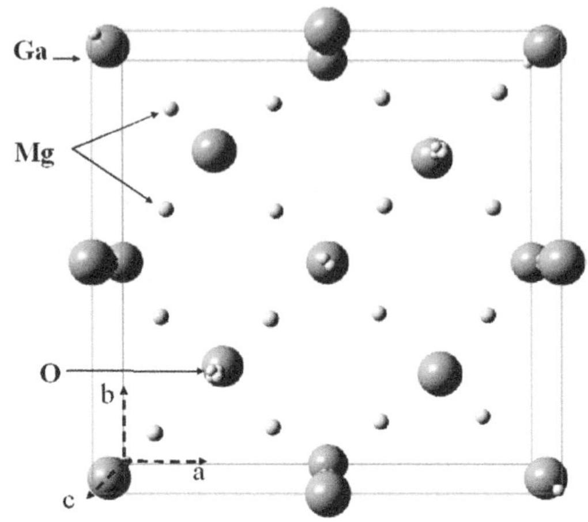

Figure 4.147 Crystal structure of $MgGa_2O_4$.

The unit cell parameters of ICSD-46019 ($SrGa_2S_4$) are a = 20.932 Å, b = 20.549 Å, c = 12.227 Å, very similar to synthesized sample as seen from Table 4.30. The little differences in cell parameters can be explained by a slight difference in atomic positions. The positions of Sr atoms in $SrGa_2S_4$ are the same as the reference, but the positions of Ga atoms and S atoms are a little different. Crystallographic data for synthesized $SrGa_2S_4$ and multiphase [$SrGa_2S_4$+$MgGa_2O_4$] samples are given in Tables 4.23a, 4.24, and 4.31.

The thiogallate compounds of the type $Sr^{I}Ga_2S_4$ belong to the orthorhombic crystal class with the space group D^{24}_{2h} (*Fddd*) [390], and $MgGa_2O_4$ to cubic structure with the space group O^7_h (*Fd3m*). The detailed analysis of orthorhombic crystal class was given in previous paragraph and Fig. 4.146 and the inset figure in Table 4.24 illustrate the symmetry of the S and Sr environments in orthorhombic crystals.

Figure 4.147 illustrates the symmetry of the S and Sr environments in cubic crystals.

The $MgGa_2O_4$ crystal has a spinel structure with a partially inverted nature, and belongs to the cubic space group *Fd3m*. In this compound, the populations of Mg^{2+} ions in tetrahedral and octahedral sites are 0.16 and 0.84, respectively. The site symmetry of each atom can be obtained from the correlation tables [391].

According to data in Table 4.24, the shortest distance between Sr and S is estimated as 3.30 Å. For $MgGa_2O_4$, the bond Mg–O is shorter

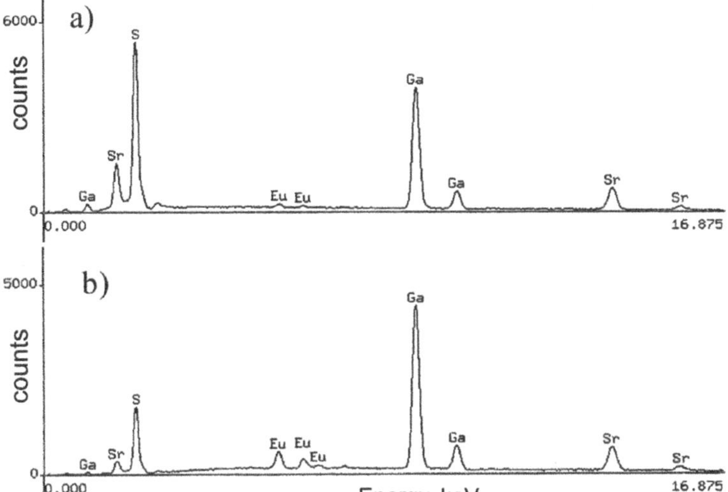

Figure 4.148 EDS patterns corresponding to $SrGa_2S_4$ (a) and multiphase [$SrGa_2S_4$+$MgGa_2O_4$] sample (b).

and it was found as 1.94 Å. (Table 4.31). (The measurements were made for all interatomic distance less than 5 Å).

The compositional analysis by energy dispersive X-ray spectrometry on a JEOL 2010 CX electron microscope was carried out. EDS spectra from one single phase $SrGa_2S_4$ phosphor and multiphase [$SrGa_2S_4$+$MgGa_2O_4$] sample are shown in Fig. 4.148.

The distribution of elements is analyzed in Table 4.32.

From Fig. 4.148 and Table 4.32, it is clearly seen that at the initial 6 mol% Eu^{2+} concentration in single and multiphase samples, the

Table 4.32 Compositional EDS analysis

	Atomic%		Weight%		Wt % Error	
Element	Single (a)	Multi (b)	Single (a)	Multi (b)	Single (a)	Multi (b)
Sr–Lα	6.79	2.29	9.89	2.95	0.16	0.10
S - K	29.85	11.96	15.90	5.64	0.11	0.07
Eu - Lα	0.59	2.89	1.50	6.45	0.17	0.22
Ga - K	62.76	82.86	72.71	84.96	0.59	0.65
Total	100.00	100.00	100.00	100.00		

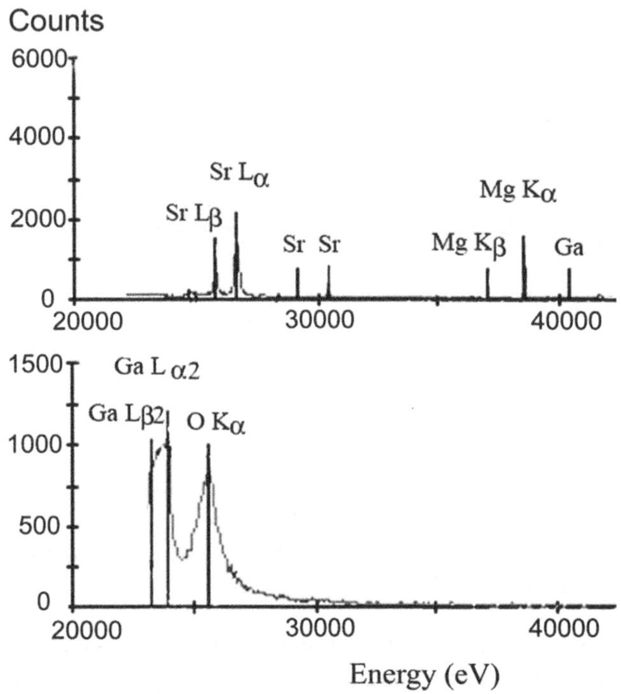

Figure 4.149 WDS spectra of multiphase phosphor [$SrGa_2S_4$+$MgGa_2O_4$].

activator incorporation degrees in two samples differ about 4 times. This AID parameter is very important because not only the initial activator concentration, but also the activator incorporation degree is responsible for luminescence properties [401]. Unfortunately, EDS analysis could not detect the light elements such as magnesium and oxygen. Thus, the compositions of the samples should be also examined by EPMA.

The EPMA technique allows us to use wavelength dispersive spectroscopy, and the sample composition can be easily identified. Magnesium and oxygen, analyzed by EPMA, were detected in the multiphase phosphor and shown in Fig. 4.149. The quantitative data were found to be in good agreement with EDS and XRD data.

The Stokes Raman spectra of the polycrystalline $SrGa_2S_4$ and [$SrGa_2S_4$+$MgGa_2O_4$] are given from 50 cm^{-1} to 500 cm^{-1} in Fig. 4.150.

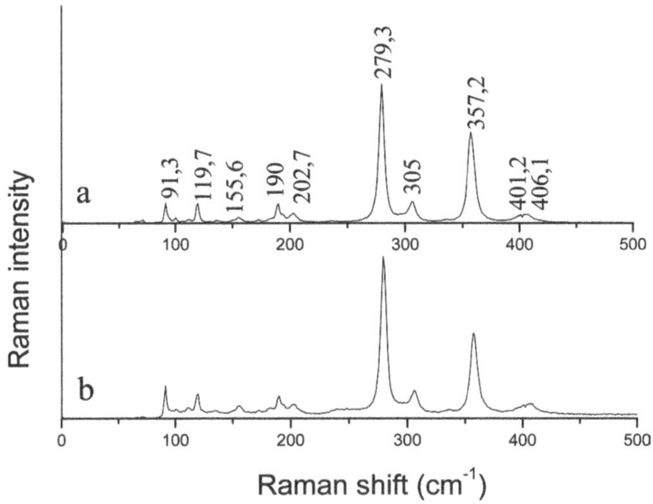

Figure 4.150 Raman spectra of $SrGa_2S_4$ (a) and [$SrGa_2S_4$ + $MgGa_2O_4$] (b), excited with the 785 nm laser line.

The Stokes spectra for both samples are very similar and consist of 19 vibration lines positioned between 68 cm^{-1} and 408 cm^{-1}. Two dominant peaks at 279 cm^{-1} (34.5 meV) and at 357 cm^{-1} (44.2 meV)

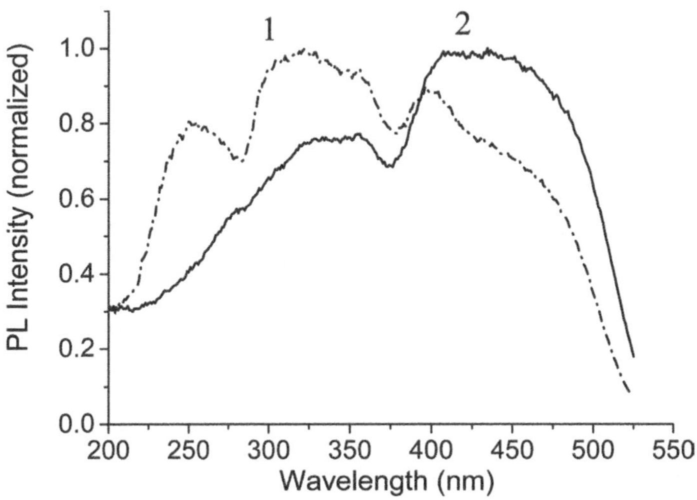

Figure 4.151 Excitation (PLE) spectra of $SrGa_2S_4:Eu^{2+}$ (1) and [$SrGa_2S_4$ + $MgGa_2O_4$]:Eu^{2+}(2).

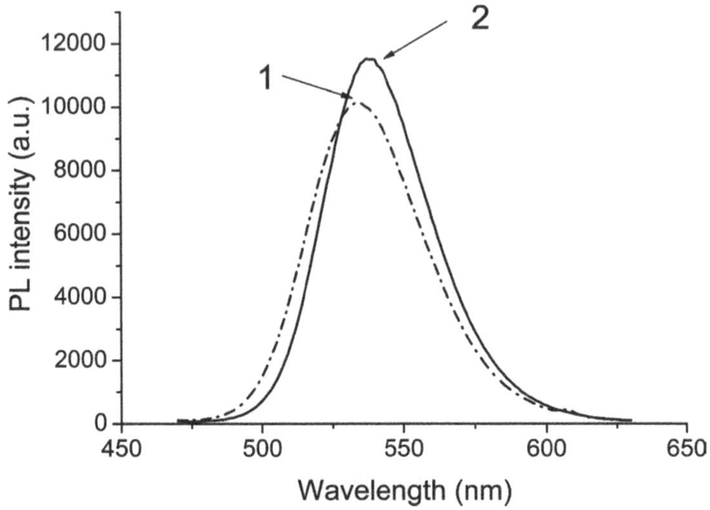

Figure 4.152 Emission (PL) spectra of $SrGa_2S_4$:Eu^{2+} (1), a well-known commercial green phosphor (Sarnoff Co.) and [$SrGa_2S_4$+$MgGa_2O_4$]:Eu^{2+}(2) excited at 460 nm.

for $SrGa_2S_4$ and [$SrGa_2S_4$+$MgGa_2O_4$] are registered. The intensity ratios of these peaks are 1.5 for $SrGa_2S_4$ and 1.9 for [$SrGa_2S_4$+$MgGa_2O_4$]. All other peak intensities and peak positions coincide very well. These Raman spectra are in good agreement with those reported in the literature for $SrGa_2S_4$ [395, 396] and could be used for the fitting of emission bands.

The photoluminescence excitation and emission spectra of the investigated compounds doped with 6 mol% of Eu^{2+} at 300 K are presented in Figs. 4.151 and 4.152.

For both samples, the PL spectra under excitation at 460 nm consist of a broad band in the visible range. The PL peak for multiphase phosphor (2) is shifted to 539 nm in comparison with 535 nm for $SrGa_2S_4$ (1). The spectral bandwidths at the FWHM are 44 nm (2) and 49 nm (1), correspondingly. No additional band due to the emission of Eu^{2+} is observed. For comparison purpose, a commercial sample of $SrGa_2S_4$ (Sarnoff Co.) was used for PL measurement in Fig. 4.152, which is known as the best green emitter of this type and has almost identical structure and optical properties with synthesized $SrGa_2S_4$ sample. The new [$SrGa_2S_4$+$MgGa_2O_4$] sample clearly shows a better emission than the best commercial green emitter.

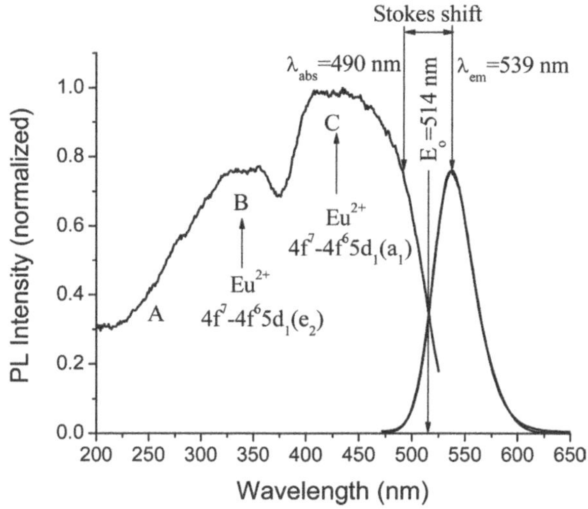

Figure 4.153 PLE and PL spectra of multiphase phosphor $[SrGa_2S_4+MgGa_2O_4]:Eu^{2+}$ at 300 K.

The PLE and PL spectra of multiphase phosphor $[SrGa_2S_4+MgGa_2O_4]:Eu^{2+}$, as well as its detailed analysis for 539 nm wavelength, are presented in Fig. 4.153.

This PLE spectrum is similar to the spectrum of $SrGa_2S_4:Eu^{2+}$ described in previous paragraph. Figure 4.153 shows that the PLE spectrum of $[SrGa_2S_4+MgGa_2O_4]:Eu^{2+}$ is also composed of three main bands in the UV–visible range. We think that it would be useful to repeat the same method of photoluminescence analysis for multiphase phosphor. Though the similar way, the results will be a little different.

PLE band *A* is ascribed to the transitions between the valence band and the conduction band of the host matrix. The two other PLE bands, *B* and *C* are ascribed to the $4f^7(^8S_{7/2})\rightarrow 4f^6\,(^7F)5d$ transitions in Eu^{2+} electronic levels. In multiphase compounds, the Sr^{2+} ions, as well as Mg^{2+}, are substituted by the Eu^{2+} ions. They occupy three different sites in C_2 and D_2 symmetry [376]. Because their shapes have minor differences, the emission bands of Eu^{2+} in the three available sites are expected to lie near each other. In the first order approximation, they are located at the center of square anti-prisms formed by the 8 S^{2-} anions (Tables 4.23a and 4.24). In this D_{4d} symmetry, the 5*d* orbitals are split in three levels (Chartier et al. [382]) as was illustrated in Fig. 4.144.

The band *C* is ascribed to the $4f^7$ ($^8S_{7/2}$)→$4f^6$ (^7F)$5d$ (a_1) transitions and the band *B* to the $4f^7$ ($^8S_{7/2}$)→$4f^6$ (^7F)$5d$ (e_2) transitions. By evaluating the energy gap between *B* and *C* bands to 8000 cm^{-1} (1 eV), the crystal field strength, $\Delta_{a-p} = 8.9Dq$, can be estimated at about 16,000 cm^{-1} (2 eV).

An almost mirror image relationship seems to hold between the PL and the PLE spectra in the energy region between the two maxima (Fig. 4.153). This relation is characteristic of a phonon-broadened emission and confirms the hypothesis made previously; the mean phonon energy hν is the same for the $4f^7$ ($^8S_{7/2}$) ground state as for the $4f^6$ (^7F)$5d$ (a_1) excited state. We can determine the energy of the zero-phonon line E_0 at the intersection of the PL and PLE spectra. From Fig. 4.153, $E_0 = 2.41$ eV, which corresponds to a wavelength of 514 nm (comparable with that given by Eichenauer et al.[398] and Chartier et al. [382]). Unfortunately, the fine structure due to the spin–orbit coupling in the $4f^6$ configuration leading to the splitting of the ^7F levels into seven levels (^7F_J with $J = 0 \rightarrow 6$) was not observable at room temperature. The energy E_{abs} of the $f \rightarrow d$ transition from the ground state to the lowest level of the $4f^6 5d$ (a_1) corresponds to the $^7F_0$ level. Because this level is not resolved, we can only give an estimation of E_{abs}. By using the mirror image relationship between the PL and the PLE spectra, we have found $E_{abs} = 2.53$ eV (490 nm). The energy of the transition from the $4f^7$ ($^8S_{7/2}$) ground state to the $4f^6$ ($^7F_0$)$5d$ excited state decreases from the free ion value when the Eu^{2+} ion is brought into the crystal environment. The energy of the $f \rightarrow d$ absorption and of the $d \rightarrow f$ emission can be written according to the formalism of Dorenbos [388]: $E_{abs} = E_{free} - D$; $E_{em} = E_{free} - D - \Delta S$, where E_{free} is the energy difference between the lowest $4f^7$ level and the $4f^6$ ($^7F_0$)$5d$ level for the free or gaseous ion, *D* is the energy decreased also called a redshift, and ΔS is the Stokes shift. The redshift energy *D* depends on the crystal environment and can be represented by a term called the centroid shift or nephelauxetic (covalence) effect (ε_c), concerning the shift of the average of the five $5d$ levels relative to the free ion value, and another term associated with the crystal field splitting (ε_{cfs}) as seen in Fig. 4.154. For Eu^{2+} ions, E_{free} is known to be 4.19 eV [399]. The PLE and PL spectra

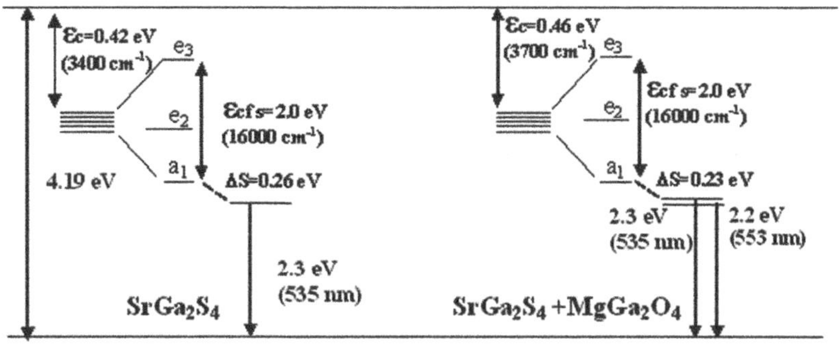

Figure 4.154 Nephelauxetic effect in polycrystalline single phase $SrGa_2S_4:Eu^{2+}$ phosphor and multiphase $[SrGa_2S_4 + MgGa_2O_4]:Eu^{2+}$ sample.

provide the values of absorption and emission energies, $E_{abs} = 2.53$ eV and $E_{em} = 2.30$ eV, accordingly. The Stokes shift is then $\Delta S = E_{abs} - E_{em} = 0.23$ eV (1900 cm^{-1}) and the redshift $D = E_{free} - E_{abs} = 1.66$ eV (13,400 cm^{-1}).

The smaller redshift in $SrGa_2S_4:Eu^{2+}$ ($D = 13{,}000$ cm^{-1}) compared with $[SrGa_2S_4+MgGa_2O_4]:Eu^{2+}$, (13400 cm^{-1}) is mainly explained by the smaller nephelauxetic effect or centroid shift. This shift ε_c is calculated and schematically shown in Fig. 4.154.

The Sr element is less electronegative than the Mg element and the cation–anion interatomic distance Sr–S in strontium thiogallate is larger (3.30 Å) than Mg–O in magnesium digallium oxide (1.94 Å). The Sr–S bonds are then less covalent than the Mg–O ones and the nephelauxetic effect is weaker in $SrGa_2S_4$ than in $MgGa_2O_4$. Stronger nephelauxetic effect (or centroid shift) replaces the peak wavelength from 535 nm to 539 nm, which is very important for improving the color rendering index.

Table 4.33 Luminescence data for $SrGa_2S_4$ and $[SrGa_2S_4+MgGa_2O_4]$ doped with 6 mol% Eu^{2+}

Formula	λ(abs) nm/ (eV)	*Eo* nm/ (eV)	λ(em) nm/ (eV)	FWHM nm	*S*	hv (meV)	ΔS cm^{-1}/ (eV)	*D* cm^{-1}/ (ev)	ε(cfs) cm^{-1}/ (eV)	ε(centr) cm^{-1}/ (eV)
$SrGa_2S_4$	4822.57	5082.44	5352.31	49	4	34.5	20000.26	13000 1.62	16000 2.0	3400 0.42
$\{SrGa_2S_4+$ $MgGa_2O_2\}$	4902.53	5142.41	5392.30	44	4	35.5	19000.23	13400 1.66	16000 2.00	3700 0.46

Table 4.34 Literature data on the $4f^7 \rightarrow 4f^6$ [$^7F_0$]$5d_1$ absorption wavelength λ_{abs}, zero-phonon line E_0, $d \rightarrow f$ emission λ_{em}, FWHM emission band width, and Stokes shift ΔS with corresponding references. The first column compiles the type of compound and the second one the Eu^{2+} concentration.

Formula	Eu (%)	T (K)	λ_{exc} (nm)	λ_{abs} (nm)	Eo (nm)	λ_{em} (nm)	FWHM (nm)	ΔS (cm^{-1})	Ref.
$BaGa_2S_4$	low	300	mean	453		497	60	1954±390	[388]
	1	90	337	450	–	507	47	5000–6000	[378]
	0.5	300	254			495	58		[375]
	2	300	361			500	52		[372]
	1	77	360	420±10	459	505	38	4400±1000	[381]
	1	300	360			493	60		[385]
$SrGa_2S_4$	low	300	mean	503		532	45	1084±395	[388]
	2	300	351			535	48		[372]
	2	300	254	360–510		538	52	2400	[375]
	1	300	400		506	535	50	2273	[398]
	1	30	400		506	535	38		[398]
	1	90	337	502	–	538	39	2000–3000	[378]
	0.1	77	420	486	508	537	35	1700±200	[382]
$CaGa_2S_4$	low	300	mean	530		558	56	947±356	[388]
	0.2	300	254			560	57		[375]
	2	300	368			555	51		[372]
	7	300	465			553	50		[383], [389]
	0.5	300	470			556	54		[386]
	1	77	500	512	537	561	40	1700	[402], [403]
	2	300				555	50		[400]
	1	300	325	496		562		2368	[404]
	10	300	325	525		562		1254	[404]
	1	300	441.6	500		555	52	1982	[405]
	2	300	365			556	50		[379]
		300				562	51		[406]

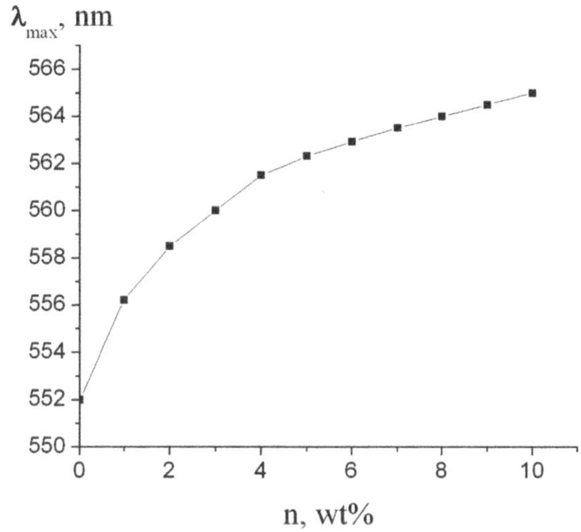

Figure 4.155 Peak intensity wavelength λ_{max} of the emission of $CaGa_2S_4:Eu^{2+}$ phosphor as a function of the activator concentration (n). $T = 300$ K.

Luminescence and crystallographic data for the investigated thiogallates doped with 6 mol% Eu^{2+} are summarized in Table 4.33.

4.5.4 *Electron-Vibrational Interaction in 4f–5d Optical Transitions*

The calculation of the Stokes shift (ΔE) is related to the difference between the absorption and emission energies. First of all, the detailed analysis of literature data about the emission and absorption of thiogallate phosphors was made. Data presented in the literature on the $f \to d$ spectroscopy of Ca, Sr, and Ba thiogallates doped with Eu^{2+} ions have been re-analyzed, and the information derived is shown in the Table 4.34. The cited literature can be consulted for more detailed information.

Occasionally, different publications yielded different results and then average values are compiled as in Ref. [388], but in other references the original data are summarized. Whenever both $f \to d$ absorption and $d \to f$ emission wavelength are known, the Stokes shift ΔS has been determined. From the analysis of literature data one can see that the emission peak position for the same materials and

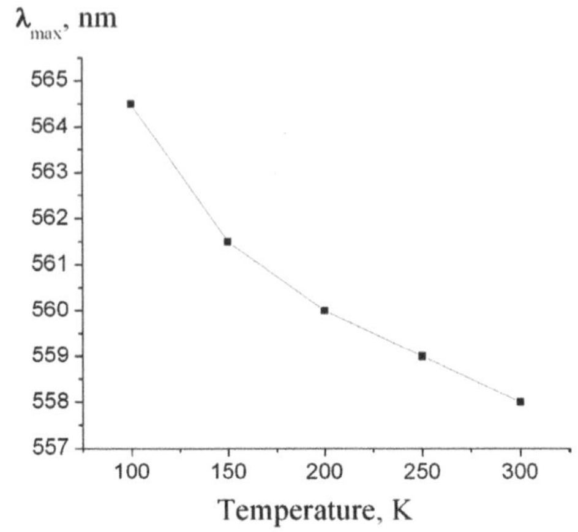

Figure 4.156 The dependence of the emission peak intensity wavelength λ_{max} of $CaGa_2S_4:Eu^{2+}$ (6 mol%) on temperature.

FWHM are identical from different cited results. Band positions may shift by typically 5–10 nm depending on the Eu^{2+} concentration and the temperature. Some of experiments for $CaGa_2S_4:Eu^{2+}$ (Figs.4.155 and 4.156) confirm this conclusion.

The emission spectra of the $BaGa_2S_4:Eu^{2+}$ (1%) powder as a function of the temperature from 77 to 450 K for excitation wavelength of 360 nm are presented by Jabbarov et al. [381]. The wavelength of the emission maximum decreases from 503 to 488 nm and the FWHM of the band is rising from 42 to 70 nm with increasing temperature.

It can be also seen in Table 4.34 that the main differences in data are in absorption wavelength. Many authors did not indicate this, or gave different values. For example, Donohue et al. have found for $Sr_{0.98}Ga_2S_4:0.02Eu^{2+}$ that the absorption is between 360 and 510 nm which they associated with Eu^{2+} $4f^7 \rightarrow 4f^6 5d$ transitions [375]. These data are in satisfactory agreement with 500 nm value derived from fluorescence decay lifetime data and quite reasonable for a parity allowed transition in a multielectron transition metal ion. λ_{abs} pertains to that $f \rightarrow d$ transition that leaves the remaining six $4f$ electrons in the lowest, i.e. $^7F_0$ level of the $4f^6$ configuration and the $5d$ electron in the lowest crystal field $5d$ level. Usually, $^7F_0$ level is not resolved in excitation spectra and in those cases its location had to be estimated. That's why the

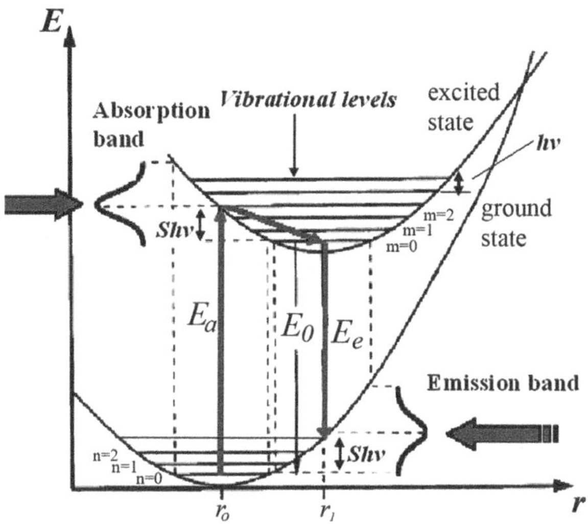

Figure 4.157 Configurational diagram.

Stokes shift cannot be determined accurately, since the first maximum in the *f*→*d* excitation band is often difficult to identify. There is one other problem concerning the excitation spectra. In many articles the excitation spectra differ for the same material with different activator concentration. Hidaka et al. paid attention to dependence the excitation spectra on the concentration [404]. The excitation spectrum (1 Eu mol%) has a maximum near the absorption edge at 2.5 eV (496 nm) and two peaks at 3.1 eV (400 nm) and 3.7 eV (335 nm) which correspond to those observed in the absorption spectrum, suggesting that the excited states consist of three levels being ascribed to the splitting of the 5*d* multiplets of Eu ion. For larger value of Eu concentration (10 Eu mol%) it seems that a band structure related absorption takes places and shape of spectrum changed. The steep in excitation spectrum decreases and the two maxima at 3.1 and 3.7 eV became smoother.

The classical description cannot satisfactorily explain emission and excitation spectral shapes and non-radiative transition probabilities. The single configuration coordinate model based on quantum mechanics was used to fit the emission band of considered phosphors Fig. 4.157.

Since the 4*f*-shell is well shielded the electron-vibrational (vibronic) interaction within this shell is negligible so that the intensities of the vibronic satellites corresponding to $4f \to 4f$ transitions are small. On the contrary, $4f \to 5d$ transitions give rise to the broad vibronic bands whose shapes are closely related to a significant vibronic interaction in $5d$-shells. We will use the quantum-mechanical descriptions based on the single configuration coordinate model based also on the assumption that the frequency of the active vibrational mode remains the same in both electronic states and the anharmonicity is neglected. We will neglect also the Jahn–Teller interaction that can be active in the d-shell and can lead to an essential complication of the vibronic band (see review article [407]). Within this framework the form function of the band was found in the pioneering papers [408, 409] and was called by Dexter [410] Pekarian (or Pekar–Huang–Rhys formula) [411]. Very extended model of configuration diagrams was considered in Ref. [412]. In the seventies to nineties the radiative and non-radiative processes were extensively studied [413–415] and the basic model (linear vibronic coupling) was extended with due account for anharmonicity [416] and shift of the vibrational frequencies [417].

The broad vibronic bands whose shape-function can be described by the normalized "Pekarian" type spectral distribution that represents a series of the discrete lines [407, 411] each related to an individual electron-vibrational transition:

$$F(\Omega) = exp\left[-\frac{a}{2}Coth\left(\frac{p\hbar\omega}{2kT}\right)\right]\sum_{-\infty}^{\infty} exp\left(\frac{p\hbar\omega}{2kT}\right)I_p\left(\frac{a}{2Sinh\left(\frac{\hbar\omega}{2kT}\right)}\right)\delta\left(\Omega-\Omega_0-p\omega\right) \tag{4.41}$$

In Eq.(4.41) $I_p(x)$is the modified Bessel function that represents a series of the discrete lines, a is the so-called heat release parameter (shift of the dimensionless vibrational coordinate after photon absorption), $\delta(\Omega - \Omega_0 - p\omega)$ is the Dirac δ-function that describes an individual vibrational line of the dimensional frequency $p = (\Omega - \Omega_0)/\omega$, Ω_0 is the frequency of the zero-phonon line, ω is the frequency of the active vibrational mode. If the dispersion zone of the lattice vibrations is taken into account the shape function is smoothed that is achieved by replacing the δ-function by the Gaussian or Lorentzian bands. In the case of low temperature Pekarian can be simplified. In fact, in this case

$$Sinh(\hbar\omega/2kT) \approx (1/2)exp(\hbar\omega/kT),$$

$$exp\left(-\tfrac{a}{2}Coth\tfrac{\hbar\omega}{2kT}\right) \approx exp\left(\tfrac{a}{2}\right)$$

and the modified Bessel function can be represented as a series in terms of the argument $ae^{\frac{p\hbar\omega}{2kT}}$:

$$I_p\left(a\,e^{-\frac{\hbar\omega}{2kT}}\right) \approx \left[\left(\frac{a}{2}\right)^p \frac{1}{p!} + \left(\frac{a}{2}\right)^{p+2} \frac{1}{(p+1)!} e^{-\frac{\hbar\omega}{kT}}\right] e^{\frac{p\hbar\omega}{2kT}} \quad (4.42)$$

In the decomposition (4.42) small terms $o\left[ae^{-\frac{\hbar\omega}{2kT}}\right]^4$ (and high order terms) are neglected that is justified when the temperature is low enough $(kT \leq \hbar\omega)$ and the vibronic coupling is not too strong. Combining Eqs (4.41) and (4.42) one can obtain the following low-temperature approximation for the band profile of the vibronic band:

$$F(\Omega) = e^{-\frac{a}{2}}\left[\sum_{p=0}^{\infty}\left(\frac{a}{2}\right)^p \frac{1}{p!} + \sum_{p=-1}^{\infty}\left(\frac{a}{2}\right)^{p+2} \frac{1}{(p+1)!} e^{-\frac{\hbar\omega}{kT}}\right]\delta\left(\Omega - \Omega_0 - p\omega\right) \quad (4.43)$$

This expression describes a discrete set of the vibronic lines (each represented by a corresponding δ-function) and therefore reflects quantum properties of the vibrations. The dispersion of the vibrations makes smooth the discrete structure. Therefore, one can omit the δ-functions and the required summations and consider that intensity of the band in Eq. (4.43) as a function of the continuous parameter p:

$$F(p) = \frac{e^{-s} \cdot S^P}{p!}\left[1 + S^2 \frac{1}{(p+1)} e^{-\frac{\hbar\omega}{kT}}\right] \quad (4.44)$$

where $S = a/2$ is the Huang–Rhys–Pekar factor – the number of emitted phonons accompanying the optical transition. At temperature $T = 0$ K from this equation we obtain a well-known transition probability given by a simple relation [418]:

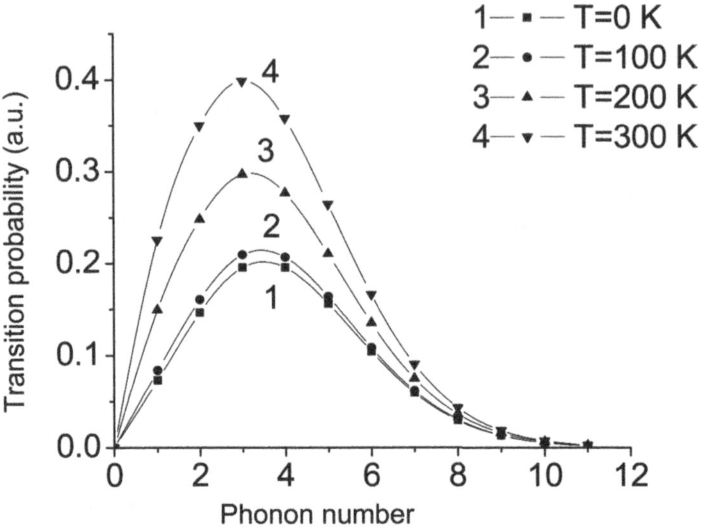

Figure 4.158 Influence of temperature on transition probability ($S = 4$).

$$F(p) = \frac{e^{-S} \cdot S^{P}}{p!} \qquad (4.45)$$

Usually, only the first term in Eq. (4.44) that is the Poisson type distribution is taken into account in consideration of the band shapes at low temperatures. From the physical point of view this means absorption (emission) originating from the fundamental vibronic level in the initial electron-vibrational state. This approximation fails when the temperature is high enough, for example, room temperature, when the most PL measurements are carried out. Under this condition the second term in Eq. (4.44) corresponding to the emission (absorption) from both ground and first excited levels are important and the second term in Eq. (4.44) plays important role.

The temperature dependence of Eq. (4.44) is shown in Fig. 4.158.

From Fig. 4.158 one can see that the temperature T = 0 K corresponds to Eq. (4.45) or the first term in Eq. (4.44). The influence of temperature becomes significant at T = 200 K and higher.

A numerical solutions of Eq. (4.44) for T = 0 K and T = 300 K and for varies S parameters are presented in Fig. 4.159.

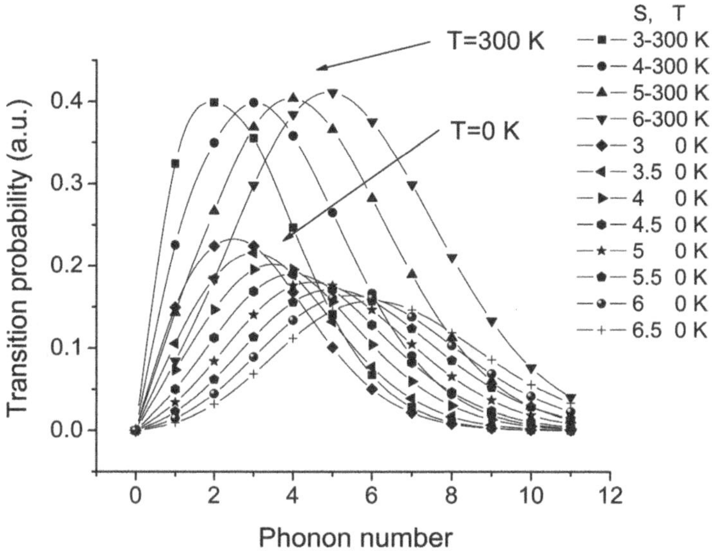

Figure 4.159 The spectral shape calculated for $T = 300$ K (upper curves) and $T = 0$ K (lower curves) from Eq. 4.44 at different parameter S.

It is clearly seen from Figs. 4.160 and 4.161 that the temperature influence is very important and must be taken into account. By comparing the family of curves at low and high temperatures one can see that at $T = 0$ K the maximum likelihood decreases with increasing S, whereas for $T = 300$ K it practically does not change. Using Stirling's approximations for factorial [419] in Eq. (4.45), it can be shown that the maximum likelihood at $T = 0$ K satisfies the next equation:

$$F(p) = \left.\frac{e^{-S} \cdot S^{P}}{\sqrt{2\pi p} \cdot p^{p} \cdot e^{-p}}\right|_{p=p^*} = \frac{1}{\sqrt{2\pi p}} \tag{4.46}$$

as since the peak is located at $p^* = S$ at $T = 0$ K.

At $T \neq 0$ K we can find the peak position from a numerical solution Eq. (4.44) (Fig. 4.161) or from analytical consideration:

$$\begin{aligned} F(p^*) &> F(p^* - 1) \\ F(p^*) &> F(p^* + 1) \end{aligned} \tag{4.47}$$

One can see that the peak position is shifted from $p^* = S$ to $p^* = S - 1$.

From the practical point of view the application of Eq. (4.44) instead of Eq. (4.45) gives us the more correct and more precise fitting of experimental results. In turn, the knowledge of Huang–Rhys–Pekar factor allows us to determine the Stokes shift:

$$\Delta S = (2S - 1) \cdot \hbar\omega \tag{4.48}$$

The temperature variation of the FWHM can be described by using the configuration coordinate model (Fig. 4.159) and the Boltzmann distribution according to the following equation proposed by Henderson [406]:

$$\mathrm{FWHM}(T) = \sqrt{8 \cdot \mathrm{In}2} \cdot \hbar\omega \cdot \sqrt{S} \cdot \sqrt{\coth\frac{\hbar\omega}{2kT}} \tag{4.49}$$

where $\hbar\omega$ is the mean phonon energy, S the Huang–Rhys–Pekar parameter, and k the Boltzmann constant. This equation can be used only when the parabolas of the ground state and the excited state have the same curvature. Combining Eq. (4.44) with (4.49) we can find the Huang–Rhys–Pekar parameter S and phonon energy $\hbar\omega$ for investigated materials and calculate the Stokes shift.

This method was applied for studying Ba, Ca, and Sr thiogallates prepared under the same synthesis conditions with the same activator concentration. 6 mol% of Eu ions was used, because this concentration shows the best emission in thiogallate powders. The proposed phosphors are usually used in LEDs at room and higher temperatures, and we have to take into account the temperature influence on transition probability (Eq. 4.44).

To determine the energy of an absorbed photon $\hbar\omega$ the Raman scattering spectra of $SrGa_2S_4$, $CaGa_2S_4$, and $BaGa_2S_4$ were measured. The main Raman peaks were given in Fig. 4.139.

Two dominant peaks at 279 cm^{-1} (34.5 meV) and at 357 cm^{-1} (44.2 meV) for $SrGa_2S_4$ and similar peaks at 284 cm^{-1} (35.2 meV) and 361 cm^{-1} (44.8 meV) for $CaGa_2S_4$ are registered. Only one intense vibration line at 302 cm^{-1} (37.5 meV) is found for $BaGa_2S_4$. (Fig. 4.139). These Raman spectra are in good agreement with calculated data from Eqs.

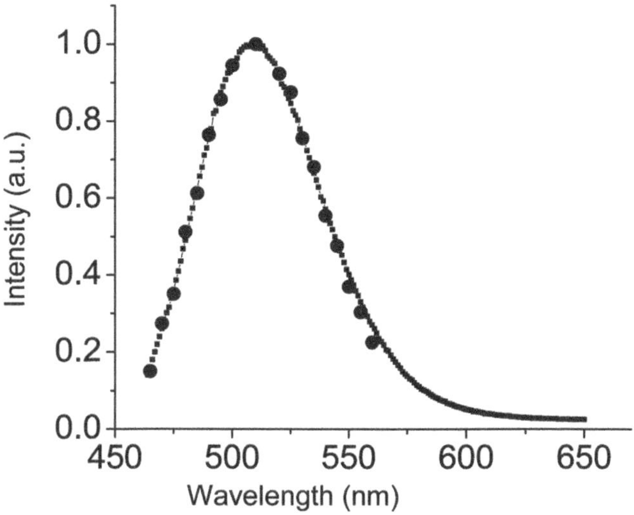

Figure 4.160 Experimental (broken curves) and calculated (circles) emission spectrum at 300 K of the Ba thiogallate doped with 6 mol% of Eu^{2+} under excitation at 460 nm. The fitting was carried out using Eq. (4.44): $\hbar\omega = 37.5$ meV, $S = 7$.

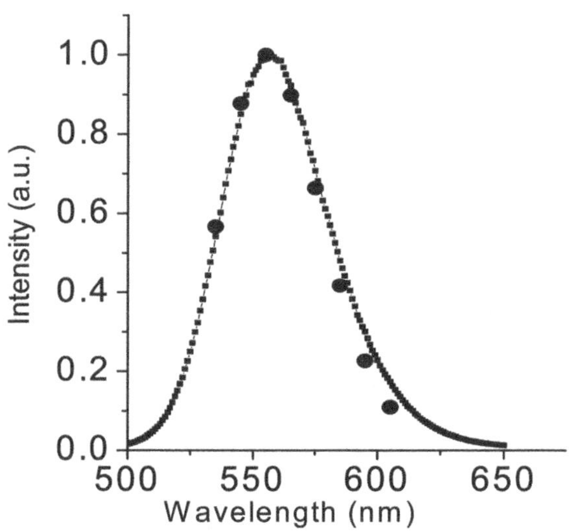

Figure 4.161 Experimental (broken curves) and calculated (circles) emission spectrum at 300 K of the Ca thiogallate doped with 6 mol% of Eu^{2+} under excitation at 460 nm. The fitting was carried out using Eq. (4.44): $\hbar\omega = 35.5$ meV, $S = 4$.

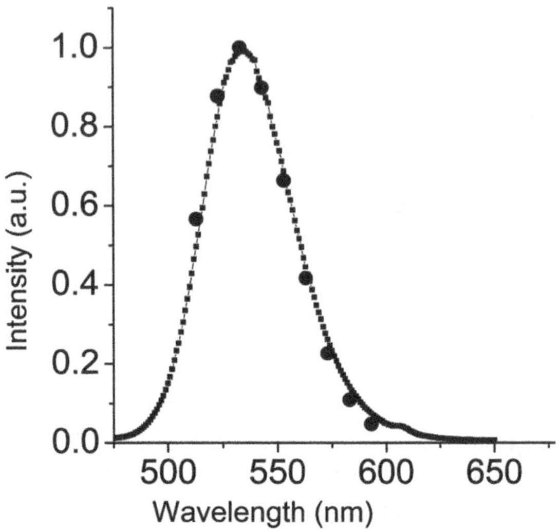

Figure 4.162 Experimental (broken curves) and calculated (circles) emission spectrum at 300 K of the Sr thiogallate doped with 6 mol% of Eu^{2+} under excitation at 460 nm. The fitting was carried out using Eq. (4.44): $\hbar\omega = 34.5$ meV, $S = 4$.

4.44 and 4.49, and those reported in the literature [385, 395]. These phonon energies could be used in fits of the emission bands.

The emission spectra for $BaGa_2S_4$, $CaGa_2S_4$, and $SrGa_2S_4$ phosphors are presented in Figs. 4.160–4.162, respectively.

The best fits of the emission bands for $BaGa_2S_4:Eu^{2+}$, $CaGa_2S_4:Eu^{2+}$ and $SrGa_2S_4:Eu^{2+}$ were obtained with Huang–Rhys parameter $S = 7$, 4, 4 correspondingly and shown in Figs. 4.162–4.164. The Stokes shift was calculated by Eq. (4.48). At room temperature we found $\Delta S = (3927 \pm 100)$ cm^{-1} for $BaGa_2S_4:Eu^{2+}$, $\Delta S = (1975 \pm 100)$ cm^{-1} for $CaGa_2S_4:Eu^{2+}$, and $\Delta S = (1919 \pm 100)$ cm^{-1} for $SrGa_2S_4:Eu^{2+}$.

These data were compared with experimentally measured excitation and emission spectra. The main problem at room and higher temperature is to determine correctly the λ_{abs}, because usually this wavelength is not resolved at room temperature. An almost mirror image relationship seems to hold between the emission and the excitation spectra in the energy region between the two maxima for all investigated thiogallates (Figs. 4.163–4.165).

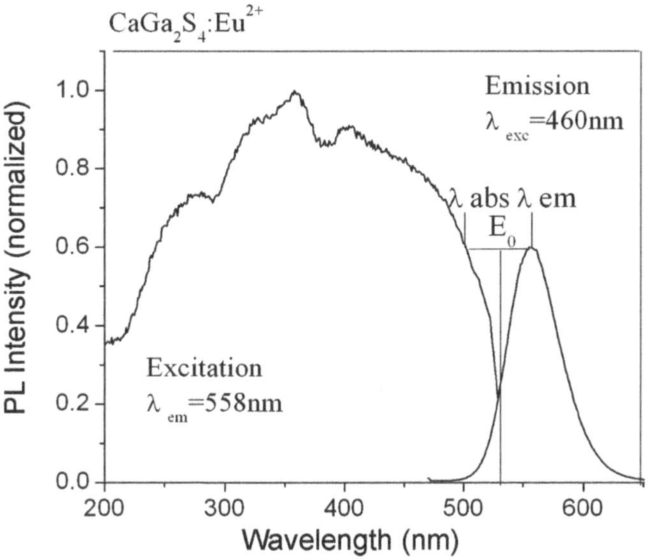

Figure 4.163 Excitation and emission spectra for $CaGa_2S_4:Eu^{2+}$ phosphors at 300 K.

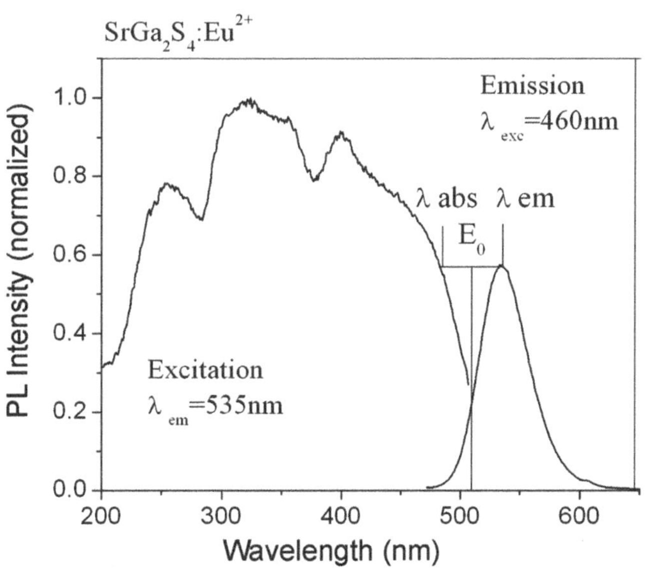

Figure 4.164 Excitation and emission spectra for $SrGa_2S_4:Eu^{2+}$ phosphors at 300 K.

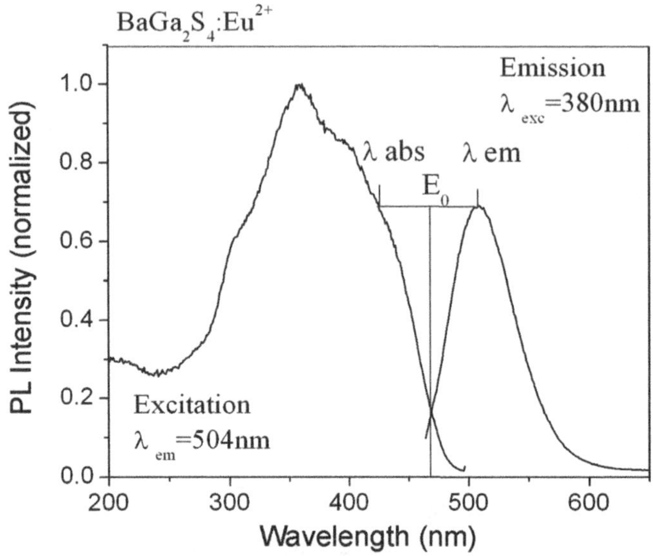

Figure 4.165 Excitation and emission spectra for $BaGa_2S_4:Eu^{2+}$ phosphors at 300 K.

The excitation spectra consist of extremely broad bands which extend well into the visible region and overlap the emission bands to a considerable extent. The excitation bands in the visible range are due to Eu^{2+} absorption since the unactivated host compounds do not absorb in this region. As shown in Figs. 4.163–4.165, the emission bands depend on the choice of host cation and they have peaks in the blue-green-yellow regions of the spectrum. The isostructural calcium and strontium thiogallates form a complete series of solid solutions whose spectral energy distribution vary uniformly from the green of $Sr_{1-x}Eu_xGa_2S_4$ ($\lambda = 535$ nm) to the yellow of $Ca_{1-x}Eu_xGa_2S_4$ ($\lambda = 558$ nm). Solid solutions in the barium-strontium system are limited. The radiant efficiency of the Eu^{2+} activated thiogallates decreases in order Ca > Sr > Ba, but their stability decreases in the opposite order: Ba > Sr > Ca.

The mirror image relationship in Figs.4.163–4.165 is characteristic of a phonon-broadened emission and confirms the hypothesis that the mean phonon energy $\hbar\omega$ is the same for the $4f^7$ ($^8S_{7/2}$) ground state as for the $4f^6$ (^7F)$5d$ excited state. We can determine the energy of the zero-phonon line E_0 at the intersection of the emission and ex-

Table.4.35 Luminescence data for Ba, Sr, Ca thiogallate phosphors doped with 6 mol% of Eu^{2+}. The spectra measurements were carried out at 300 K.

Phosphor		λ_{abs}	E_0	λ_{em}	FWHM	ΔS_{math}	ΔS_{exp}
$BaGa_2S_4:Eu^{2+}$	nm	424	464	504	62	–	80
	eV	2.92	2.67	2.46	–	0.49	0.46
	cm^{-1}	23584	21551	19841	2450	3927 ± 100	3743 ± 200
$SrGa_2S_4:Eu^{2+}$	nm	482	508	535	49	–	53
	eV	2.57	2.44	2.31	–	0.24	0.26
	cm^{-1}	20746	1968	18691	1600	1919 ± 100	2055 ± 200
$CaGa_2S_4:Eu^{2+}$	nm	500	529	558	50	–	58
	eV	2.48	2.34	2.22	–	0.25	0.26
	cm^{-1}	20000	18903	17921	1620	1975 ± 100	2079 ± 200

citation spectra. As an example, for $BaGa_2S_4:Eu^{2+}$ zero-phonon line E_0 and corresponding wavelengths were determined: $E_0 = 464$ nm, $\lambda_{abs} = 424$ nm (23584 cm^{-1}) and $\lambda_{em} = 504$ nm (19841cm^{-1}) (Fig. 4.165). The Stokes shift was calculated as $\Delta S = \lambda_{abs} - \lambda_{em} = (3927 \pm 100)$ cm^{-1}. The calculated absorption and emission energies and related Stokes shift agree well with the experimental results.

The similar analysis was carried out for other thiogallate phosphors. Summarized luminescence data for synthesized phosphors are given in the Table 4.35.

One can see that in all considered systems we are dealing with case of strong vibronic interaction giving rise to the broad structureless optical bands. The small difference in the Stokes shift between $SrGa_2S_4:Eu^{2+}$ and $CaGa_2S_4:Eu^{2+}$ can be explained as the Eu ion is situated in the center of decahedron composed from eight sulfur atoms in $SrGa_2S_4$ as well as in $CaGa_2S_4$. In the both cases the Eu ion is surrounded by theses sulfur atoms and the circumstance is similar. This is the main reason why we cannot find any essential difference of the energy levels and the Stokes shift.

The calculated values of the Stokes shift are in good agreement with experimental PL results at 300 K and differ from those presented in literature at low-temperature measurements [381, 382, 402]. It means that the temperature dependence is essential and must be taken into account.

4.5.5 *Concluding Remarks*

To summarize the results of this section the following is to be mentioned.

The luminescence properties in barium, strontium and calcium thiogallate phosphors were revisited. This study shows the comparison of the properties of ternary sulfide compounds prepared at the same synthesis conditions with the same 6 mol% of Eu^{2+} concentration. The luminescence and crystallographic data are summarized and listed in Tables 4.28, 4.29, 4.33, and 4.35. These data will be useful to evaluate the quality of the powders prepared for different electronic devices

A new green multiphase phosphor [$SrGa_2S_4$+$MgGa_2O_4$]: Eu^{2+} with improved luminescence properties was synthesized. Experimental data and detailed analysis show that the luminescence intensity of this new compound is about 15% higher than that of the best commercial strontium thiogallate. The FWHM of the emission band is also narrower while the luminescence peak is shifted from 535 nm to 539 nm.

A model of $4f \rightarrow 5d$ optical transitions, which takes into account the electron-vibrational interaction, was discussed. It was shown that the temperature influence on the fitting of experimental emission spectra becomes significant at room and higher temperatures. The analysis of the experimental results with proposed model can serve as an efficient tool for better understanding the luminescent mechanism in rare earth activated phosphors. At the same time the data so far obtained do not allow to analyze the contribution of the Jahn–Teller coupling due to the fact the optical bands are represented by the broad structureless distributions. Under this condition the inclusion of the additional fitting parameters would lead to an excessive flexibility in the fitting procedure. For this reason we have restricted our consideration to a simple basic model of vibronic interaction. More detailed study of the polarization dichroism effects [407] is expected to reveal the character of Jahn–Teller coupling in *d*-shell.

Presented results provide a deeper knowledge of the Eu luminescence in barium, strontium, and calcium thiogallates and confirm that these materials are good candidates for solid state lighting, LED and display devices.

4.6 Multiexcited Phosphors for LED Application

4.6.1 *Introductory Remarks*

Since the commercial blue light emitting diodes was first reported by Nichia chemical Co. in 1994, much progress has been achieved on GaN-based LEDs [344, 420–424]. Blue and near-UV light GaN-based LEDs can be used to excite longer wavelength emitting phosphor to realize white light emission. Nakamura pioneered this approach at Nichia Chemical Co. and the first white light emitting diodes became commercially available in 1997 [425–427].

White light-emitting diodes, with the advantages of long lifetime, saving energy consumption, and environmental-friendly characteristics, are thought to be the most important solid state light sources for substituting the widely used incandescent lamps and fluorescent lamps [428]. Great interest has been focused on these devices and tremendous progress has been achieved since the first commercial white-emitting LED was fabricated with about 450 nm blue-emitting GaN-based LED in 1997 [426]. At present, the most common and convenient method to obtain the white light is combining a yellow-emitting phosphor, for example, trivalent cerium activated yttrium aluminum garnet (YAG:Ce^{3+}) wavelength converter or terbium aluminum garnet (TAG:Ce^{3+}) with a GaN blue LED chip.

However, this light combination has exposed some drawbacks: firstly, the degradation rates of LEDs and phosphor are not isochronous, thus the white light point in the color coordinates scheme shifts with working time. Secondly, the "white" light from this combination route has an undesirable color balance, the deficiency of the red light in the visible spectra (around 600 nm) results in a low color rendering index [429]. So a novel red-emitting phosphor for compensating the red component is needed in order to obtain white light with higher color rendering index. To improve the color properties of the device there are two different ways. First of them is based on adding another component to create a white LED with three emission bands (a three-band white LED). Blue LEDs have been combined with phosphors to convert part of the blue light to red and green light, to produce white light (Fig. 4.166) [430].

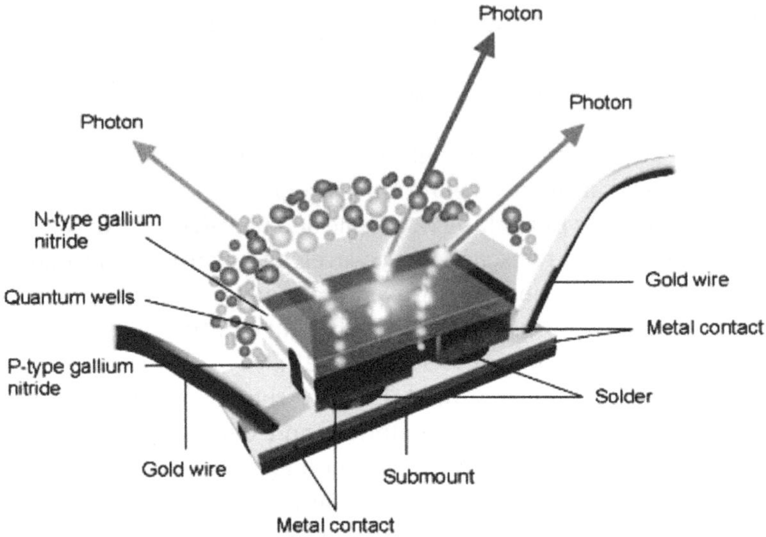

Figure 4.166 Artist's view of two phosphors converted LED: On a copper slug and underneath a plastic lens, a "flip-chip," is soldered to metal contacts; flip-chip meaning the substrate on which the stack of GaN and InGaN layers has been deposited is used as light exit, the (bottom) p-contacts being highly reflective. The color converting phosphors (original red and green) are placed on top of the chip, embedded in silicone. (Reprinted with permission from [430]. Copyright: 2005 WILEY-VCH.)

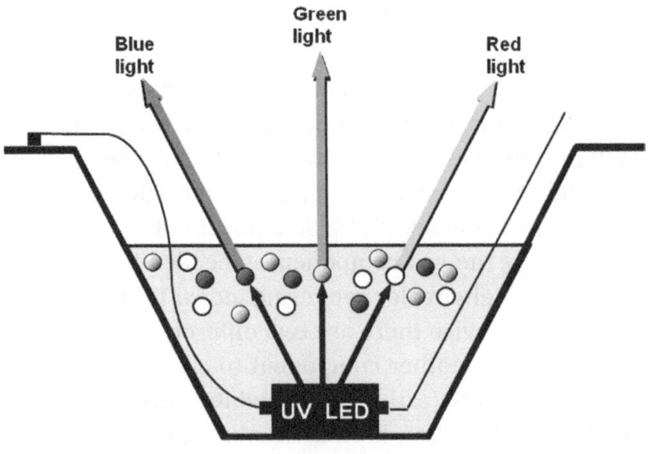

Figure 4.167 Three phosphors converted UV LED.

Another approach for obtaining white light is to combine a near-UV LED (around 370–410 nm) with three color phosphors. Fig. 4.167

The white light consists of the red, green, and blue light emitted from the three phosphors while excited by the near-UV LEDs [431, 432]. In this case, the float of white color point can be eliminated because all the visible lights come from the three color phosphors. In order to realize this type of white LEDs, novel phosphors are required, since most of commercially applicable phosphors were developed on different purposes of applications and do not meet the white LEDs applications. The phosphors for LED applications that needs to be excited effectively by the near-UV light is lacking, especially the red one.

To improve the color properties of the LED devices one new non traditional way – to excite the red phosphor not only by UV source, but also by the emission of blue phosphor (double excitation) as well as by green phosphor (triple excitation) is proposed in this section. Nowadays, the commercially applicable red emitting phosphors for white LED are divalent Eu ion activated sulfides. These sulfide phosphors are chemically unstable when exposed in the moisture environment. Therefore, there is an urgent demand upon the superior red emitting phosphors for white LEDs applications. Eu^{3+}-doped materials, especially in which the Eu^{3+} ions occupy a non-centrosymmetric site in the host, have been widely used as the red emitting phosphors due to their intense $^5D_0 \rightarrow {}^7F_2$ emission in the red spectral region. Previous investigations [429, 433] have showed that Eu^{3+}-doped phosphors exhibited relatively strong absorption in the near-UV region and intense red emission with good color purity. The search for stable, inorganic rare earth based red phosphors with high absorption in the near UV/blue spectral region is therefore an attractive research task. The requirement of a suitable red-emitting UV/blue phosphor is that it should have a stable host, exhibit strong and broad absorption around 400 nm (LED emission wavelength) and good absorption around 440–480 nm (emission of blue phosphor) with the chromaticity coordinates near the National Television Standard Committee (NTSC) values. The blue, green and red phosphors that satisfy these criteria were developed and synthesized on the basis of the recent patents [434–436] and are presented in this chapter. Some new ideas to use an additional scattering layer around the chip and special reflecting medium, consisting from powder with high reflectance and UV reflecting film are proposed.

4.6.2 *Requirements of Multiexcitation*

A luminescent material usually absorbs energy in one region of the electromagnetic spectrum and emits radiation energy in another region of the spectrum. Typically, the energy of the photons emitted is lower than the energy of the photons absorbed. Most useful phosphors emit radiation in the visible portion of the spectrum in response to the absorption of radiation, which is outside the visible portion of the spectrum. Thus, the phosphor converts electromagnetic radiation to which the human eye is not sensitive into electromagnetic radiation to which the human eye is sensitive.

However, it is very difficult to find suitable three phosphors equally excited by UV in the green, blue and red regions to obtain really a white light. Usually, the quantum yield of red phosphor is very weak in comparison with green and blue phosphors. The present work is directed to overcoming or at least reducing the problem set forth above.

To resolve this problem a white light illumination system comprising a radiation source and a special combination of three phosphors is proposed. Three phosphors have to meet the next requirements: a first luminescent material (blue phosphor) possesses a peak emission wavelength of about 420 to about 480 nm, a second luminescent material (green phosphor), has a peak emission wavelength of about 490 to about 550 nm and a peak absorption wavelength of about 350 to about 480 nm, which is different from the first luminescent material, and a third luminescent material (red phosphor), has a peak emission wavelength of about 580 to about 620 nm and a peak absorption wavelength of about 350 to about 550 nm, which is different from the first and second luminescent material. A first blue phosphor can be selected from at least one of

(a) $(Sr,Ca)_5(PO_4)_3Cl:Eu^{2+}$
(b) $La_{2.99}Ce_{0.01}(SiS_4)_2I$ [437]
(c) $Ce_3(SiS_4)_2X$ (X = Cl, Br, I) [437]

a second green phosphor can be selected from at least one of

(a) $SrGa_2S_4:Eu^{2+}$

(b) $(Ba\ Sr)_2SiO_4:Eu^{2+}$

(c) $Sr_2Si_3O_82SrCl_2:Eu^{2+}$ [8]

(d) $Sr_4Al_{14}O_{25}$: Eu^{2+} [8]

(e) $Y_2SiO_5:Ce^{3+},Tb^{3+}$ [8]

(f) $BaYSi_4N_7:Eu^{2+}$ [438]

and a third red phosphor:

(a) $K_5Eu_{2.5}(WO_4)_{6.25}$:Sm

(b) $Li(WO_4)_{1.25}$:Eu,Sm

(c) $Ca_{0.76}MoO_4:Eu_{0.24}{}^{3+}$ [429]

(d) $Ca(Eu_{0.5}La_{0.5})Si_3O_{13}$ [439]

(e) $Sr_xCa_{1-x}S:Eu^{2+}$,Cl

In such combination of proposed phosphors we can achieve a triple excitation of the red phosphors by UV LED, a first phosphor and a second phosphor. The second phosphor can be double excited by UV LED and first phosphor.

Some of these cited phosphors are used as commercial phosphors and others were synthesized by traditional or original methods.

The synthesis of $(Sr,Ca)_5(PO_4)_3Cl:Eu^{2+}$, $(Ba,Sr)_2SiO_4$: Eu^{2+}, $K_5Eu_{2.5}(WO_4)_{6.25}$:Sm, and $Li(WO_4)_{1.25}$:Eu,Sm phosphors was performed by traditional solid state reaction method. Polycrystalline $SrGa_2S_4:Eu^{2+}$ and some improved modification of thiogallate samples were also synthesized by solid state reaction, but without toxic H_2S. SrS and Ga_2S_3 sulfide powders mixed in stoichiometric composition and annealed at 900–1000°C with a carbon reduction atmosphere for 4 h (instead of toxic gas atmosphere). The doping ions were introduced in the form of EuS with 6 mol% Eu^{2+} concentration. This method provides powder samples with good crystalline properties as shown by X-ray diffraction measurements. The thiogallate samples exhibit a deep green color with peak maximum at 535 nm. This peak's maximum and the FWHM can be changed on a large scale.

The red emitting phosphor $Sr_xCa_{1-x}S:Eu^{2+}$ was prepared by solid state reaction by combining the starting materials: $Sr(NO_3)_2$, $Ca(NO_3)_2$·

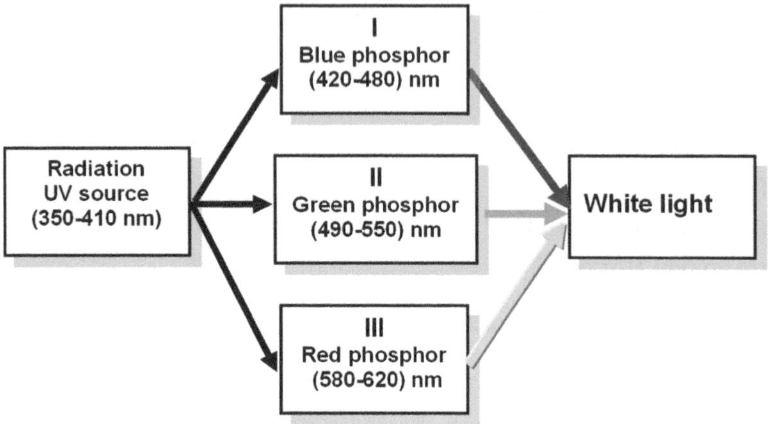

Figure 4.168 Classical UV LED scheme.

$4H_2O$, and Eu_2O_3. The element S of the phosphor is introduced from SrS/CaS. In alternative method we used CaS, SrS, and EuS as raw materials and additional amount of NH_4Cl was applied as a flux. The annealing was carried out at 1000°C about 3 h with a carbon reduction atmosphere. In this case the luminescence intensity of $Sr_xCa_{1-x}S:Eu^{2+},Cl$ was about 5–10% higher in comparison with $Sr_xCa_{1-x}S:Eu^{2+}$.

All phosphor samples were characterized by crystalline structure and luminescence properties. XRD measurements, EDS and EPMA qualitative and quantitative non-destructive elemental analysis, optical spectroscopy and PL characterization, morphology and size measurements were carried out. All these techniques are described in detail in Chapter 3.

Classical UV LED scheme of a white light illumination system is presented in Fig. 4. 168.

Disadvantage of this scheme is in very weak quantum yield of red phosphor in comparison with green and blue phosphors. To increase the quantum yield of red phosphor, the double and triple excitation of red phosphor is proposed.

4.6.3 *Double Excitation in UV LED*

A schematic illustration of double excitation of red phosphor is shown in Fig. 4.169.

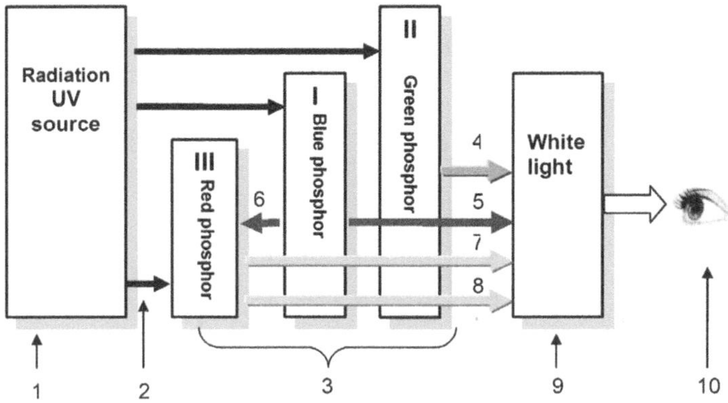

Figure 4.169 Schematic illustration of a white light illumination system with a double excitation of red (III) phosphor.

In Fig. 4.169, a radiation source (1), for example, LED emits radiation (2) incident on three luminescent materials (3), such as a first phosphor I, second phosphor II, and third phosphor III. These three phosphors I, II, and III may be blended together or separated and may comprise discrete overlying layers, as shown in Fig. 4.169.

The radiation (2) may have a wavelength to which the human eye is not sensitive, such as 420 nm and below. After absorbing the incident radiation (2), the second phosphor II emits green light (4) having a peak emission wavelength between 490 and 550 nm. The first phosphor I emits blue light (5) and (6) having a peak emission wavelength between 420 and 480 nm, while the third phosphor III emits red light (7) and (8) having a peak emission wavelength between 580 and 620 nm and a peak absorption wavelength of about 350 to about 480 nm. The red light (7) is the light emitting by the third phosphor III due to absorption of the radiation (6) of the second phosphor II and light (8) is due to absorbing the incident radiation of the LED. The human observer (10) perceives the combination of green (4), blue (5) and red (7, 8) light as white light (9). Figure 4.169 schematically shows that green light, blue light, and red light emanates from discrete phosphor areas to illustrate the concept of color mixing. However, it should be understood that the green, blue, and red light may be emitted from the same area if the first, second, and

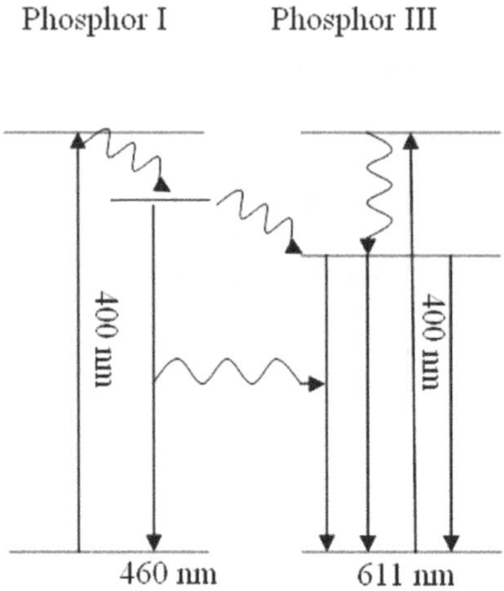

Figure 4.170 Energy transfer between the first (blue) and third (red) phosphors.

third phosphors are blended together to form a single blended phosphor layer.

To realize in practice this double excitation of red phosphor by UV source and blue phosphor let us consider the energy transfer between the first (blue) and third (red) phosphors, presented in Fig. 4.170.

The first blue phosphor I is excited by UV source (400 nm, for example) and emits a blue light at 460 nm. If the third red phosphor III is capable to absorb this wavelength as well as 400 nm from UV source, it will be double excited. Quantum yield and brightness will increase. All proposed phosphors from above mentioned list fit these criteria. Two high efficient candidates as phosphor III may be $Ca_{0.76}MoO_4{:}Eu_{0.24}{}^{3+}$ and $K_5Eu_{2.5}(WO_4)_{6.25}{:}Sm$.

The excitation and absorption spectra of these phosphors are shown in Figs. 4.171 and 4.172.

From these figures one can see that both phosphors can be excited by UV LED (400 nm) and blue light (460 nm).

As example of blue phosphor we can take $La_{2.99}Ce_{0.01}(SiS_4)_2I$ and $Ce_3(SiS_4)_2X$ (X = Cl, Br, I). Their spectra are shown in Figs. 4.173 and 4.174. All these phosphors are good excited by UV source (400 nm) and emit a blue light with high intensity at 460 nm.

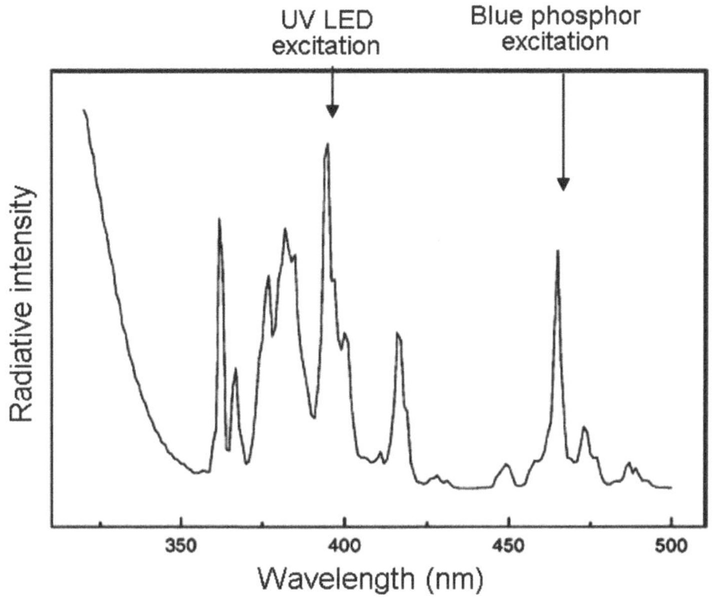

Figure 4.171 Excitation spectrum of $Ca_{0.76}MoO_4:Eu_{0.24}$ $^{3+}$ phosphor (monitored in wavelength 616 nm).

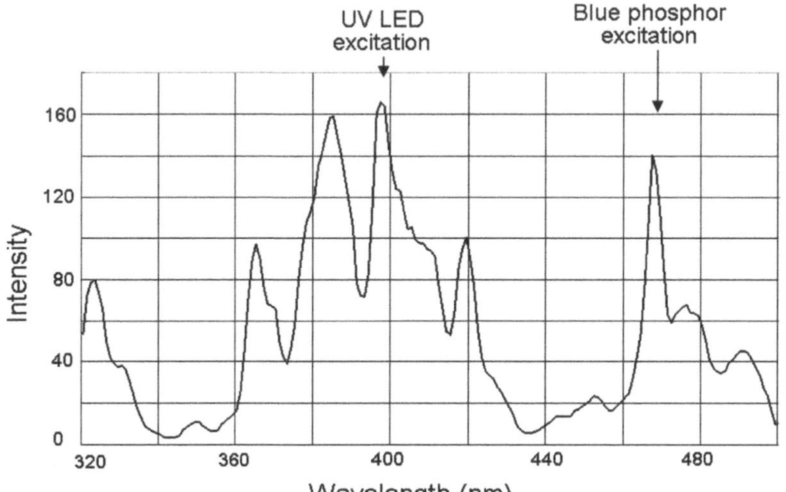

Figure 4.172 Absorption spectra of $K_5Eu_{2.5}(WO_4)_{6.25}$:Sm.

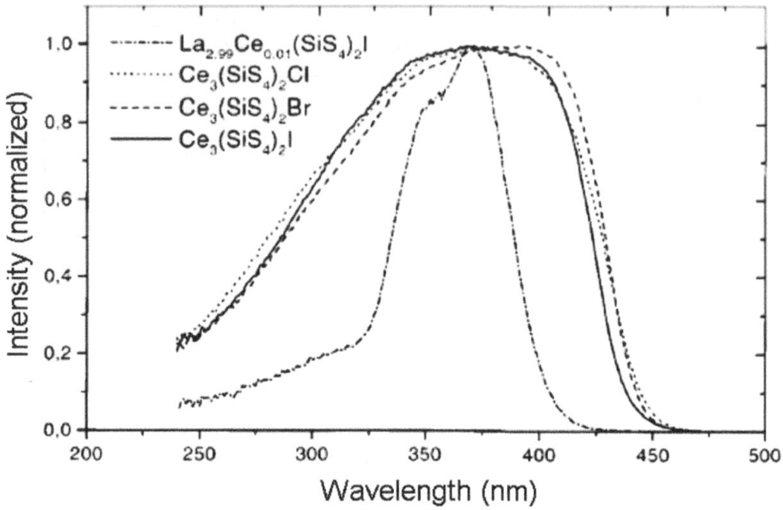

Figure 4.173 Excitation spectra at 300 K (λ_{em} = 500 nm) of La $_{2.99}$Ce $_{0.01}$ $(SiS_4)_2I$ and $Ce_3(SiS_4)_2X$ (X = Cl, Br, I).

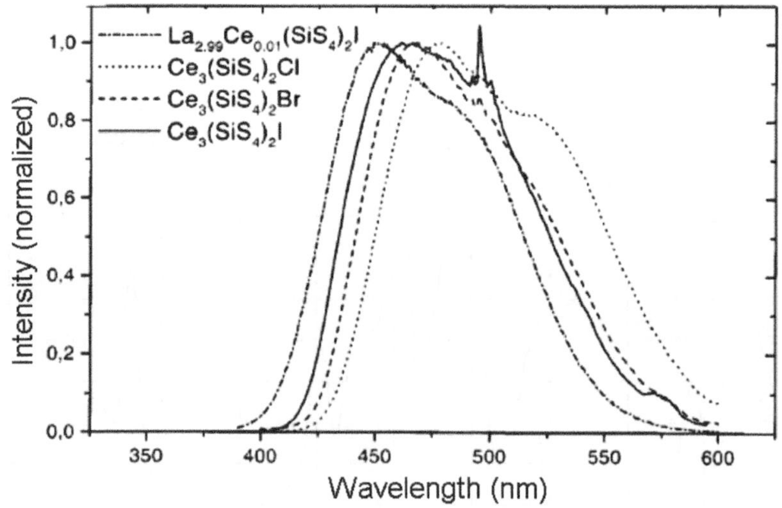

Figure 4.174 Emission spectra (λ_{exc} = 372 nm) of $La_{2.99}Ce_{0.01}(SiS_4)_2I$ and $Ce_3(SiS_4)_2X$ (X = Cl, Br, I) at 300 K. The lines around 495 nm are due to traces of praseodymium in cerium starting materials.

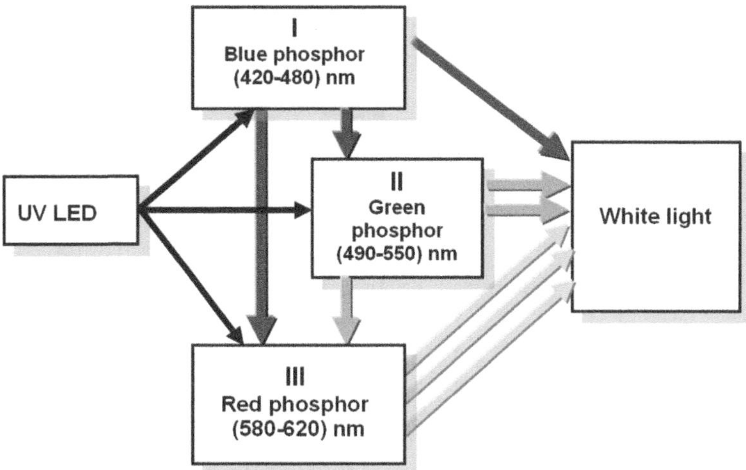

Figure 4.175 Schematic illustration of a white light illumination system with a double excitation of green phosphor and a triple excitation of red phosphor.

4.6.4 *Triple Excitation in UV LED*

The model of a white light illumination system with a double excitation of second (green) phosphor and a triple excitation of third (red) phosphor is shown in Fig. 4.175.

Figure 4.175 schematically illustrates a double excitation of green phosphor and a triple excitation of red phosphor. In Fig. 4.175, a radiation source (1), such as an LED, emits radiation (2) incident on three luminescent materials (3), such as a first (blue) phosphor I, second (green) phosphor II, and third (red) phosphor III. The principal difference of this model in comparison with Fig. 4.169 consists in important choice of green phosphor. It must be excited by UV LED and blue phosphor and excite the red phosphor. The energy transfer scheme between the first, second, and third phosphor is given in Fig. 4.176.

The first luminescent material may be the same as in previous case, for example, $La_{2.99}Ce_{0.01}(SiS_4)_2I$ or $Ce_3(SiS_4)_2X$ (X = Cl, Br, I). Its spectra are shown in Figs. 4.173 and 4.174.

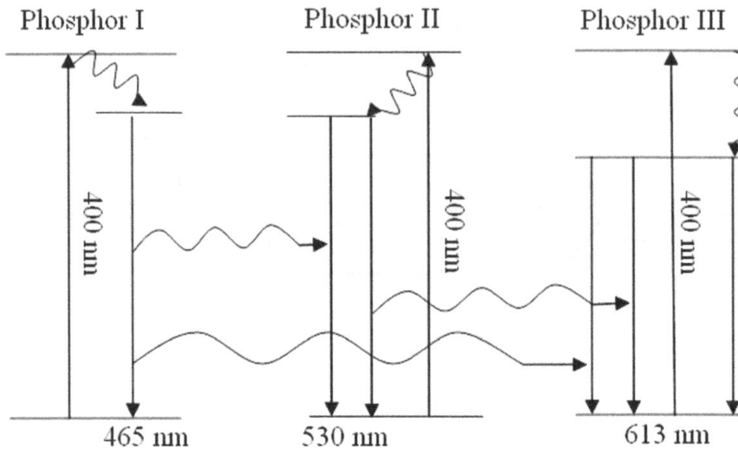

Figure 4.176 Energy transfer scheme between the first, second, and third phosphor.

The second luminescent material may be any phosphor II, which in response to the incident radiation from the radiation source emits visible light having a peak emission wavelength between 490 and 550 nm and a peak absorption wavelength of about 350 to about 480 nm. For example, the following phosphors fit these criteria:

(a) $SrGa_2S_4:Eu^{2+}$

(b) $(Ba\ Sr)_2SiO_4:Eu^{2+}$

The $SrGa_2S_4:Eu^{2+}$ phosphor is preferred because it has a quantum efficiency of at least 90% for incident radiation having a wavelength of 350–500 nm, has little or no selective absorption of the visible light and strong excitation at 465 nm. That's why the Sr thiogallate with improved properties was carefully prepared by modified chemical route. PL characteristics of $SrGa_2S_4:Eu^{2+}$ are shown previously in Fig. 4.141.

The high-efficiency red phosphor $Ca(Eu_{0.5}La_{0.5})Si_3O_{13}$ may be proposed as candidate of phosphor III with a triple excitation. Excitation and emission spectra of $Ca(Eu_{0.5}La_{0.5})Si_3O_{13}$ are shown in Fig. 4.177.

This phosphor may be tripling excited by the incident radiation (2) from the radiation source (1) and by incident radiation (6)

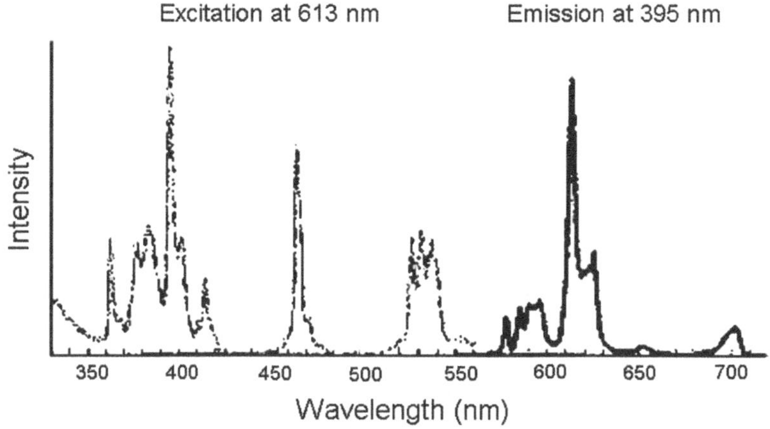

Figure 4.177 Excitation at 613 nm (left) and emission at 395 nm (right) spectra of $Ca(Eu_{0.5}La_{0.5})Si_3O_{13}$ phosphor (monitored in wavelength 613 nm).

from the first phosphor I and radiation (8) from the second phosphor II. (Figs. 4.175, 4.176, and 4.177). At that quantum yield of phosphor III, brightness of device and color rendering index increase.

4.6.5 *Double Excitation in Blue LED*

The white light in blue LED is obtained usually by combining a 465 nm blue light from the GaN-based LED and yellow light from the phosphor YAG:Ce. However, the "white" light from this combination route has an undesirable color balance, the deficiency of the red light in the visible spectra (around 600 nm) results in a low color rendering index. Another approach for obtaining white light is to combine a blue LED (around 460 nm) with green and red phosphors. The white light consists of the blue light of the chip and red and green light, emitted from the two-color phosphors while excited by the blue LEDs. So, a novel red emitting phosphor with double excitation by blue LED and by green phosphor for compensating the red component is needed in order to obtain white light with higher color rendering index.

The same scheme, proposed in Fig. 4.169, can be used for a white light illumination system with a double excitation of red phosphor.

The $SrGa_2S_4:Eu^{2+}$ as a green phosphor is preferred in this case. This phosphor is very good excited as by UV (400 nm) as well as by blue light (460 nm) (Fig. 4.141, excitation spectra). Moreover, it has highest possible quantum efficiency about 90% for incident radiation having a wavelength of 400 to 480 nm and has little or no selective absorption of the visible light.

The red phosphor used in this scheme is capable simultaneously to absorb one part of blue light from the diode at 460 nm, and an other part of green light from green phosphor ($SrGa_2S_4:Eu^{2+}$, for example). Figure 4.178 schematically illustrates the above principle.

One of the most efficient candidates as red phosphor may be proposed $Sr_xCa_{1-x}S:Eu^{2+}$,Y (wherein x is a number of from about 0.3 to 0.8, and Y is one or more halogens, in either their atomic or ionic forms). This proposed phosphor may be double excited by the incident radiation from the radiation source and by incident radiation from the green phosphor. Quantum yield and brightness will increase. The synthesis route and luminescence properties of $Sr_xCa_{1-x}S:Eu^{2+}$,Cl red phosphor were examined in detail in previous works [440, 441]. The excitation and emission spectra of $(Ca_{1-x},Sr_x)S:Eu^{2+}$ with a different Sr/Ca ratio, satisfying the above principle, are shown in Fig. 4.179.

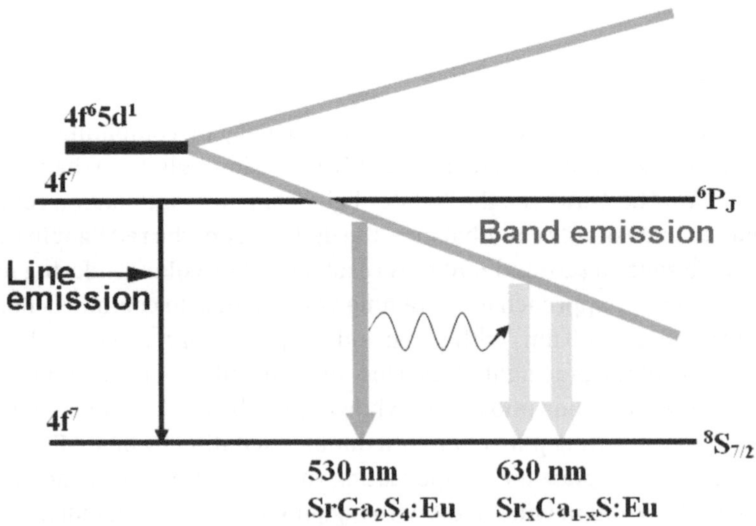

Figure 4.178 Energy transfer between the first (green) and second (red) phosphor.

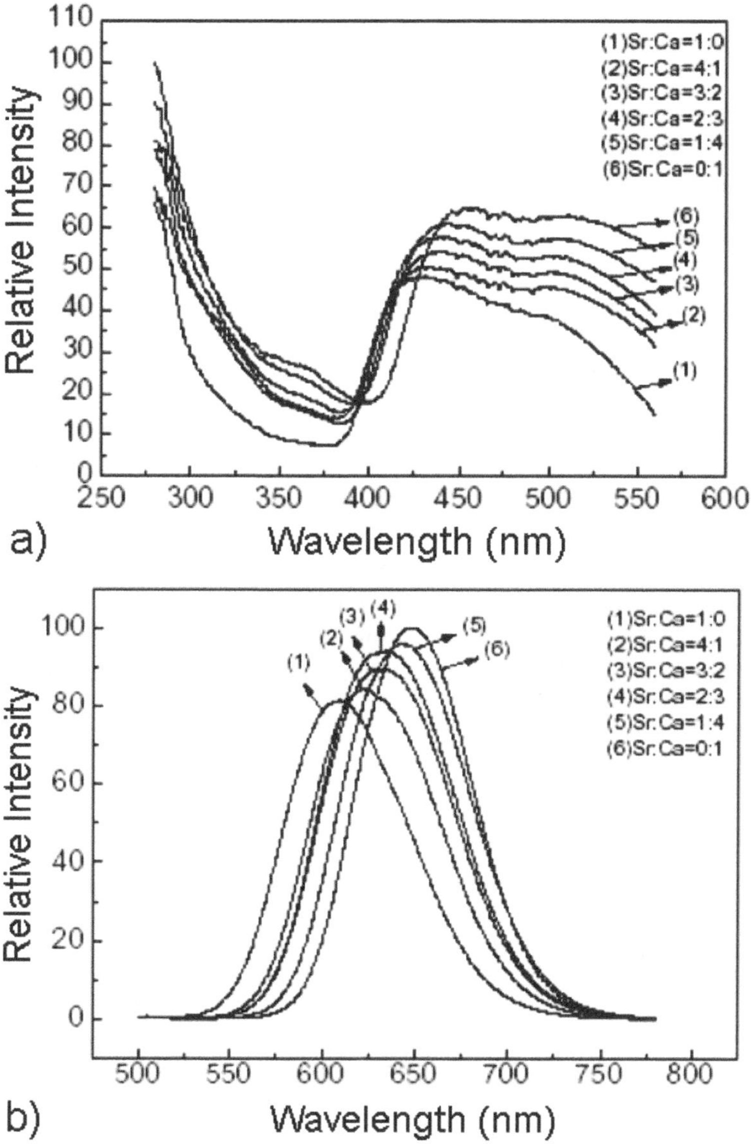

Figure 4.179 The excitation (a) and emission (b) spectra of $(Ca_{1-x},Sr_x)S:Eu^{2+}$ with a different Sr/Ca ratio (excitation made by 460nm light source).

The proposed multiexcited phosphors were combined in a LED device with an nUV Ga(In)N chip and blue InGaN chip and tested. For this purpose, a 395 nm UV Ga(In)N chip and 460 nm blue InGaN chip were used to fabricate the white-color LEDs based on different combinations of multiexcited phosphors. Compared with previous results, these phosphors show in experiments higher emission intensity.

4.6.6 *LED with Scattering Layer and Reflecting Medium*

4.6.6.1 UV LED

A schematic view of a scattering layer-coated LED with reflecting medium is presented in Fig. 4.180. In this figure is shown a light-emitting device (1), which comprises a UV-emitting LED (2), covered with a special scattering layer (3), a phosphor composition (4), comprising a phosphor blend containing green-emitting phosphors, blue-emitting phosphors, and red-emitting phosphors, being excitable by UV-emitting LED, the reflecting medium, consisting of powder (5) with high reflectance and/or UV reflecting film (6)

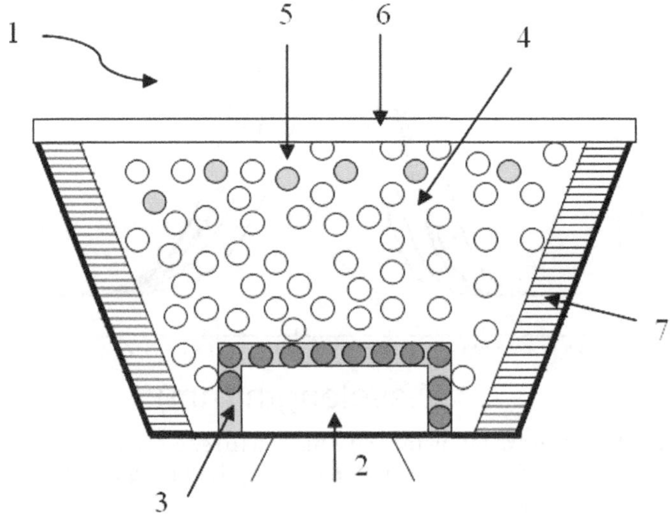

Figure 4.180 A schematic view in cross-section of a scattering layer-coated LED with reflecting medium.

and diffusive or specular reflector cup (7), reflecting light from the shell.

GaN and InGaN UV-emitting LEDs (2) are generally known in the art. UV-emitting LED has an emission peak at 360–420, optionally about 380–400 nm. Usually, a combination of three phosphors – blue, green, and red, excited by the UV emission from LED – is applied to generate a white light.

The different possibilities of phosphor packaging around the chip are shown in Fig. 4.181.

The best results from numerous experiments have been achieved in the case of uniform distribution of phosphors in reflector cup (b). To improve the color properties of the LED devices and the quantum yield of phosphor blend some new solutions are proposed: (a)

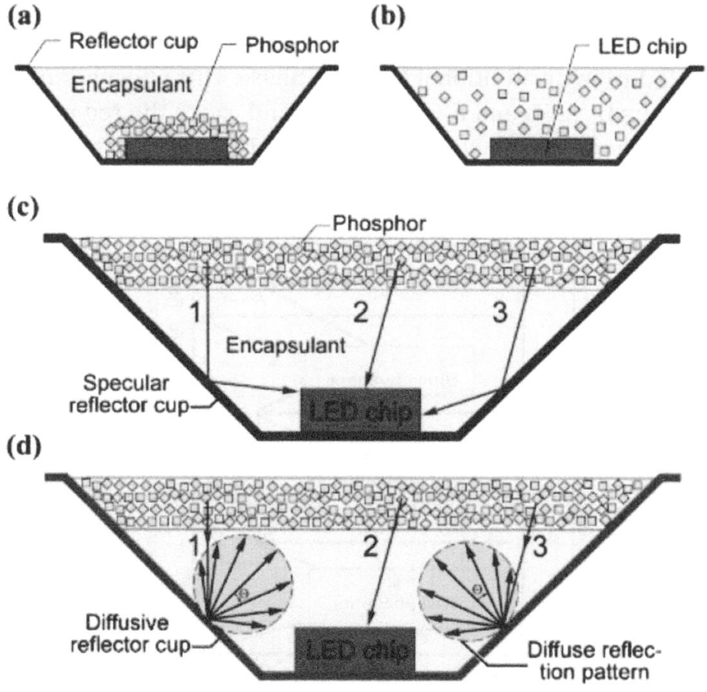

Figure 4.181 Arrangements of phosphor in white LED: (a) Conformal distribution directly on LED chip. (b) Uniform distribution in reflector cup (phosphor-in-cup). (c) Uniform distribution thin layer above LED chip (remote phosphor). (d) Remote phosphor distribution in diffuse reflector cup.

multiexcitation of phosphors; (b) scattering layer around the chip; (c) reflecting medium, consisting from powder with high reflectance and/or UV reflecting film. This idea was described in detail in some patents [442–444] and realized in practice.

Figure 4.182 schematically illustrates a double excitation of green phosphor (II) and a triple excitation of red phosphor (III). The multiexcitation model of phosphors was described above, but we think it is useful to repeat some details in combination with reflecting medium. A modified green phosphor is recommended also in this scheme.

In Fig. 4.182 a radiation source UV LED emits radiation incident on three luminescent materials, such as a first (blue) phosphor I, second (green) phosphor II, and third (red) phosphor III. The main principal difference of this model in comparison with traditional LED consists in important choice of green phosphor. The best candidate of green phosphor was synthesized on the basis of strontium thiogallate $SrGa_2S_4:Eu^{2+}$ with additional $MgGa_2O_4$ phase. This phosphor must be excited by UV LED and blue phosphor and excite the red phosphor. For higher quantum yield of red phosphor and for better color

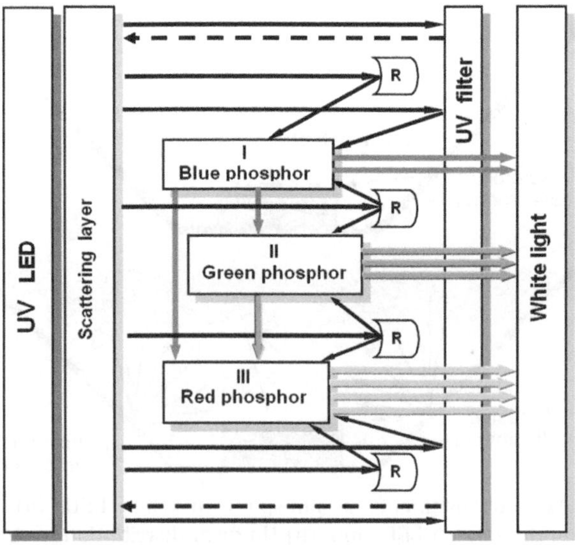

Figure 4.182 Proposed scheme of a white light illumination system with a reflecting medium and a multiexcitation of phosphors.

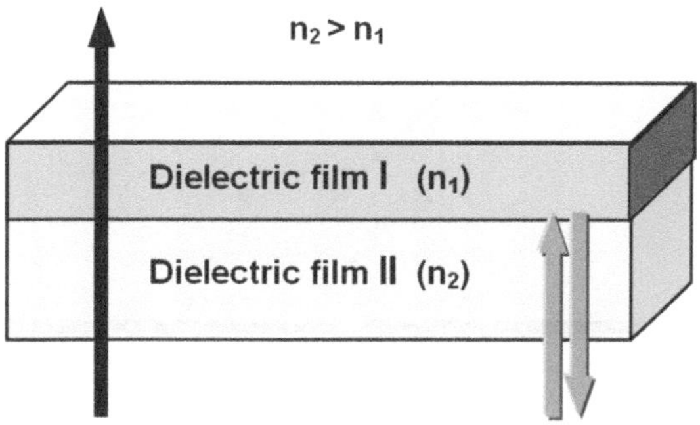

Figure 4.183 A schematic view of a multilayer UV reflecting filter.

rendering index and brightness of device it is desirable to carefully select a combination of red, green and blue phosphors capable absorb UV and visible blue-green range of spectra. The detailed analysis of these phosphors and their possible combinations were given in sited article [445].

A reflecting powder medium (R) comprises at least one powder selected from the group of MgO, $BaSO_4$, Al_2O_3, ZrO_2 , TiO_2, or any other with high reflecting index. When the incident UV light falls into these particles, it is reflected and excites again one of the phosphors.

The other type of reflecting medium may be a UV reflecting film. Multilayer dielectric film (band pass filter) serves for total UV reflectance and maximum visible light transmission. The scheme of such filter is shown in Fig. 4.183.

This filter covers the outside optical surface of the device and is placed over the phosphors. The multilayer film consists from alternate layers with high refractive index (about 1.82–2.5) and low (about 1.35–1.45) refractive index. Usually, TiO_2 ($n = 2.38$), ZrO_2 ($n = 1.99$), MgF_2 ($n = 1.38$), and LiF_2 ($n = 1.39$) are applied. The number of layers is about 15–40 and the thickness of one layer is 50–500 nm. Using such filter, the intensity of LED can be increased essentially, because reflected UV light excites repeatedly the phosphor blend.

To increase a quantum yield of LED and to increase, as consequence, the total brightness and light intensity of device, a UV-emitting LED is covered by a scattering layer. With reference to Figs. 4.184–4.186, there is shown a detailed scattering layers coated LED

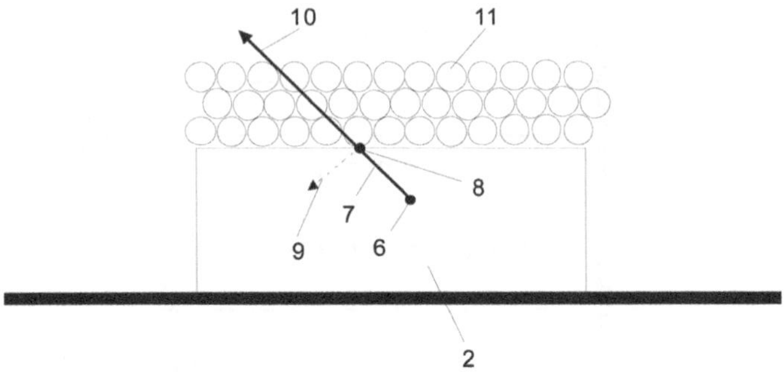

Figure 4.184 A schematic view in cross-section of a scattering layer, which coats the LED and consists of single material coating with identical particle size (11).

with a single material coating with identical particle sizes (Fig. 4.184), single material coating with different particle sizes (Fig. 4.185) and different materials coating with different particle sizes (Fig. 4.186).

Usually, a refractive index of chip substrate exceeds one of an encapsulant. It leads to significant light losses because of the total internal reflections (TIR) (9) and Fresnel reflections (FR). Index mismatch between LED/Epoxy and LED/phosphors reduces extraction/out-coupling efficiency of LED light due to TIR and FR losses. Losses become more significant when phosphors are mixed with a lower index epoxy, further reducing effective index below 1.8. To increase the extraction and efficiency of emitting light a special

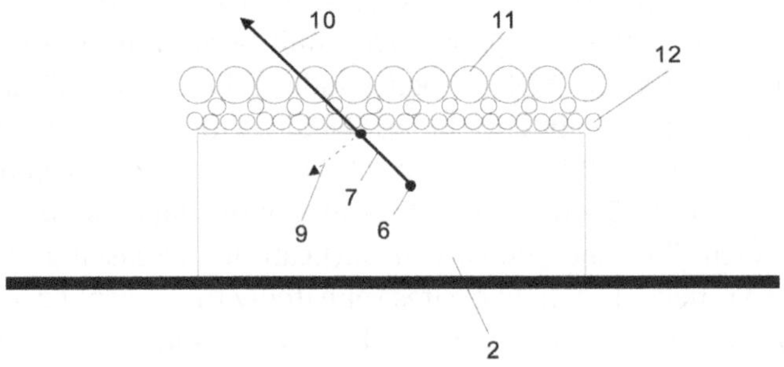

Figure 4.185 A schematic view in cross-section of a scattering layer, which coats the LED and consists of single material coating with different particle size (11 and 12).

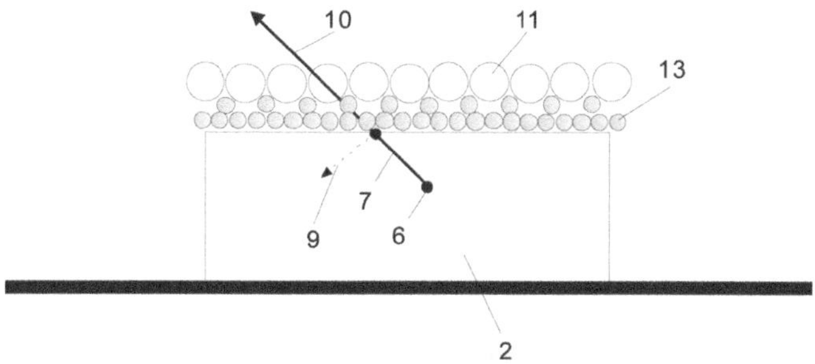

Figure 4.186 A schematic view in cross-section of a scattering layer, which coats the LED and consists of different materials coating with different particle size (11 and 13).

scattering cover layer around the chip is proposed. The main goal of this layer is a modification of the mechanism of TIR and symmetry of ray paths from the chip emitting areas. When the distance between particles in the scattering layer and the chip surface is comparable or less the wavelength of emitting light, the tunneling process of the light into the scattering particles appears. The light can go out from this layer (10) or reflect again into the chip. In this case the symmetry of light rays in the chip is changed because of not specular reflection, and the efficiency of emitted light increases. The refraction index in the scattering layer has to be comparable or more than the refraction index of the chip. The particle sizes have to be enough for the light scattering ($d_{\text{particles}} \geq \lambda_{\text{emission}}$) and optimized for the best tunneling process. The number of the particle monolayers in the scattering layer is optimized also and does not exceed usually 5 monolayers. A special additional subcoating layers (with particle size less than in the basic scattering layer and not more than a wavelength of emitting light) can be applied for better penetration of the light into the basic scattering layer (Figs.4.185 and 4.186).

The optimized parameters for the particle sizes and refraction index of the coating layer are presented in Figs. 4.187 and 4.188. The silicon encapsulant and subcoating particles with size 250 nm from phosphor with refraction index 1.82 were used in theoretical ray-tracing simulation.

As a result of multiexcitation of phosphors and scattering layer around the chip and applying the reflecting medium, the quantum

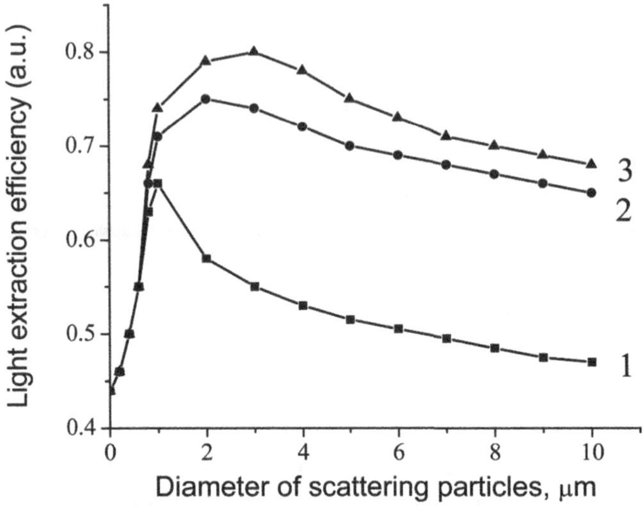

Figure 4.187 Efficiency of chip light extraction in dependence on particle size in the scattering layer. 1 – sapphire particles without subcoating, 2 – sapphire particles with sapphire subcoating, and 3– sapphire particles with additional phosphor subcoating.

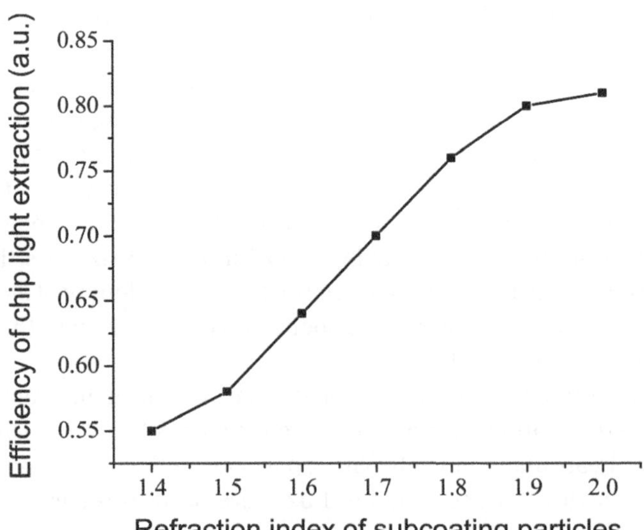

Figure 4.188 Efficiency of chip light extraction versus a refraction index of subcoating particles.

yield of green phosphor increases about 5%, the quantum yield of red phosphor increases about 15%, the total brightness and light intensity of device increases about 20%. The light, emitted from the chip and phosphors, has the same spatial and angular distribution in the output of device and CRI is improved about 15%.

4.6.6.2 Blue LED

The white light in blue LED is obtained usually by combining a 465 nm blue light from the GaN-based LED and yellow light from the phosphor YAG:Ce or TAG:Ce. As a result, a slightly bluish white light with a color rendering index of ~70 is produced that may, in principle, be improved by adding red photons from a complementary phosphor. The second generation of white LEDs is still based on blue InGaN LEDs but with a mix of green and red phosphors that increases the CRI up to 90 and makes the phosphor deposition process potentially easier. In fact, new red or orange phosphors are urgently needed for the development of efficient white LEDs, regardless of the type of LED used, and they need to be very bright in order to compensate for the low red sensitivity of the human eye. So, we can apply here the same method (as in UV LED) of double excitation of a red emitting phosphor by the blue LED and by the green phosphor in order to obtain white light with higher color rendering index. The detailed analysis of phosphor combination was made in previous article [445]. The scattering layer around the chip, considered above, and all conclusions for UV LED are also available for the blue LED.

In this paragraph we pay attention only to the influence of reflecting powder on the efficiency of LED device. The simplified scheme of a white light illumination system with a phosphor blend containing green-emitting phosphor and red-emitting phosphor and a reflecting medium is presented in Fig. 4.189. As in previous case, a reflecting powder medium comprises at least one powder selected from the group of MgO, $BaSO_4$, Al_2O_3, ZrO_2 , TiO_2, or any other with high reflecting index.

In Fig. 4.189, a radiation source (blue LED) emits radiation (2) incident on the blend of phosphors comprising a green-emitting phosphor (G), a red-emitting phosphor (R), and a reflecting medium

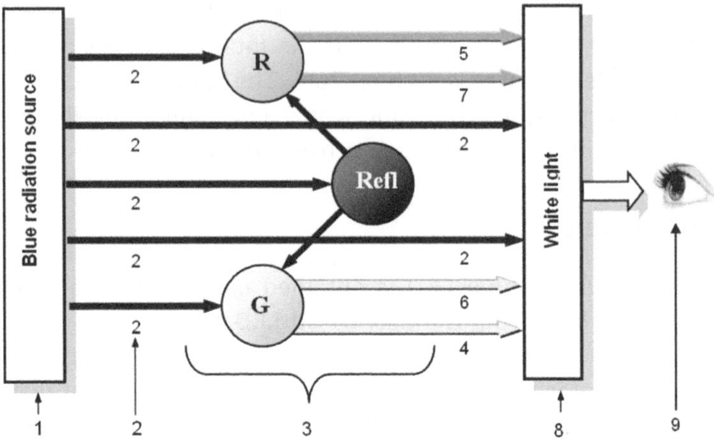

Figure 4.189 Schematic illustration of a white light illumination system with a phosphor blend comprising a reflecting medium. R – red phosphor, G – green phosphor, and Refl – reflecting powder (MgO, for example).

from powder samples with high reflectance. The human observer (9) perceives the combination of blue (2) and green (4, 6) and red (5, 7) light as white light (8). The two phosphors R and G and the reflecting powder may be blended together (3) or separated (as shown in Fig. 4.175) and may comprise discrete overlying layers. Most preferably, the reflecting medium powder is over the green-emitting phosphor and the red-emitting phosphor. In this case, the additional excitation from the reflecting medium could be absorbed by the green-emitting phosphor and the red-emitting phosphor and additional light (6) and (7) increases the total brightness of the device. At that, the CRI also increases. The purity of reflecting powders, exclusion of true absorption and extensive reflecting medium are the main factors that increase the total brightness. When the refractive index of reflecting medium is higher than environment index, the intensity also increases because of the total internal reflection.

The optimized parameters for the particle sizes for different reflecting powders are presented in Fig. 4.190. The silicon encapsulant and the most popular reflecting materials in powder form were used in theoretical ray-tracing simulation.

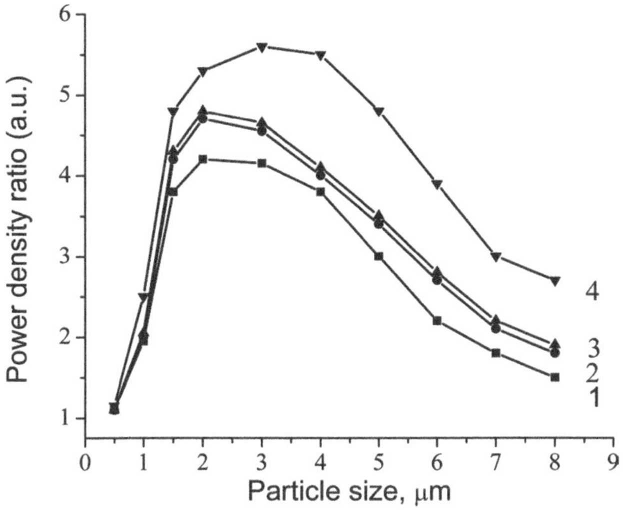

Figure 4.190 Efficiency of LED intensity in dependence on particle size and materials in the reflecting medium. 1 – $BaSO_4$, 2 – MgO, 3 – Sapphire, and 4 – ZrO_2.

For the most reflecting materials the optimal particle size is between 2 and 4 μm.

The ratio of reflecting powder amount and phosphor is very important in the blend composition. Their is an optimal concentration of reflecting powder amount and phosphors and it depends on reflectance and absorption in reflecting medium as well as geometry of chip packaging. The optimized results were achieved when the reflecting medium blend accounts about 50% of phosphors.

4.6.7 *Concluding Remarks*

A new idea of multiexcitation phosphors in LED with additional scattering layer around the chip and reflecting medium was proposed and discussed. These LEDs show better quantum yield of green and red phosphors, higher total brightness and improved color rendering index.

4.7 Other Luminescent Materials

4.7.1 *Luminescent Properties of Ca-α-SiAlON:Eu^{2+}*

White light-emitting diodes are a promising new-generation light source which can replace conventional incandescent and fluorescent lamps due to their reliability and low energy consumption [446–448]. At the present time, the commercial white LEDs comprising a blue LED chip and YAG:Ce^{3+} yellow phosphor are widely used as outdoor lighting sources [426]. However, these white LEDs are not applicable for indoor lighting purpose due to their less red luminescence and relatively lower color rendering index value (CRI ≈ 85) [42]. Improvement of CRI is possible by doping ions that can emit in orange/red region or by mixing of another orange/red phosphor [449, 450]. Nevertheless, a more creative solution lies in developing a new orange-yellow phosphor and combining it with a blue LED.

Inorganic phosphors based on cation doped oxynitrides are attracting much attention in recent years because of their non-toxicity, thermal stability, and interesting luminescence properties [451–453]. Among these oxynitrides, Ca^{2+} doped α-SiAlON has been regarded as one of the candidate host material for yellow or orange phosphors. The α-SiAlON, isostructural with α-Si_3N_4 has a solid solution composition $M_x^{v+}Si_{12-(m+n)}Al_{m+n}O_nN_{16-n}$ ($x = m/v$, x and v are solubility and valence of the M ion, respectively. M is one of the cations Li, Mg, Ca, Y, or some rare-earths) in which $m+n$(Si–N) bonds are replaced by m(Al–N), and n(Al–O) bonds [454].

The crystal structure of M-α-SiAlON has been widely studied by many research groups [455–458]. It is well known that in contrast to the purely substitutional solubility in β-SiAlON ($Si_{6-z}Al_zO_zN_{8-z}$), the α-SiAlON crystal structure is derived from α-$Si_{12}N_{16}$ by partial replacement of Si^{4+} by Al^{3+} and stabilized by modifying cations such as Li, Mg, Ca, or rare earths in large, closed interstitial sties of the (Si,Al)–(O,N) network [455], as shown in inset of Fig. 4.191. Izumi et al. [456, 457] have already reported highly accurate atomic position and occupation factors of Ca and rare earth cations by the Rietveld method. In M-α-SiAlON, the M cation is coordinated by seven (N,O) anions at three different M–(N,O) distances [458]. The size of the M cation site is dominated by the $(Si,Al)_{12}(O,N)_{16}$ network [458], as demonstrated by almost the same average M–(N,O) distances found in M-α-SiAlON [456–458], independent of the difference in ionic radius between Ca^{2+} (1.14 Å [459]) and the rare earth cation (Eu^{2+} = 1.31 Å [459]).

For the Ca-α-SiAlON based phosphors, much efforts have been given to verifying the effects of the activators or the host composition on their luminescent properties [460–462]. Ca-α-SiAlON:Eu^{2+} is thought to be a promising phosphor for high-temperature applications due to its thermal and chemical stability. However, its luminescent properties in elevated temperature ranges were not systematically studied. In this work, Eu^{2+} activated yellow and orange Ca-α-SiAlON phosphors which are suitable for application in white LEDs have been prepared by gas pressured sintering (GPS) method. The photoluminescence emission and excitation spectra for the prepared Ca-α-SiAlON:Eu^{2+} were observed in a temperature range from room temperature to 150°C, and their temperature dependence was analyzed.

Two compositions of Ca-α-SiAlON:Eu^{2+} phosphors were synthesized from α-Si_3N_4 (Ube Industries LTD., Japan), AlN (Tokuyama Corp., Japan), $CaCO_3$ (Kojundo Chemical Co., Ltd., Japan) and Eu_2O_3 (Shin-Etsu Chemical Co., Ltd., Japan) powders. The emission wavelength of Ca-α-SiAlON:Eu^{2+} was tailored by changing the concentration of Eu^{2+} or the composition of the α-SiAlON host with a small change of the emission intensity [463, 464]. In this work, the compositions of $Ca_{0.485}Eu_{0.015}Si_{9.2}Al_{2.8}O_{1.8}N_{14.2}$ and $Ca_{0.94}Eu_{0.06}Si_3Al_3ON_{15}$ were selected for the study of yellow and orange Ca-α-SiAlON:Eu^{2+} phosphor, respectively.

The raw powder mixtures were prepared using a planetary ball mill. After drying in a vacuum oven, the powder mixture was granulated using a test sieve, and then loaded into a boron nitride (BN) crucible. GPS was carried out at 1750°C for 2 h in 0.4 MPa of N_2 atmosphere. The crystalline phase of the synthesized powders was identified by X-ray powder diffraction (XRD, Rigaku, Japan), operating at 40 kV using Cu-Kα radiation (λ = 1.5406 Å). The data were collected in the continuous scan mode at the speed of 3° at 2θ/min with step size of 0.014° from 15 to 70°. The powder morphology was investigated by scanning electron microscopy (JEOL, JSM 5900 LV, Japan). The photoluminescence properties of the synthesized

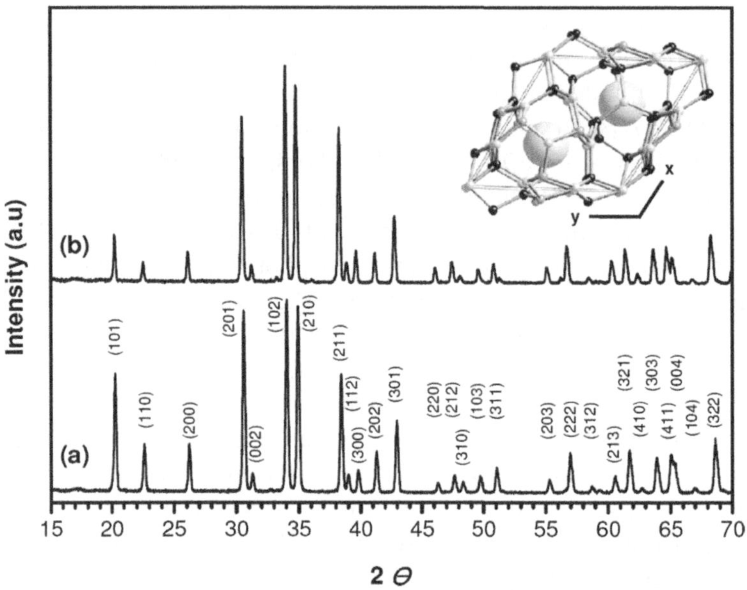

Figure 4.191 XRD patterns of (a) yellow and (b) orange Ca-α-SiAlON:Eu^{2+} phosphors prepared by GPS at 1750°C for 2 h under 0.4 MPa of N_2 atmosphere. The inserted figure shows the Ca-α-SiAlON structure viewed along a direction nearly parallel to the *c* axis. Modifying cations (Ca or luminescent ions) are represented by pink spheres. (Si,Al) atoms are shown as blue or red, and (N,O) atoms are shown as white spheres, respectively.

phosphor powders were measured using a spectrofluorometer (Fluorolog Tau-3, Horiba, USA) in a temperature range of 25 and 150°C with a 450 W xenon lamp as an excitation source.

Figure 4.191 shows XRD patterns of GPS Ca-α-SiAlON:Eu^{2+} phosphors for the compositions of (a) yellow and (b) orange. The peak positions of the XRD patterns in Fig. 4.191 confirm that the prepared samples have a single phase of Ca-α-SiAlON (JSPDS card No. 33-0261). No secondary phase such as β-SiAlON, $CaAlSiN_3$, or AlN-polytype was detected [465, 466].

Figure 4.192 shows SEM micrographs of (a) yellow and (b) orange Ca-α-SiAlON:Eu^{2+} phosphors. At high magnification, compared with those in the yellow phosphor, individual particles of orange phosphors are connected to each other more frequently. The differences in the particle morphology can be attributed to

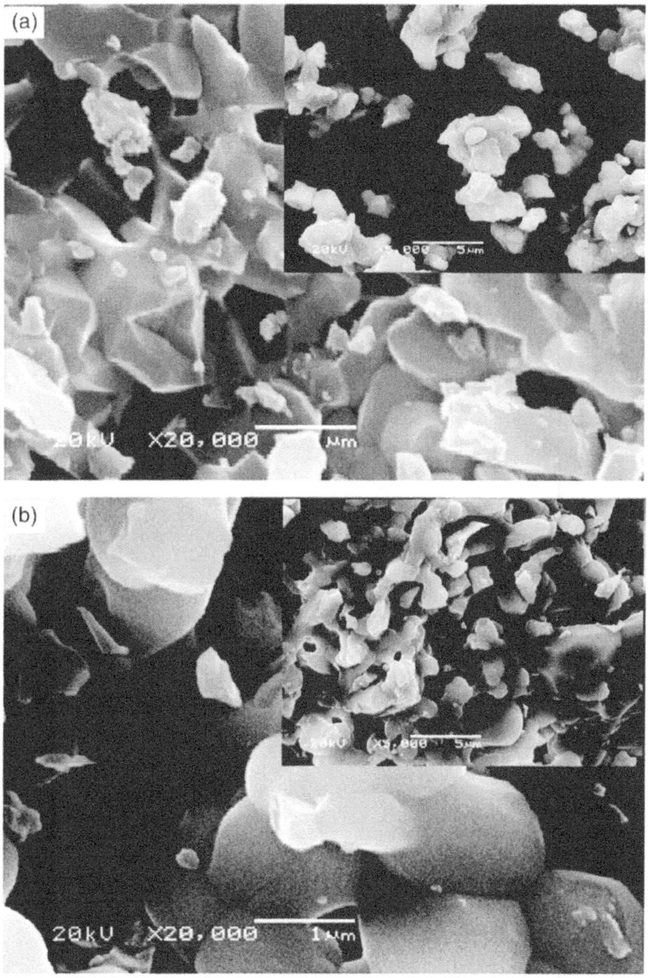

Figure 4.192 SEM photographs of (a) yellow and (b) orange Ca-α-SiAlON:Eu^{2+} phosphors prepared by GPS at 1750°C for 2 h under 0.4 MPa of N_2 atmosphere. Low magnification photographs are shown in the inset.

the different sinterability of powders with different compositions in the Ca-α-SiAlON system. The volume of transient liquid phase in the $M_x^{v+}Si_{12-(m+n)}Al_{m+n}O_nN_{16-n}$ system increases with increasing *m*, which would accelerate the materials transport and promote the solution-diffusion-precipitation process [467]. The residual liquid phase would act as a binder to form large agglomerates in the orange phosphor samples.

Figure 4.193 shows typical PL spectra of the (a) yellow and (b) orange Ca-α-SiAlON:Eu^{2+} phosphors measured in a temperature range from 25 to 150°C. The excitation spectra which were measured

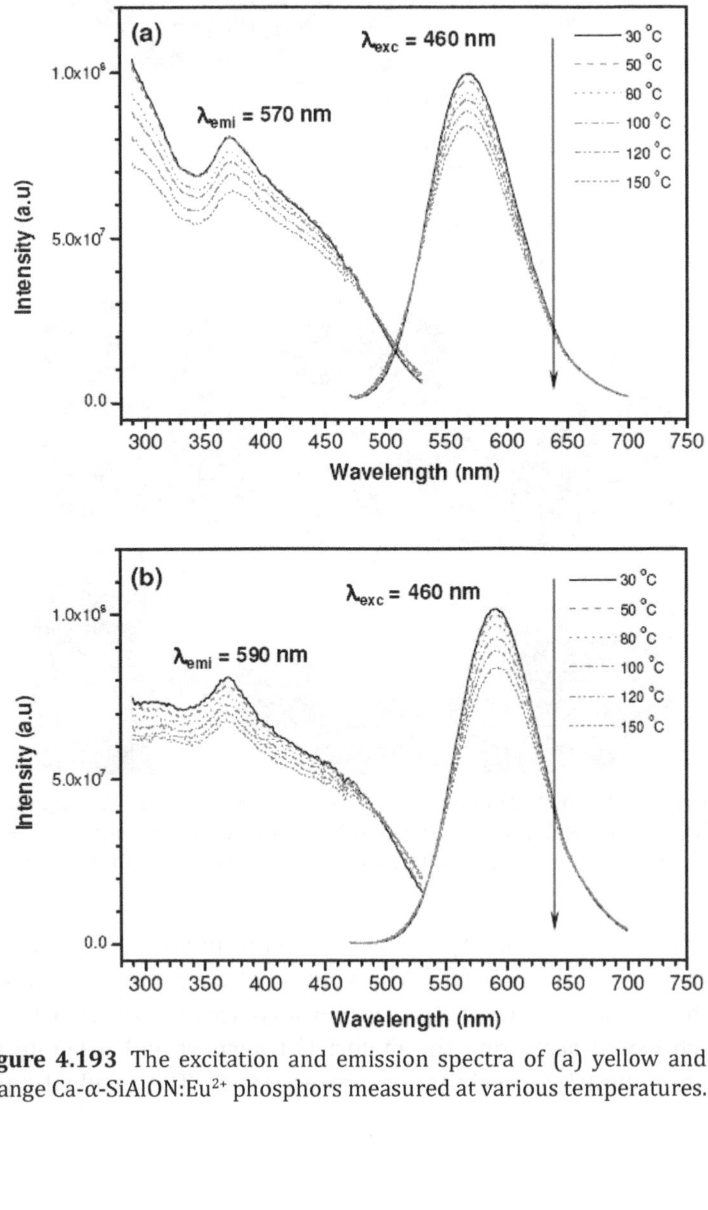

Figure 4.193 The excitation and emission spectra of (a) yellow and (b) orange Ca-α-SiAlON:Eu^{2+} phosphors measured at various temperatures.

at emission maxima are broad and structureless at all measuring temperatures. They cover the spectral region from the UV to the visible range. A broad band is observed in the excitation spectra with maxima at 370 nm, which corresponds to the $4f^7 \rightarrow 4f^65d^1$ transition of Eu^{2+} [20]. The emission spectra show a single intense broad emission band at 570 (yellow) and 590 nm (orange) under excitation at 460 nm at all measuring temperatures. These emission bands are attributable to the allowed $4f^65d^1 \rightarrow 4f^7$ transition of Eu^{2+} [20]. The characteristic emission spectra for Eu^{3+} are not observed, which exhibits sharp lines between 580 and 650 nm [468, 469]. This suggests that the Eu ion in the Ca-α-SiAlON host exists as the divalent state.

The excitation in the long wavelength region can be explained based on the nitrogen-rich surrounding of Eu^{2+} ions in Ca-α-SiAlON host [458, 470]. Fig. 4.194a represents the splitting of energy levels of Eu^{2+} due to the crystal field in the Ca-α-SiAlON host.

Both the higher formal charge of N^{3-} compared with O^{2-} and the nephelauxetic effect induced by the difference in electro-negativity of nitrogen (3.04) with oxygen (3.44) would lead to larger ligand-field splitting of the 5*d* levels and the center of gravity of the 5*d* states at lower energy as shown in Fig. 4.194a. It is also suggested that the preferential orientation of the *d*-orbital or low-lying state of the conduction band could account for the long emission wavelength [471].

The Stokes shift of the prepared Ca-α-SiAlON:Eu^{2+} yellow and orange phosphors was calculated mathematically and estimated from the PL spectral data.

The classical description cannot satisfactorily explain emission spectral shapes and non-radiative transition probabilities. The single configuration coordinate model [8] based on quantum mechanics was used to fit the emission band of considered phosphors as shown in Fig. 4.194b. Because the 4*f*-shell is well shielded, the electron-vibrational interaction within this shell is negligible so that the intensities of the vibronic satellites corresponding to $4f \rightarrow 4f$ transitions are small. On the contrary, $4f \rightarrow 5d$ transitions give rise to the broad vibronic bands whose shape-function can be described by the normalized Pekarian type spectral distribution [8, 406]. We used electron-vibrational interaction in $4f \rightarrow 5d$ optical transitions (see paragraph 4.5.4, Eqs. (4.41–4.45)).

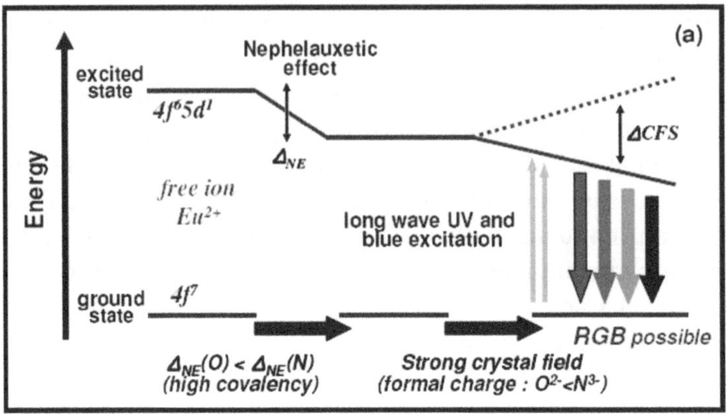

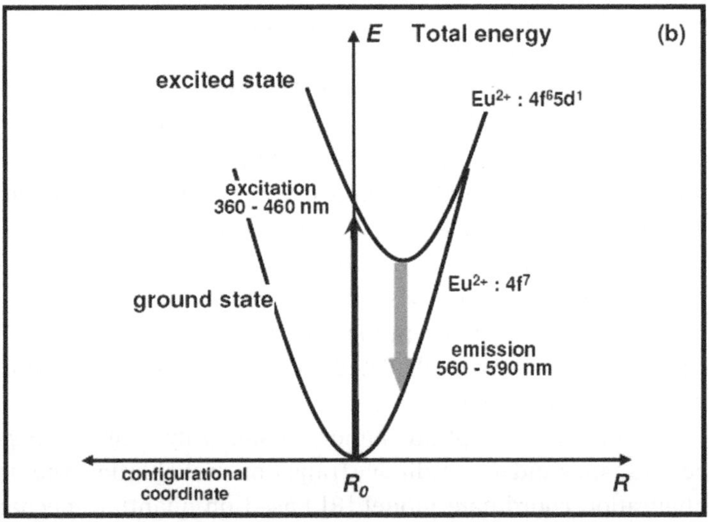

Figure 4.194 (a) Schematic diagram for splitting of energy levels of Eu^{2+} due to the crystal field in Ca-α-SiAlON host. (b) Schematic illustration of a configurational coordinate model for excitation by blue light. The two vertical arrows indicate the absorption of blue light and the emission of yellow or orange light, respectively.

The application of Eq. (4.44) instead of Eq. (4.45) gives us more correct and more precise fitting of experimental results. We obtained the best fits of the emission bands for the yellow phosphor with

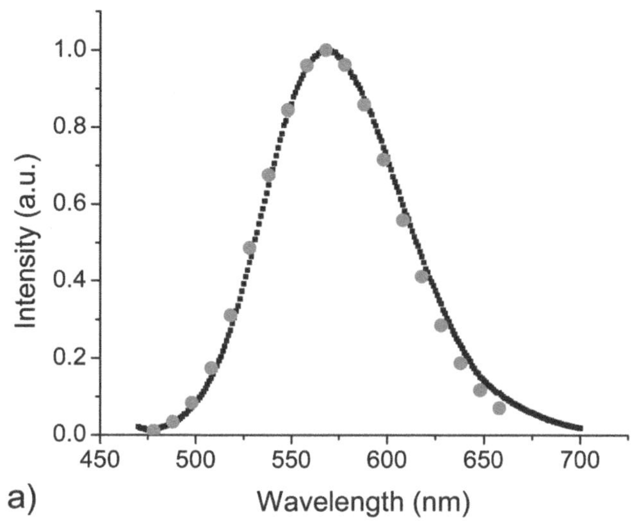

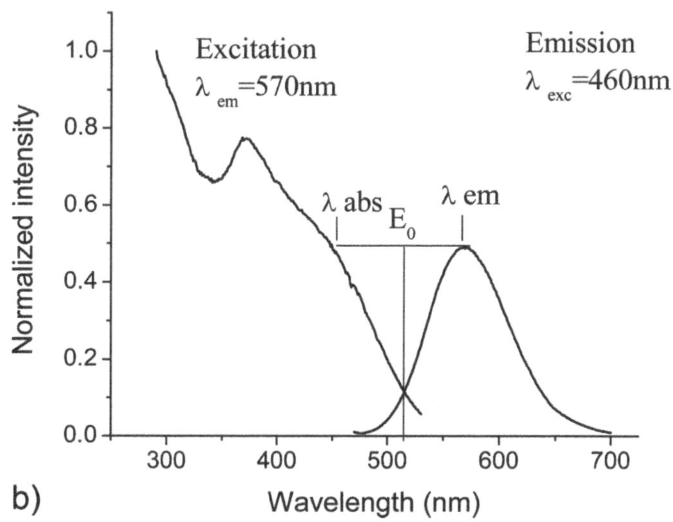

Figure 4.195 (a) Experimental (broken curves) and calculated (circles) emission spectrum at 25°C of the yellow Ca-α-SiAlON:Eu^{2+} under excitation at 460 nm. (S = 12.5 and phonon energy $\hbar\omega$ = 24.1 meV). (b) Excitation and emission spectrum of the yellow phosphor providing values of λ_{abs} and λ_{em} at 25°C.

Huang–Rhys–Pekar parameter S = 12.5 and phonon energy $\hbar\omega$ = 24.1 meV. These data are shown in Fig. 4.195a. For the orange Ca-α-SiAlON:Eu^{2+} phosphor the best fitting was obtained with S = 12.5 and phonon energy $\hbar\omega$ = 22 meV.

In turn, the knowledge of Huang–Rhys–Pekar factor and phonon energy allows us to determine the Stokes shift. By supposing that the ground state parabola of the configuration coordinate model presents the same curvature as the excited state parabola, i.e. the phonon energy is the same for the $4f^7(^8S_{7/2})$ ground state as for the $4f^6(7F)5d$ excited state, we can determinate the Stokes shift (ΔS) (Eq.4.48). ΔS is related to the offset of the parabolas in the configuration coordinate diagram and ΔS is equal to: $\Delta S = (2S - 1) \cdot \hbar\omega$.

This method was applied for study synthesized Ca-α-SiAlON:Eu^{2+} phosphors.

For the yellow phosphor, the Stokes shift is 4700 cm^{-1}, and for the orange phosphor is 4300 cm^{-1}. These data were compared with experimentally excitation and emission spectra. The main problem at room or higher temperatures is to determine correctly the λ_{abs}, because usually this wavelength is not resolved at room or higher temperatures. An almost mirror image relationship seems to hold between the emission and the excitation spectra in the energy region between the two maxima as shown in Fig. 4.196b. This relation is characteristic of a phonon-broadened emission and confirms the hypothesis that the mean phonon energy $\hbar\omega$ is the same for the $4f^7(^8S_{7/2})$ ground state as for the $4f^6(^7F)5d$ excited state. We can determine the energy of the zero-phonon line E_0 at the intersection of the emission and excitation spectra. From Fig. 4.195b, E_0 = 508 nm was found and Stokes shift $\Delta S = \lambda_{em} - \lambda_{abs}$ = 4716 cm^{-1}. The similar analysis for the orange phosphor gives ΔS = 4295 cm^{-1}. Summarized luminescence data for both synthesized phosphors are given in Table 4.36.

As one can see from Table 4.36, the mathematical calculations with consideration of the temperature dependence are in good agreement with experimental PL results. A little difference between the Stokes shift for yellow and orange phosphors can be explained by the different centroidal shift, because the nephelauxetic effect is stronger for the orange Ca-α-SiAlON:Eu^{2+} phosphor in comparison

Table 4.36 Mathematically calculated PL data for the yellow and orange Ca-α-SiAlON:Eu^{2+} phosphors

Phosphor		λ_{abs}	E_0	λ_{em}	FWHM	ΔS_{math}	ΔS_{exp}
	nm	448	508	568	82		
Yellow	eV	2.76	2.44	2.18			
	cm^{-1}	22.321		17.605		4700	4716
	nm	472	532	592	80		
Orange	eV	2.62	2.33	2.09			
	cm^{-1}	21.186		16.891		4300	4295

with the yellow phosphor. Presented here data for synthesized phosphors generally are in agreement with previous reports [458, 470].

The thermal quenching property is one of the most important technological parameters for phosphors applied in white LEDs. The temperature-dependent PL properties of the prepared Ca-α-SiAlON:Eu^{2+} phosphors were investigated in a temperature range from 25 to 150°C, and the results are shown in Figs. 4.193 and 4.196.

Upon heating the phosphor samples, the decrease in emission intensity and broadening of FWHM are apparent as shown in Fig. 4.193, which can be explained by the thermal quenching in the configurational coordinate model for Eu^{2+} in the Ca-α-SiAlON host structure [8]. The lower potential curve and the higher one represent the total energy of the ground state of $4f^7(^8S_{7/2})$ and the excited state of $4f^65d^1$, respectively. The equilibrium positions of the two states are different from each other because of the spatial distribution of the electron orbitals. The excited PL center is thermally activated through phonon interaction, and then thermally released through the crossing point between the excited state and the ground state in a configurational coordinate diagram. This non-radiative transition induced by thermal activation is strongly dependent on temperature and results in thermal quenching of the PL intensity [472]. The electron–phonon interaction resulted from increased population density of the phonon, which broadened the FWHM of the PL spectra at high temperature [472].

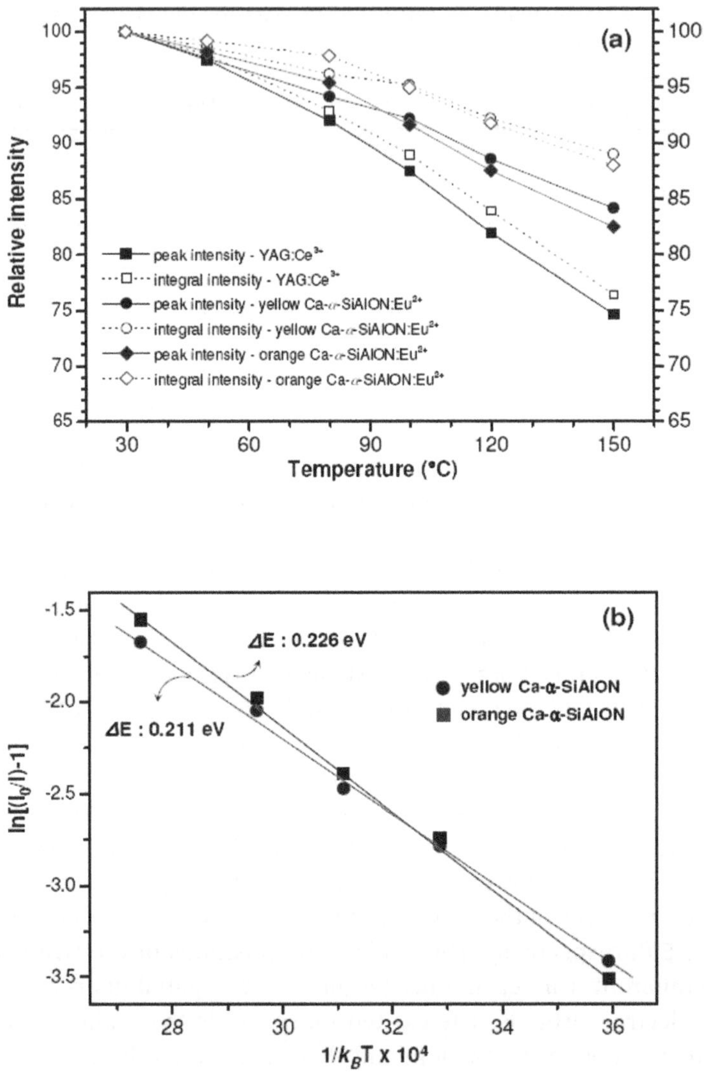

Figure 4.196 (a) Temperature dependence of the relative PL peak intensity and integral PL intensity of the prepared yellow and orange Ca-α-SiAlON:Eu^{2+} phosphors in a temperature range from 25 to 150°C. (b) Arrhenius fitting of the PL peak intensity in the measuring temperature range. Activation energies for thermal quenching (ΔE) of the prepared Ca-α-SiAlON:Eu^{2+} phosphor samples are shown.

Figure 4.196a displays a decrease of the relative PL peak intensity and integral PL intensity for the prepared yellow and orange Ca-α-SiAlON:Eu^{2+} phosphors with increasing temperature. For comparison, decrease of the PL for a commercially used YAG:Ce^{3+} phosphor was measured and shown with in Fig. 4.196a. The integral PL emission intensity which is proportional to the quantum efficiency decreased slightly less than the peak intensity as shown in Fig. 4.196a. At 150°C, the integral emission intensity of the yellow and orange Ca-α-SiAlON:Eu^{2+} phosphors were about 89.0% and 87.9%, respectively, of those measured at room temperature, which is higher than that of the YAG:Ce^{3+} phosphor by about 13%. This result confirms the Ca-α-SiAlON:Eu^{2+} phosphors are promising for application in elevated temperatures or high-power LEDs.

To further understand the temperature dependence of PL intensity and to determine the activation energy for thermal quenching, the Arrhenius equation was fitted to the thermal quenching data of the yellow and orange Ca-α-SiAlON:Eu^{2+} phosphors, as shown in Fig. 4.196b. According to the classical theory of thermal quenching, the temperature-dependent PL intensity can be described by the expression [473].

$$I(T) = \frac{I_0}{1 + A\exp\left(\frac{-\Delta E}{k_B T}\right)}$$

where I_0 is the initial intensity, $I(T)$ is the intensity at a given temperature T, A is a constant, ΔE is the activation energy for thermal quenching, and k_B is Boltzmann's constant. Figure 4.196b plots $\ln[(I_0/I) - 1]$ vs. $1/(k_B T)$ and gives a straight line up to T = 150°C. The best fit gives comparable activation energy (ΔE) of 0.211 and 0.226 eV for yellow and orange Ca-α-SiAlON:Eu^{2+} phosphors, respectively, which is consistent with the reported value (0.23 eV) in Ca-α-SiAlON:Yb^{2+} phosphor [474].

4.7.2 *$(Ca_{1-x}Sr_x)(S_{1-y}Se_y)$:$Eu^{2+}$ Phosphors with Controlled Properties*

Rare earth calcium and strontium sulfides have been the object of increased interest in recent years [475–479]. One reason has been the low cost of these red phosphors and the good linear behavior. RE phosphors can be separated into two types: broad band emitting owing to the $5d \rightarrow 4f$ transition (Eu^{2+}, Ce^{3+}) or narrow band emitting owing to the transition between the 4f levels (Eu^{3+}, Tb^{3+}, Sm^{3+}). For white light in blue LED the broad band emitting phosphors are preferred. Eu^{2+}-doped CaS and SrS as well as CaSe and SrSe are currently under investigation as efficient phosphors for the emission of red-orange light for the LED application. To improve the luminescent properties and to receive the necessary spectra shift, the multicompound phosphors are used. Some formulas can be considered: $M_xM:Eu^{2+}$, where M_x is variable Ca/Sr ratio as $Ca_{1-x}Sr_x$ and M is constant, for example, M is only sulfur [429, 480, 481] or only selenium [482]. Other combination $MM_y:Eu^{2+}$, is also known from literature. Here M is constant, for example, M is Ca or Sr and M_y is variable $S_{1-y}Se_y$ [483].

In the present work we propose a new high efficiency phosphor described by general formula with two variable parameters $M_xM_y:Eu^{2+}$, where M_x can be $Ca_{1-x}Sr_x$, and M_y can be, for example, $S_{1-y}Se_y$. This phosphor $(Ca_{1-x}Sr_x)(S_{1-y}Se_y)$: Eu^{2+} emits in the deep red, orange or yellow range of spectra in dependence of x and y values. A mathematical diagram and algorithm is proposed to predict the luminescence properties. White light has also been achieved by combining a blue LED (460 nm) with $SrGa_2S_4$:Eu (green) and $(Ca_{1-x}Sr_x)(S_{1-y}Se_y)$: Eu^{2+} (red) phosphor.

Classical inorganic phosphors usually consist of a host lattice with activator ions doped into it in small concentrations, typically a few mole percent or less. Selection of optimum phosphor material (host, dopant and codopant) plus optimization the composition leads to color purity, high brightness, high efficiency and long-term stability. Right synthesis rout and synthesis conditions give suitable powder characteristics.

One example of a phosphor composition for a three-band white LED comprises the green-emitting phosphor $SrGa_2S_4:Eu^{2+}$ in combination with at least one of the following red-emitting phosphors consisting of $(Ca_{1-x}Sr_x)(S_{1-y}Se_y)$: Eu^{2+}, wherein x and y is a number from 0 to 1. All these phosphors can be synthesized by solid state reaction. By combining the present red phosphor with green phosphor and blue LEDs, full color white light can be obtained.

The red-emitting phosphor can be prepared by combining the starting materials: $Sr(NO_3)_2$, $Ca(NO_3)_2 \cdot 4H_2O$, SeO_2, and Eu_2O_3. The element S of the phosphor is introduced from SrS/CaS and Se from SrSe/CaSe. The host materials, CaSe and SrSe, can be prepared from calcium selenate and strontium selenate, respectively. The phosphors can be synthesized by high-temperature solid state reaction. The stoichiometric amounts of the corresponding raw materials (CaSe, SrSe, and EuF_3, which could be prepared by hydrothermal synthesis) have to be thoroughly mixed by grinding in an agate mortar and then heated at 1100°C for 2 h in a stream of reductive atmosphere (N_2+H_2) or CO.

Usually blue-emitting LED has an emission peak at 420–480, optionally about 450–470 nm. A combination of two blue-excited phosphors (excited by the blue emission from LED), one emitting in the green and one emitting in the red, is used in place of the yellow emitting yttrium aluminum garnet. Because these new emissions are close to peaks in the tristimulus functions in combination with the blue from the LED they produce light with higher luminous efficacy and better color rendering than the phosphors used in prior devices. These green and red phosphors are selected so that they are excited by the blue-emitting LED. The green color-emitting phosphor typically has an emission peak at 510–550 nm. The red color-emitting phosphor typically has an emission peak at 600–650 nm. To predict the properties and to receive the desirable color of emission, the careful analysis between x and y parameters needs to be carried out. From the literature data it is well known that there are no interstitials when strontium substitutes calcium in a rock-salt crystalline structure and selenium substitutes sulfur. Vegard's law states that there is a linear relationship between substitional element concentration and the lattice parameter in cases where the solute and solvent have similar bonding properties. In these cases the lattice parameter is influenced only by the relative sizes of the atoms and has been shown to be valid in ionic salts and compounds such as CaSe or SrSe where sulfur substitutes for the selenium [484]. In conformity with the Vegard's law and the experimental linear dependence of wavelength peak versus compositional parameters (Figs. 4.197 and 4.198), a 3D diagram for multicompound composition can be created (Fig. 4.199).

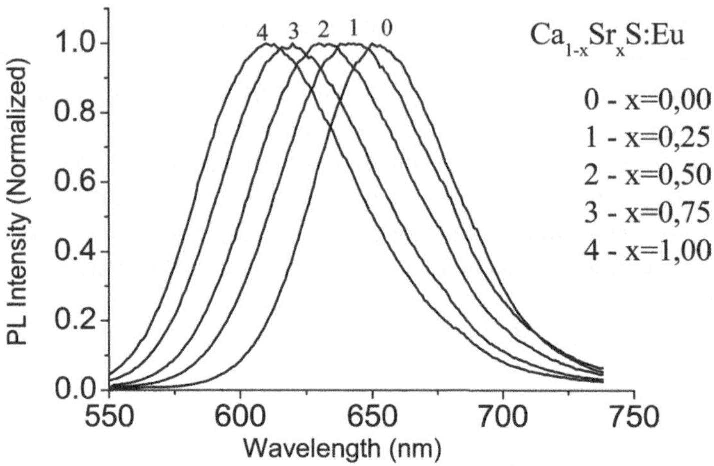

Figure 4.197 Emission spectra of $Ca_{1-x}Sr_x$:Eu^{2+}. Excitation made by 460 nm light source.

Similar linear dependences for other three-component compositions were obtained. Computer simulation for $(Ca_{1-x}Sr_x)(S_{1-y}Se_y)$: Eu^{2+} gives the plane surface for the peak wavelength versus compositional parameters (Fig. 4.199). Experimental data (black

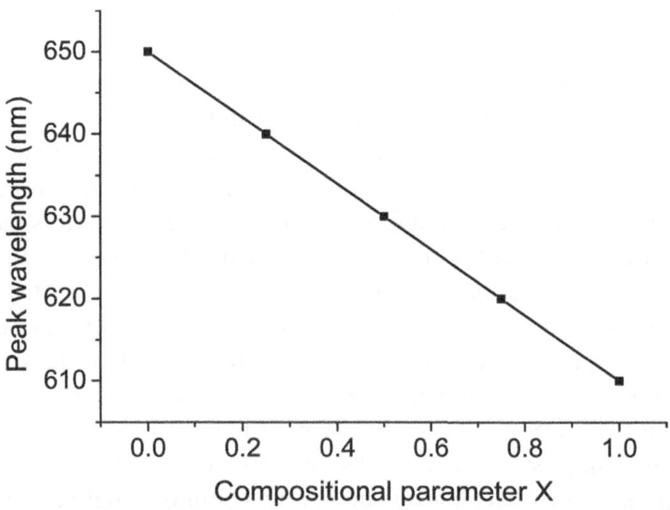

Figure 4.198 The dependence of peak wavelength on the composition parameter X for $Ca_{1-x}Sr_xS$:Eu^{2+}.

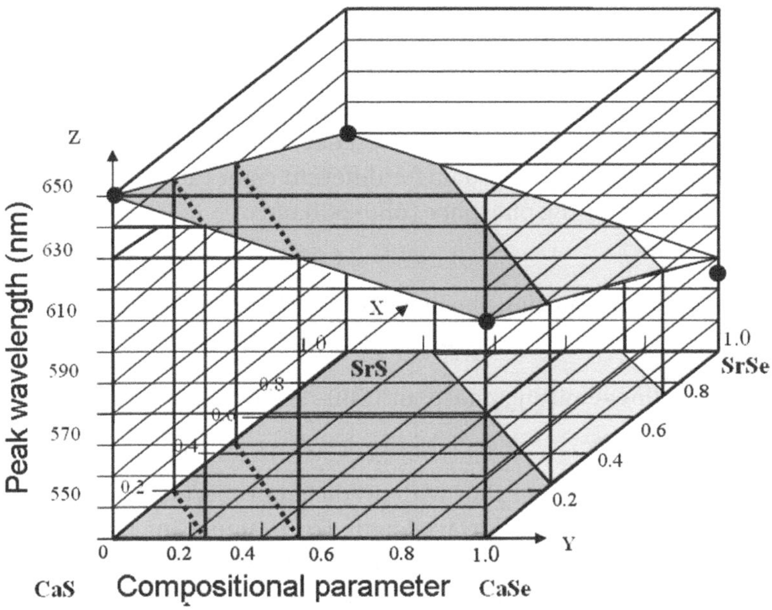

Figure 4.199 Peak wavelength diagram for $(Ca_{1-x}Sr_x)(S_{1-y}Se_y)$: Eu^{2+}.

points) from different publications for CaS [480, 481], SrS [429], CaSe [482], and SrSe [482] show very good agreement with presented data and proposed model.

From this diagram it is evident that any peak shift from yellow to red color can be achieved in this system. Equation of the plane in space may be represented by

$$Z = Ax + By + C$$

where Z means the peak wavelength, x and y are independent variable compositional parameters and A, B, C are coefficients, that can be determined from experimental data and boundary conditions $(x, y = 0, 1)$.

For our range

$$0 \leq x \leq 1$$

$$0 \leq y \leq 1$$

the peak position Z can be found as

$$Z = -40x - 40y + 650$$

Intersection of this plane with any Z = Const gives a line, which projection determines the relation between compositional parameters x and y for every predetermined color. Using this equation, we find x and y data for different color ranges.

For red-emitting phosphor (650–600 nm):

$$0 \le x + y < 1.25$$

For orange-emitting phosphor (600–580 nm):

$$1.25 < x + y < 1.75$$

For yellow-emitting phosphor (580–570 nm):

$$1.75 < x + y \le 2$$

To receive the phosphor with predicted properties, first we have to fix the desirable peak wavelength. For example, for red-emitting phosphor with maximum emission at 630 nm we obtain from diagram $x + y = 0.5$. It means that if we want to substitute only Ca by Sr, ($y = 0$, $x = 0.5$) the phosphor composition will be the next:

$$Ca_{0.5}Sr_{0.5}S{:}Eu^{2+}$$

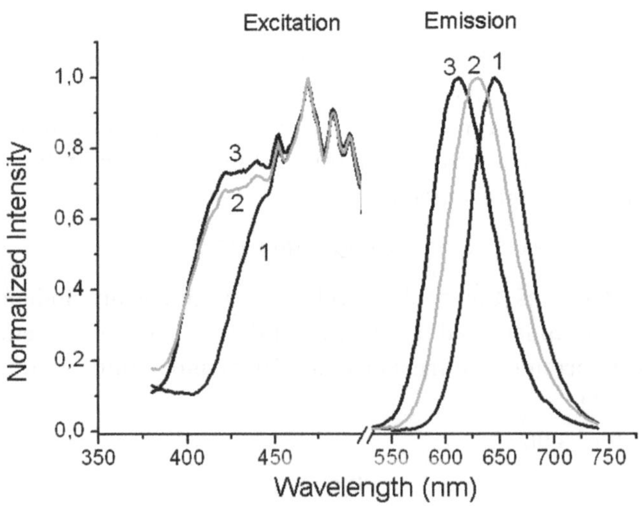

Figure 4.200 The excitation and emission spectra of $(Ca_{1-x},Sr_x)S{:}Eu^{2+}$ with a different Sr/Ca ratio: (1) $x = 0$, (2) $x = 0.5$, and (3) $x = 1$. Excitation made by 460 nm light source.

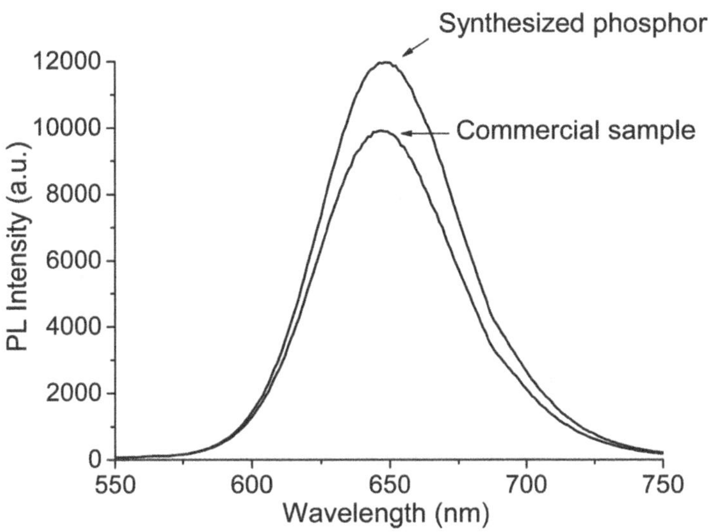

Figure 4.201 Relative PL emission spectra of synthesized and commercial phosphors.

This phosphor was synthesized by above mentioned solid state reaction and excitation and emission spectra are presented in Fig. 4.200.

Really, one can see the spectra shift from 650 nm ($CaS:Eu^{2+}$) spectrum 1 to 630 nm ($Ca_{0.5}Sr_{0.5}S:Eu^{2+}$) (Figs. 4.197 and 4.200). Further substitution Ca by Sr leads to subsequent spectra shift to 610 nm ($SrS:Eu^{2+}$). All these results are in good agreement with proposed diagram (Fig. 4.199).

The brightness of all our phosphors exceeds the commercial samples about 20% (Fig. 4.201).

As shown in Fig. 4.200, the excitation spectra of phosphors are very broad; this means the phosphors can be well excited by the visible light from 430 to 500 nm, fitting in with blue LED chip. The lowest excited state of 4*f* levels is located at about 28×10^3 cm^{-1} [8] and is higher than the $4f^65d^1$ level in most crystals, so that Eu^{2+} usually gives characteristic broadband emission due to transitions between the crystal field components of the $4f^65d^1$ excited state configuration and the $^8S_{7/2}$ ($4f^7$) ground state. The wavelength positions of the emission bands depend very much on hosts, changing from the

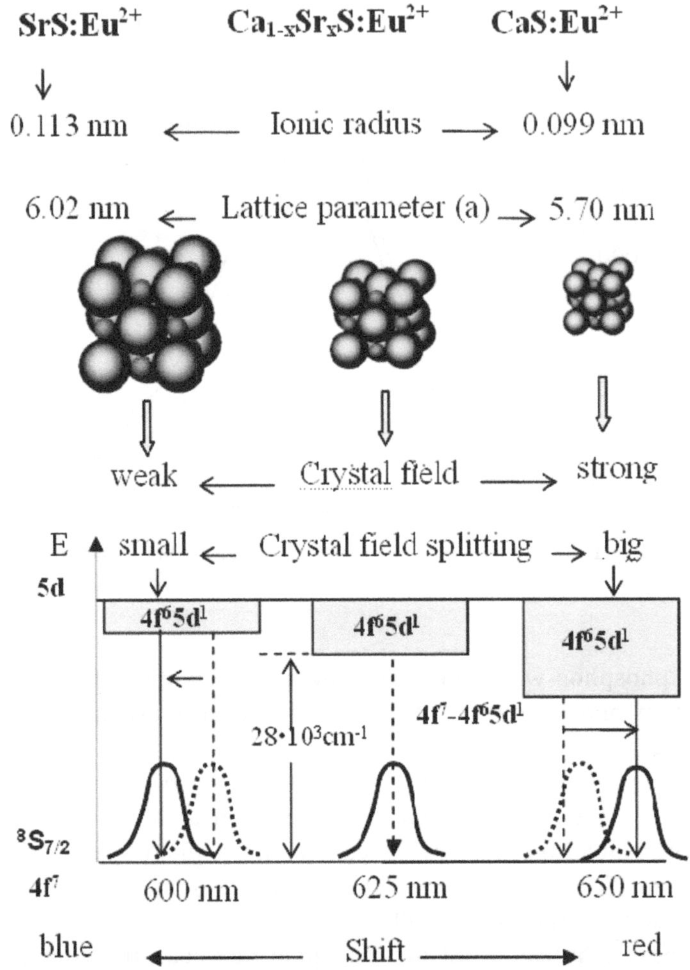

Figure 4.202 Schematic illustration of crystal field splitting and spectra shift in $(Ca_{1-x},Sr_x)S:Eu^{2+}$ with a different Sr/Ca ratio.

near-UV to the red of the electromagnetic spectrum. This dependence is interpreted as the crystal-field splitting of the 5*d* level, as shown schematically in Fig. 4.202.

With the increment of crystal-field strength, the emission bands shift to longer wavelength. As noted above, the $(Ca_{1-x},Sr_x)S:Eu^{2+}$ is a cubic rock-salt structure where Eu^{2+} is on Sr^{2+}/Ca^{2+} sites and is O_h symmetry and small, which result in the strong crystal field and lead to the red emission. In addition, as illustrated in Fig. 4.198, the emission of the phosphor shifts from 610 nm up to 650 nm by degrees with the Sr concentration (x) from 1 down to 0. Besides the dissimilarity in ionic radius between Sr^{2+} and Ca^{2+}, the energy gaps are 4.41 and 4.30 eV, respectively [485], therefore, the mixing of SrS and CaS makes for the change of the crystal field and energy gap, and accounts for the red shift of the emission wavelength. On this basis, one can vary the Sr/Ca ratio to obtain the desired wavelength.

4.7.3 *Concluding Remarks*

In this paragraph some new phosphors for solid state lighting and LED with improved and predicted properties were proposed and discussed.

A model of $4f \to 5d$ optical transitions, taking into account the electron-vibrational interaction, was applied to these materials. It was shown that the influence of temperature on the fitting of experimental emission spectra becomes significant at room and higher temperatures. The analysis of the experimental results with the proposed model can serve as an efficient tool for better understanding the luminescent mechanism in rare earth activated phosphors. Using the electron-vibrational interaction of $4f \to 5d$ optical transitions in Ca-α-SiAlON:Eu^{2+} phosphors, as well as the temperature influence on the fitting of experimental emission spectra, the Stokes shift was calculated and the good agreement of the theory and experiment were realized. The observed long-emission wavelengths were explained based on the nephelauxetic effect induced by nitrogen-rich surroundings of Eu^{2+} and on large ligand-field splitting of Eu^{2+} $5d$ energy levels. The thermal quenching properties of the prepared Ca-α-SiAlON:Eu^{2+} phosphors were superior to the YAG:Ce^{3+} phosphor, which means Ca-α-SiAlON:Eu^{2+} phosphors are promising for high-power LED applications. By fitting the temperature dependent PL data based on the classical thermal quenching theory, the activation energy (ΔE) for thermal quenching was determined to be 0.211 and 0.226 eV for the yellow and orange Ca-α-SiAlON:Eu^{2+} phosphors, respectively.

As we gain a more detailed knowledge of the absolute placement of substitional ions within large band gap materials, we can better predict the behavior of these materials from different points of view including the consideration of the overall performance efficiency of luminescent materials.

The $(Ca_{1-x},Sr_x)S:Eu^{2+}$, $(Ca_{1-x},Sr_x)Se:Eu^{2+}$, as well as $(Ca_{1-x}\,Sr_x)(S_{1-y}Se_y):Eu^{2+}$ phosphors can be excited efficiently by visible light from 430 to 490 nm and emit red, orange, and yellow light with broadband, which is in agreement with the blue LED chips.

The luminescence originates from the doped Eu^{2+} that replaces the Sr/Ca sites in crystal and acts as the luminescence center.

With the Sr/Ca ratio decreasing, the lattice parameters get lower, and the emission wavelength shows red shift. A suitable wavelength can be got by adjusting the Sr/Ca ratio.

A mathematical peak wavelength diagram is in good agreement with experimental data.

References

[1] White H.E. *Introduction to Atomic Spectra*. McGraw-Hill, New York, 1934.

[2] Liu G., Jacquier B., eds. *Spectroscopic Properties of Rare Earths in Optical Materials*. Springer Series in Materials Science, Springer, Berlin, 2005.

[3] Tsukerblat B.S. *Group Theory in Chemistry and Spectroscopy*. Dover Pub., Mineola, New York, 2006.

[4] Bünzli J.-C.G., Piguet C. Taking advantage of luminescent lanthanide ions. *Chem. Soc. Rev.*, 2005, **34**: 1048

[5] Russell H.N., Saunders F.A. New regularities in the spectra of the alkaline earths. *Astrophys. J.*, 1925, **61**: 38.

[6] Harvey E.N. *A History of Luminescence*. American Philosophical Society, Philadelphia, PA, 1957.

[7] Feldmann C., Justel T., Ronda C.R., Schmidt P.J. Inorganic luminescent materials: 100 years of research and application. *Adv. Funct. Mater.*, 2003, **13**(7): 511.

[8] Shionoya S., Yen W.M., eds. *Phosphor Handbook*. 1 ed., CRC Press, Boca Raton, 1999.

[9] Kim C.-H., Kwon I.-E., Park C.-H., Hwang Y.-J., Bae H.-S., Yu B.-Y., Pyun C.-H., Hong G.-Y. Phosphors for plasma display panels. *J. Alloys Compd.*, 2000, **311**(1): 33.

[10] Rao R.P., Devine D.J. RE-activated lanthanide phosphate phosphors for PDP applications. *J. Lumin.*, 2000, **87–89**: 1260.

[11] Ronda C.R. Recent achievements in research on phosphors for lamps and displays. *J. Lumin.*, 1997, **72–74**: 49.

[12] Brixner L.H. New X-ray phosphors. *Mater. Chem. Phys.*, 1987, **16**(3–4): 253.

[13] Martin N. Scintillation detectors for X-rays. *Meas. Sci. Technol.*, 2006, **17**(4): R37.

[14] Mills A. Solid state lighting – a world of expanding opportunities at LED 2002. *III-Vs Rev.*, 2003, **16**(1): 30.

[15] Rohwer L.S., Srivastava A.M. Development of phosphors for LEDS. *Electrochem. Soc. Interface*, 2003, **12**(2): 36.

[16] Mills A. Phosphors development for LED lighting. *III-Vs Rev.*, 2005, **18**(3): 32.

[17] Mills A. Phosphors to brighten your future. *III-Vs Rev.*, 2006, **19**(5): 36.

[18] Xie R.-J., Hirosaki N. Silicon-based oxynitride and nitride phosphors for white LEDs: a review. *Sci. Technol. Adv. Mat.*, 2007, **8**(7–8): 588.

[19] Mikami M., Watanabe H., Uheda K., Shimooka S., Shimomura Y., Kurushima T., Kijima N. New phosphors for white LEDs: material design concepts. *IOP Conf. Series: Mat. Sci. Eng.*, 2009, **1**: 012002.

[20] Blasse G., Grabmeyer B.C. *Luminescent Materials*. Springer-Verlag, Berlin, 1994.

[21] Kroger F.A. *Some Aspects of the Luminescence of Solids*. Elsevier, Amsterdam, 1948.

[22] West A.R. *Solid State Chemistry and Its Applications*. John Wiley & Sons, New York, 1990.

[23] Blasse G. Luminescence of inorganic solids: from isolated centres to concentrated systems. *Progr. Solid State Chem.*, 1988, **18**(2): 79.

[24] Butler K.H. *Fluorescent Lamp Phosphors*. The Pensylvania State University Press, University Park, 1980.

[25] Vij D.R. *Luminescence of Solids*. Plenum Press, New York, 1988.

[26] Ozawa L. *Cathodolurninescence: Theory and Applications*. VCH Verlagsgesellschaft, Weinheim/VCH Publishers, New York/Kodansha, Tokyo, 1990.

[27] Kitai A.H. *Solid State Luminescence: Theory, Materials and Devices*. Chapman and Hall, London, 1993.

[28] Singh-Nalwa H., Rohwer L.S., eds. *Handbook of Luminescence, Display Materials, and Devices*. American Scientific Publishers, 2003.

[29] Reineker P., Stefan V., Vitukhnovsky A., eds. *Select Topics in Luminescent Materials*. Stefan University Press, 2004.

[30] Ropp R.C. *Luminescence and the Solid State*, 2nd ed. Elsevier Science, Amsterdam, 2004.

[31] Nazarov M.V., Tsukerblat B.S. Luminescent performance of europium and terbium activated phosphors, in *Focus on Materials Science Research*, Rosse D.V., ed. Nova Science Publishers, New York, 2007.

[32] Gaft M., Reisfeld R., Panczer G. *Modern Luminescence Spectroscopy of Minerals and Materials*, 1st ed. Springer Leipzig, 2005.

[33] Yen W.M., Shionoya S., Yamamoto H., eds. *Measurements of Phosphor Properties*. CRC Press, Boca Raton, 2006.

[34] Yen W.M., Shionoya S., Yamamoto H., eds. *Phosphor Handbook*. 2nd ed., CRC Press, Boca Raton, 2007.

[35] Yen W.M., Shionoya S., Yamamoto H., eds. *Practical Applications of Phosphors*. CRC Press, Boca Raton, 2006.

[36] Yen W.M., Shionoya S., Yamamoto H., eds. *Fundamentals of Phosphors*. CRC Press, Boca Raton, 2006.

[37] Lakshmanan A. *Luminescence and Display Phosphors: Phenomena and Applications*. Nova Science Publishers, 2007.

[38] Ozawa L. *Cathodoluminescent Phosphors: Theories and Practical Applications*. CRC Press, Portland, 2007.

[39] Ronda C., ed. *Luminescence: From Theory to Applications*. Wiley-VCH, Darmstadt, 2007.

[40] Kitai A., ed. *Luminescent Materials and Applications*. John Wiley & Sons, Singapore, 2008.

[41] Fouassier C. Luminescent materials. *Curr. Opin. Solid State Mater. Sci.*, 1997, **2**(2): 231.

[42] Justel T., Nikol H., Ronda C. New developments in the field of luminescent materials for lighting and displays. *Angew. Chemie Int. Ed.*, 1998, **37**(22): 3084.

[43] Ronda C.R., Justel T., Nikol H. Rare earth phosphors: fundamentals and applications. *J. Alloys Compd.*, 1998, **275–277**: 669.

[44] Kenyon A.J. Recent developments in rare-earth doped materials for optoelectronics. *Prog. Quant. Electron.*, 2002, **26**(4–5): 225.

[45] Srivastava A.M., Ronda C.R. Phosphors. *Electrochem. Soc. Interface*, 2003, **12**(2): 48.

[46] Abrams B.L., Holloway P.H. Role of the surface in luminescent processes. *Chem. Rev.*, 2004, **104**(12): 5783.

[47] Chander H. Development of nanophosphors. Review. *Mat. Sci. Eng. R: Reports*, 2005, **49**(5): 113.

[48] Hase T., Kano T., Nakazawa E., Yamamoto H. Phosphor Materials for cathode-ray tubes. *Adv. Electron .Electron Phys.*, 1990, **79**: 271.

[49] Leverenz H.W. *An Introduction to Luminescence of Solids*. John Wiley & Sons, New York, 1950.

[50] Rao C.N.R. *Chemical Approaches to the Synthesis of Inorganic Materials*. Wiley, New Delhi, 1994.

[51] Yen W.M., Weber M.J., eds. *Inorganic Phosphors: Compositions, Preparation, and Optical Properties*. CRC Press, Boca Raton, 2004.

[52] Ropp R.C. *Luminescence and the Solid State*, 1st ed. Elsevier Science, Amsterdam, 1991.

[53] Segal D. *Chemical Synthesis of Advanced Ceramic Materials: Chemistry of Solid State Materials*. Cambridge University Press, 1989.

[54] Nazarov M. Luminescence mecanisme of highly efficient YAG and TAG phosphors. *J. Mold. Phys. Sci.*, 2005, **4**(3): 347.

[55] Sutton W.H. Microwave processing of ceramic materials. *Am. Ceram. Soc. Bull*, 1989, **68**(2): 376.

[56] Whittaker A.G., Mingos D.M.P. Microwave-assisted solid-state reactions involving metal powders and gases. *J. Chem. Soc. Dalton Trans*, 1993, **9**: 2541.

[57] Landry C.C., Barron A.R. Synthesis of polycrystalline chalcopyrite semiconductors by microwave irradiation. *Science*, 1993, **260**: 1653.

[58] Vaidhyanathan B., Ganguli M., Rao K.J. Fast solid state synthesis of metal vanadates and chalcogenides using microwave irradiation. *Mater. Res. Bull.*, 1995, **30**(9): 1173.

[59] Baghurst D.R., Chippindale A.M., Mingos D.M.P. Microwave syntheses for superconducting ceramics. *Nature*, 1988, **332**: 311.

[60] Kato M., Sakakibara K., Koike Y. Rapid preparation of $YBa_2Cu_3O_{7-x}$ with T_c = 90 K using a domestic microwave oven. *Jpn. J. Appl. Phys.*, 1997, **36**: L1291.

[61] Houmes J.D., Zur Loye H.-C. Microwave synthesis of ternary nitride materials. *J. Solid State Chem.*, 1997, **130**(2): 266.

[62] Whittaker A.G., Mingos D.M.P. Microwave-assisted solid-state reactions involving metal powders. *J. Chem. Soc. Dalton Trans.*, 1992: 2751.

[63] Ostorero J., Gasgnier M., Petit A. Yttrium iron garnet and Y, Fe binary oxides synthesized by microwave monomode energy transfer. *J. Alloys Compd.*, 1997, **262–263**: 275.

[64] Peelamedu R.D., Roy R., Agrawal D. Anisothermal reaction synthesis of garnets, ferrites and spinels in microwave field. *Mater. Res. Bull.*, 2001, **36**: 2723.

[65] Uematsu K., Toda K., Sato M. Preparation of YVO_4:Eu^{3+} phosphor using microwave heating method. *J. Alloys Compd.*, 2005, **389**: 209.

[66] Park W.J., Jung M.K., Masaki T., Imb S.J., Yoon D.H. Characterization of YVO_4:Eu^{3+}, Sm^{3+} red phosphor quick synthesized by microwave rapid heating method. *Mat. Sci. Eng.: B*, 2008, **146**: 95.

[67] Broadhead K.G., Heady H.H. Radiochemical precipitation studies of rare-earth oxalates. *Anal. Chem.*, 1960, **32**: 1603.

[68] Wang H.Z., Gao L., Niihara K. Synthesis of nanoscaled yttrium aluminum garnet powder by the co-precipitation method. *Mater. Sci. Eng. A*, 2000, **288**: 1.

[69] Choe J.Y., Ravichandran D., Blomquist S.M., Kirchner K.W., Forsythe E.W., Morton D.C. Cathodoluminescence study of novel sol–gel derived $Y_{3-x}Al_5O_{12}$: Tb_x phosphors. *J. Lumin.*, 2001, **95**: 119.

[70] Ruan S.K., Zhou J.G., Zhong A.M., Duan J.F., Yang X.B., Su M.Z. Synthesis of $Y_3Al_5O_{12}$:Eu^{3+} phosphor by sol-gel method and its luminescence behavior. *J. Alloys Compd.*, 1998, **275**: 72.

[71] Lo J.R., Tseng T.Y. Effect of LiCl on the crystallization behavior and luminescence of $Y_3Al_5O_{12}$:Tb. *Mater. Chem. Phys.*, 1998, **57**: 95.

[72] Veith M., Mthur S., Kareiva A., Jilavi M., Zimmer M., Huch V. Low temperature synthesis of nanocrystalline $Y_3Al_5O_{12}$ (YAG) and Ce-doped $Y_3Al_5O_{12}$ via different sol-gel methods. *J. Mater. Chem. Phys.*, 1999, **9**: 3069.

[73] Shi S., Wang J.Y. Combustion synthesis of Eu^{3+} activated $Y_3Al_5O_{12}$ phosphor nanoparticles. *J. Alloys Compd.*, 2001, **327**: 82.

[74] Johnson D.W. Nonconventional powder preparation techniques. *Am. Ceram. Soc. Bull.*, 1981, **60**(2): 221.

[75] Chen T.M., Chen S.C., Yu C.J. Preparation and Characterization of garnet phosphor nanoparticles derived from oxalate coprecipitation. *J. Solid State Chem.*, 1999, **144:** 437.

[76] Yada M., Ohya M., Machida M., Kima T. Synthesis of porous yttrium aluminium oxide templated by dodecyl sulfate assemblies. *Chem. Commun.*, 1998: 1941.

[77] Kinsman K.M., Mckittrick J., Sluzky E., Kesse K. Phase development and luminescence in chromium-doped yttrium aluminum garnet (YAG:Cr) phosphors. *J. Am. Ceram. Soc.*, 1994, **77**(11): 2866.

[78] Hoghooghi B., Healey L., Powell S., Mckittrick J., Sluzky E., Hesse K. Synthesis of YAG:Cr phosphors by precipitation from aluminum and yttrium sulfate solutions. *Mater. Chem. Phys.*, 1994, **38**(2): 175.

[79] Sordelet D.J., Akinc M., Panchula M.L., Han Y., Han M.H. Synthesis of yttrium aluminum garnet precursor powders by homogeneous precipitation. *J. Eur. Ceram. Soc.*, 1994, **14**(2): 123.

[80] Erdei S., Roy R., Harshe G., Juwhari H., Agrawal D., Ainger F.W., White W.B. The effect of powder preparation processes on the luminescent properties of yttrium oxide based phosphor materials. *Mater. Res. Bull.*, 1995, **30**(6): 745.

[81] Rao R.P. Growth and characterization of Y_2O_3:Eu^{3+} phosphor films by sol-gel process. *Solid State Commun.*, 1996, **99**(6): 439.

[82] Gowda G. Synthesis of yttrium aluminates by the sol-gel process. *J. Mater. Sci. Lett.*, 1986, **5**(10): 1029.

[83] Inoue M., Otsu H., Kominami H., Inui T. Synthesis of yttrium aluminum garnet by the glycothermal method. *J. Am. Ceram. Soc.*, 1991, **74**(6): 1452.

[84] Manalert R., Rahaman M.N. Processing and sintering of yttrium aluminum garnet. *J. Mater. Sci.*, 1996, **31**: 3453.

[85] Li Q., Gao L., Yan D.S. The crystal structure and spectra of nano-scale YAG:Ce^{3+}. *Mater. Chem. Phys.*, 2000, **64:** 41.

[86] Hakuta Y., Seino K., Ura H., Adschiri T., Takizawa H., Arai K. Production of phosphor (YAGTb) fine particles by hydrothermal synthesis in supercritical water. *J. Mater. Chem.*, 1999, **9**: 2671.

[87] Takamori T., David L.D. Controlled nucleation for hydrothermal growth of YAG powders. *Am. Ceram. Soc. Bull.*, 1986, **65**: 1282.

[88] Phillips M.L.F. Low-voltage cathodeluminescence of europium-activated yttrium orthovanadate, in *Liquid Crystal Materials, Devices, and Displays*. San Jose, CA, USA, 1995, Proc. SPIE, 2005, **2408**, 200.

[89] Pillai K.T., Kamat R.V., Vaidya V.N., Sood D.D. Synthesis of yttrium aluminium garnet by the glycerol route. *Mater. Chem. Phys.*, 1996, **44**(3): 255.

[90] Hess N.J., Maupin G.D., Chick L.A., Sunberg D.S., Mccreedy D.E., Armstrong T.R. Synthesis and crystallization of yttrium-aluminium garnet and related compounds. *J. Mater. Sci. Eng. B*, 1994, **29**(7): 1873.

[91] Shea L.E., Mckittrick J., Lopez O.A., Sluzky E. Synthesis of red-emitting, small particle size luminescent oxides used an optimized combustion process. *J. Am. Ceram. Soc.*, 1996, **79**: 3257.

[92] Roy R., Roy D.M., O'holleran T.P. Precision "doping" of simple and complex ceramic powders by DMS technique. *Powder Metal. Intl.*, 1984, **16**(6): 274.

[93] Lee M.H., Oh S.G., Yi S.C. Preparation of Eu-doped Y_2O_3 luminescent nanoparticles in nonionic reverse microemulsions. *J. Colloid. Interface Sci.*, 2000, **226**: 65.

[94] Hirai T., Asada Y., Komasawa I. Preparation of Y_2O_3:Eu^{3+} nanoparticles in reverse micellar systems and their photoluminescence properties. *J. Colloid Interface Sci.*, 2004, **276**: 339.

[95] Sharma P.K., Jilavi M.H., Nass R., Schmidt H. Tailoring the particle size from μm→nm scale by using a surface modifier and their size effect on the fluorescence properties of europium doped yttria. *J. Lumin.*, 1999, **82**: 187.

[96] Graham T. On the properties of silicic acid and other analogous colloidal substances. *J. Chem. Soc.*, 1864, **17**: 318.

[97] Ebelmen M. Sur les combinaisons des acides borique et silicique avec les ethers. *Ann. Chim. Phys*, 1864, **16**: 129.

[98] Brinker C.J., Scherer G.W. *Sol-Gel Science: The Physics and Chemistry of Sol-Gel Processing*. Academic Press, San Diego, 1990.

[99] Bescher E.P., Robson S., Mackenzie J.D., Patt B.E., S.Iwanczyk J., J.Hoffman E. New transparent polycrystalline silicate scintillators, in *Proc. of Sol-Gel Optics*, 2000, Proc. SPIE 2000, **3943**: 139.

[100] Levy D., Quintana X., Rodrigo C., Otón J.M. Gel-glass dispersed liquid crystal projection display, in *Proc. of Sol-Gel Optics III*, 1994, Proc. SPIE 1994, **2288**: 529.

[101] Pope E.J.A. Optically active sol-gel microspheres for flat-panel color displays, in *Proc. of Sol-Gel Optica III*, 1994, Proc. SPIE 2004, **2288**: 536.

[102] Lehmann W. Heterogeneous halide-silica phosphors. *J. Electrochem. Soc.*, 1975, **122**(6): 748.

[103] Costa V.C., Lochhead M.J., Bray K.L. Fluorescence line-narrowing study of Eu^{3+}-doped Sol-gel Silica: Effect of modifying cations on the clustering of Eu^{3+}. *Chem. Mater.*, 1996, **8**: 783.

[104] Lochhead M.J., Bray K.L. Rare-earth clustering and aluminum codoping in solgel silica: Investigation using europium(III) fluorescence spectroscopy. *Chem. Mater.*, 1995, **7**: 572.

[105] Jia W., Liu H., Feofilov S.P., Meltzer R., Jiao J. Spectroscopic study of Eu^{3+}doped and Eu^{3+},Y^{3+} codoped SiO_2 sol-gel glasses. *J. Alloys Compd.*, 2000, **311**: 11.

[106] Hakuta Y., Haganuma T., Sue K., Adschiri T., Arai K. Continuous production of phosphor YAG:Tb nanoparticles by hydrothermal synthesis in supercritical water. *Mater. Res. Bull.*, 2003, **38**: 1257.

[107] Somiya S., Roy R. Hydrothermal synthesis of fine oxide powders. *Bull. Mater. Sci.*, 2000, **23**(6): 453–460.

[108] Byrappa K., Yoshimura M. *Handbook of Hydrothermal Technology.* William Andrew Publishing LLC, NY, 2001.

[109] Byrappa K., ed. *Hydrothermal Growth of Crystals: Progress in Crystal Growth and Characterization of Materials.* Pergamon Press, Oxford, 1991.

[110] Suchanek W.L., Riman R.E. Hydrothermal synthesis of advanced ceramic powders. *Adv. Sci. Technol.*, 2006, **45**: 184.

[111] Li X., Liu H., Wang J.-Y., Cui H.-M., Han F., Zhang X.-D., Boughton R.I. Rapid synthesis of YAG nano-sized powders by a novel method. *Mater. Lett.*, 2004, **58**(19): 2377.

[112] Zhang X., Liu H., He W., Wang J., Li X., Boughton R.I. Synthesis of monodisperse and spherical YAG nanopowder by a mixed solvothermal method. *J. Alloys Compd.*, 2004, **372**(1–2): 300.

[113] Li X., Liu H., Wang J., Cui H., Yang S., Boughton I.R. Solvothermal synthesis and luminescent properties of YAG:Tb nano-sized phosphors. *J. Phys. Chem. Solids*, 2005, **66**(1): 201.

[114] Hiemenz P.C., Rajagopalan R. *Principles of Colloid and Surface Chemistry*, 3rd ed. Marcel Dekker, New York, 1997.

[115] Lamer V.K., Dinegar R.H. Theory, production and mechanism of formation of monodisperse hydrosols. *J. Am. Chem. Soc.*, 1950, **72**(11): 4847.

[116] Huignard A., Gacoin T., Boilot J.-P. Synthesis and luminescence properties of colloidal YVO_4:Eu phosphors. *Chem. Mater.*, 2000, **12**(4): 1090.

[117] Riwotzki K., Haase M. Wet-chemical synthesis of doped colloidal nanoparticles: YVO_4:Ln (Ln) Eu, Sm, Dy. *J. Phys. Chem. B*, 1998, **102**: 10129.

[118] Riwotzki K., Haase M. Colloidal YVO_4:Eu and $YP_{0.95}V_{0.05}O_4$:Eu nanoparticles: luminescence and energy transfer processes. *J. Phys. Chem. B*, 2001, **105**(51): 12709.

[119] Gacoina T. Luminescent nanoparticles: colloidal synthesis and emission properties, in *Nanostructured and Advanced Materials*, Vaseashta, ed. Springer, Netherlands, 2005.

[120] Buissette V., Moreau M., Gacoin T., Boilot J.-P., Chane-Ching J.-Y., Le Mercier T. Colloidal synthesis of luminescent rhabdophane

$LaPO_4:Ln^{3+}x\ H_2O$ (Ln = Ce, Tb, Eu) nanocrystals. *Chem. Mater.*, 2004, **16**(19): 3767.
[121] Singh S.K., Kumar K., Rai S.B. Synthesis and spectroscopy of transparent colloidal solution of Gd_2O_3: Er^{3+}, Yb^{3+} spherical nanocrystals by pulsed laser ablation. *Mat. Sci. Eng. B*, 2010, **166**(2): 180.
[122] Wakefield G., Keron H.A., Dobson P.J., Hutchison J.L. Synthesis and properties of sub-50-nm europium oxide nanoparticles. *J. Colloid Interf. Sci.*, 1999, **215**: 179.
[123] Wakefield G., Holland E., Dobson P.J., Hutchison J.L. Luminescence properties of nanocrystalline Y_2O_3:Eu. *Adv. Mater.*, 2001, **13**(20): 1557.
[124] Ridley B.K. *Quantum Processes in Semiconductors*. Oxford Science Publications, Oxford, 1993.
[125] Sau T.K., Rogach A.L. Nonspherical noble metal nanoparticles: colloid-chemical synthesis and morphology control. *Adv. Mater*, 2009, **21**: 1.
[126] Kwon S.G., Hyeon T. Colloidal chemical synthesis and formation kinetics of uniformly sized nanocrystals of metals, oxides, and chalcogenides. *Acc. Chem. Res.*, 2008, **41**(12): 1696.
[127] Chander H. A Review on synthesis of nanophosphors: future luminescent materials, in *Proc. of ASID '06*. New Delhi, 2006.
[128] Kang Y.C., Park S.B., Lenggoro I.W., Okuyama K. Preparation of non-aggregation YAG-Ce phosphor particles by spray pyrolysis. *J. Aerosol Sci.*, 1998, **29**(suppl. 2): S911.
[129] Lenggoro I.W., Xia B., Mizushima H., Okuyama K., Kijima N. Synthesis of $LaPO_4$:Ce,Tb phosphor particles by spray pyrolysis. *Mater. Lett.*, 2001, **50**: 92.
[130] Jung K.Y., Lee H.W., Kang Y.C., Park S.B., Yang Y.S. Luminescent properties of (Ba,Sr)$MgAl_{10}O_{17}$:Mn,Eu green phosphor prepared by spray pyrolysis under VUV excitation. *Chem. Mater.*, 2005, **17**(10): 2729.
[131] Cho J.S., Jung D.S., Lee S.H., Kang Y.C. Morphological and optical characteristics of $BaMg_{0.9}Al_{10}O_{19}:Mn^{2+}$ phosphor powders prepared by flame spray pyrolysis. *J. Ceram. Process. Res.*, 2008, **9**(1): 83~87.
[132] Joffin N., Dexpert-Ghys J., Verelst M., Baret G., Garcia A. The influence of microstructure on luminescent properties of Y_2O_3:Eu prepared by spray pyrolysis. *J. Lumin.*, 2005, **113**(3–4): 249.
[133] Kingsley J.J., Patil K.C. A novel combustion process for the synthesis of fine particle alpha-alumina and related oxide materials. *Mater. Lett*, 1988, **6**: 427.
[134] Patil K.C., Aruna S.T., Ekambaram S. Combustion synthesis. *Curr. Opin. Solid State Mater. Sci.*, 1997, **2**(2): 158.
[135] Kingsley J.J., Suresh K., Patil K.C. Combustion synthesis of fine-particle metal aluminates. *J. Mater. Sci.*, 1990, **25**: 1305.
[136] Chick L.A., Pederson L.R., Maupin G.D., Bates J.L., Thomas L.E., Exarhos G.J. Glycine nitrate combustion synthesis of oxide ceramic powders. *Mater. Lett*, 1990, **10**: 6.

[137] Ekambaram S., Patil K.C. Combustion synthesis of yttria. *J. Mater. Chem*, 1995, **5**: 905.

[138] Zych E., Deren P.J., Strek W., Meijerink A., Mielcarek W., Domagala K. Preparation, X-ray analysis and spectroscopic investigation of nanostructured Lu_2O_3:Tb. *J. Alloys Compd.*, 2001, **323–324**: 8.

[139] Moore J.J., Feng H.J. Combustion synthesis of advanced materials: part I. Reaction parameters. *Prog. Mater. Sci.*, 1995, **39**: 243.

[140] Moore J.J., Feng H.J. Combustion synthesis of advanced materials: part II. Classification, applications and modelling. *Prog. Mater. Sci.*, 1995, **39**: 275.

[141] Merzhanov A.G. Theory and practice of SHS: worldwide state of the art and the newest results. *Int. J. Self-Propag. High-Temp. Synth.*, 1993, **2**: 113.

[142] Merzhanov A.G. New manifestations of an ancient process, in *Chemistry of Advanced Materials: A Chemistry for the 21st Century*, Rao C.N.R., ed. Blackwell, London, 1993.

[143] Subrahmanyam J., Vijayakumar M. Self-propagating hightemperature synthesis. *J. Mater. Sci.*, 1992, **27**: 6249.

[144] Goldstein J., Newbury D.E., Joy D.C., Lyman C.E., Echlin P., Lifshin E., Sawyer L.C., Michael J.R. Scanning electron microscopy and X-ray microanalysis. 3rd ed. Kluwer Academic/Plenum Publishers, New York, 2003.

[145] Lee R.E. *Scanning Electron Microscopy and X-ray Microanalysis.* Prentice-Hall, 1993.

[146] Reimer L. *Scanning Electron Microscopy: Physics of Image Formation and Microanalysis.* Springer-Verlag, 1985.

[147] Holt D.B., Muir M.D., Grant P.R., Boswarva I.M., eds. *Quantitative Scanning Electron Microscopy.* Academic Press, London, 1975.

[148] Goodhew P.J., Humphreys J., Beanland R. *Electron Microscopy and Analysis.* 3rd ed. Taylor & Francis, New York, 2000.

[149] Janssens K.H.A., Adams F.C.V., Rindby A., eds. *Microscopic X-ray Fluorescence Analysis.* John Wiley & Sons, Chichester, 2000.

[150] Prutton M., Gomati M.M.E., eds. *Scanning Auger Electron Microscopy.* John Wiley & Sons, Chichester, 2006.

[151] Hawkes P.W., Spence J.C.H., eds. *Science of Microscopy.* Springer, New York, 2007.

[152] Smith J.V., Stenstrom R.C. Electron-excited luminescence as a petrologic tool. *J. Geol.*, 1965, **13**: 627.

[153] Phillips M.R., Drouin D. Comparison of low voltage cathodoluminescent phosphors. *Microsc. Microanal.*, 2006, **12**(2): 1526.

[154] Saparin G.V., Obyden S.K. Colour in the microworld: real colour cathodoluminescence mode in scanning electron microscopy. *Eur. Microsc. Anal.*, 1993, **7**(3): 7.

[155] Nazarova T.A., Saparin G.V., Obyden S.K., Ivannikov P.I. Color cathodoluminescence study of oxides subjected to thermal environments. *Scanning*, 1997, **19**(1): 48.

[156] Nazarov M.V., Nazarova T.A., Saparin G.V., Obyden S. Cathodoluminescence from semiconductors and dielectrics subjected to thermal environments. *Eur. Microsc. Anal.*, 1998, **12**(4): 21.

[157] Obyden S.K., Ivannikov P.V., Saparin G.V. Color cathodoluminescence display in the scanning electron microscope of deep relief surfaces. *Scanning* 1997, **19**: 533.

[158] Yacobi B.G., Holt D.B. *Cathodoluminescence Microscopy of Inorganic Solids*. Plenum, New York, 1990.

[159] Ozawa L. *Cathodoluminescence: Theory and Applications*. VCH Publishers, 1991.

[160] Ozawa L. *Cathodoluminescence and Photoluminescence: Theories and Practical Applications*. CRC Press, Boca Raton, 2007.

[161] Reimer L. *Scanning Electron Microscopy: Physics of Image Formation and Microanalysis*. Springer, 1998.

[162] Egerton R.F. *Physical Principles of Electron Microscopy: An Introduction to TEM, SEM, and AEM*. Springer, New York, 2005.

[163] Clarke A.R. *Microscopy Techniques for Materials Science*. CRC Press, Boca Raton, 2002.

[164] Reed S.J.B. *Electron Microprobe Analysis and Scanning Electron Microscopy in Geology*, 2nd ed. Cambridge University Press, 2005.

[165] Reed S.J.B. *Electron Microprobe Analysis*. Cambridge University Press, 1993.

[166] Terence A. *Particle Size Measurement*. 4th ed. Chapman and Hall, New York, 1990.

[167] Leschonski K. Representation and evaluation of particle size analysis data. *Part. Part. Syst. Charact.*, 1984, **1**(1–4): 89.

[168] Database. Representation of results of particle size analysis. Part 1: Graphical representation, in *Int. Org. for Standartization*: *Database ISO-9276-1*, 1998.

[169] Als-Nielsen J., Mcmorrow D. *Elements of Modern X-ray Physics*. John Wiley & Sons, 2001.

[170] Warren B.E. *X-ray Diffraction*. Dover Publishing, New York, 1990.

[171] Cullity B.D. *Elements of X-ray Diffraction*. 2nd ed. Addison-Wesley, 1978.

[172] Bowen D.K., Tanner B.K. *High Resolution X-ray Diffractometry and Topography*. Taylor & Francis, 1998.

[173] Brumberger H. *Modern Aspects of Small-Angle Scattering*. Kluwer Academic Publishers, 1993.

[174] Drenth J. *Principles of Protein X-ray Crystallography*. Springer, 1994.

[175] Samson J.A.R. *Techniques of Vacuum Ultraviolet Spectroscopy*. John Wiley & Sons, New York, 1967.

[176] Arellano I., Nazarov M., Byeon C.C., Popovici E.-J., Kim H., Kang H.C., Noh D.Y. Luminescence and structural properties of Y(Ta,Nb) $O_4:Eu^{3+},Tb^{3+}$ phosphors. *Mater. Chem. Phys.*, 2010, **119**(1–2): 48.

[177] Zamoryanskaya M.V., Konnikov S.G., Zamoryanskii A.N. a high-sensitivity system for cathodoluminescent studies with the Camebax electron probe microanalyzer. *Instrum. Exp. Tech.*, 2004, **47**(4): 477.

[178] Zamoryanskaya M.V., Zamoryanskii A.N., Vainshenker I.A. Optical spectrometer for electron-probe microanalyzer. *Instrum. Exp. Tech.*, 1987, **30**(4): 932.

[179] Li B., Gu Z., Lin J., Su M.Z. X-ray luminescence properties of rare-earth doped orthotantalate. *Mater. Res. Bull.*, 2000, **35**(12): 1921.

[180] Socrates G. *Infrared and Raman Characteristic Group Frequencies Tables and Charts*. 3rd ed. John Wiley & Sons, Chichester, 2004.

[181] Bellamy L.J. *Advances in Infrared Group Frequencies*. Methuen, London, 1968.

[182] Banwell C.N., Elaine M. *Fundamentals of Molecular Spectroscopy*. 4th ed. McGraw-Hill, London, 1994.

[183] Hollas J.M. *Modern Spectroscopy*. 4th ed. John Wiley & Sons, Chichester, 2004.

[184] Lau W.S. *Infrared Characterization for Microelectronics*. World Scientific, Singapore, 1999.

[185] Brixner L.H., Chen H.Y. On the structural and luminescent properties of the M' $LnTaO_4$ rare earth tantalates. *J. Electrochem. Soc.*, 1983, **130**: 2435.

[186] Turrell G., Corset J. *Raman Microscopy: Developments and Applications*. Elsevier/Academic Press, 1996.

[187] Ferraro J.R., Nakamoto K., Brown C.W. *Introductory Raman Spectroscopy*. Academic Press, Amsterdam, 2003.

[188] Mccreery R.L. *Raman Spectroscopy for Chemical Analysis*. John Wiley & Sons, 2000.

[189] Lewis I.R., Edwards H.G.M. *Handbook of Raman Spectroscopy*. CRC Press, Boca Raton, 2001.

[190] Withnall R., Silver J. Probes of structural and electronic environments of phosphor activators: Mossbauer and Raman spectroscopy. *Chem. Rev.*, 2004, **104**(6): 2833.

[191] Nakamoto K. *Infrared and Raman Spectra of Inorganic and Coordination Compounds*. John Wiley & Sons, New York, 1997.

[192] Smith E., Dent G. *Modern Raman Spectroscopy: A Practical Approach*. John Wiley & Sons, Chichester, 2005.

[193] Schrader B., ed. *Infrared and Rarnan Spectroscopy. Methods and Applications*. VCH Verlagsgesellschaft Weinheim, 1995.

[194] Pickwell E., Wallace V.P. Biomedical applications of terahertz technology. *J. Phys. D: Appl. Phys.*, 2006, **39**: R301.

[195] Fattinger C., Grischkowsky D. Point source terahertz optics. *Appl. Phys. Lett.*, 1988, **53**: 1480.

[196] Auston D.H., Cheung K.P., Valdmanis J.A., Kleinman D.A. Cherenkov radiation from femtosecond optical pulses in electro-optic media. *Phys. Rev. Lett.*, 1984, **53**: 1555.

[197] Ferguson B., Abbott D. De-noising techniques for terahertz responses of biological samples. *Microelectron. J.*, 2001, **32**: 943.

[198] Ferguson B., Zhang X.-C. Materials for terahertz science and technology. *Nat. Mater.*, 2002, **1**: 26.

[199] Nazarov M.V., Nazarov M.M. Infrared and terahertz spectroscopy of $Y(Ta,Nb)O_4:Eu^{3+},Tb^{3+}$ phosphors. *J. Mold. Phys. Sci.*, 2009, **8**(3–4): 311.

[200] Ceperley D.M., Alder B.J. Ground state of the electron gas by a stochastic method. *Phys. Rev. Lett.*, 1980, **45**(7): 566.

[201] Assaad F.F., Werner P., Corboz P., Gull E., Troyer M. Symmetry projection schemes for Gaussian Monte Carlo methods. *Phys. Rev. B*, 2005, **72**(22): 224518.

[202] Georges A., Kotliar G., Krauth W., Rozenberg M.J. Dynamical mean-field theory of strongly correlated fermion systems and the limit of infinite dimensions. *Rev. Mod. Phys.*, 1996, **68**(1): 13.

[203] Anisimov V.I., Zaanen J., Andersen O.K. Band theory and Mott insulators: Hubbard U instead of Stoner I. *Phys. Rev. B*, 1991, **44**(3): 943.

[204] Liechtenstein A.I., Anisimov V.I., Zaanen J. Density-functional theory and strong interactions: Orbital ordering in Mott–Hubbard insulators. *Phys. Rev. B*, 1995, **52**(8): R5467.

[205] Anisimov V., Aryasetiawan F., Lichtenstein A. First-principles calculations of the electronic structure and spectra of strongly correlated systems: the LDA + U method. *J. Phys.: Condens. Matter.*, 1997, **9**(4): 767

[206] Slater J.C. Wave functions in a periodic potential. *Phys. Rev.*, 1937, **51**(10): 846.

[207] Hohenberg P., Kohn W. Inhomogeneous electron gas. *Phys. Rev.*, 1964, **136**(3B): B864.

[208] Kohn W., Sham L.J. Self-consistent equations including exchange and correlation effects. *Phys. Rev.*, 1965, **140**(4A): A1133.

[209] Kohn W. Nobel lecture: electronic structure of matter; wave functions and density functionals. *Rev. Mod. Phys.*, 1999, **71**(5): 1253.

[210] Hybertsen M.S., Louie S.G. First-principles theory of quasiparticles: calculation of band gaps in semiconductors and insulators. *Phys. Rev. Lett.*, 1985, **55**(13): 1418.

[211] Callaway J., March N. Density functional methods: theory and applications, in *Solid State Physics*, Ehrenreich H., Seitz D., Turnbull D., eds. Academic Press, New York, 1984.

[212] Jones R.O., Gunnarsson O. The density functional formalism, its applications and prospects. *Rev. Mod. Phys.*, 1989, **61**(3): 689.

[213] Trickey S.B., ed. Density functional theory of many-fermion systems, in *Advances in Quantum Chemistry*. Academic Press, San Diego, 1990.

[214] Forgaciu F., Popovici E.-J., Ciocan C., Ungur L., Vadan M. Calcium tungstate phosphors with well-defined particle sizes, in *Proc. of Sixth Symposium on Optoelectronics, SIOEL '99*. Bucharest, 1999, Proc. SPIE 2000, **4068**: 124.

[215] Grasser R., Scharmann A. The optical properties of extrinsic anionic oxo-complexes in tungstates. *Phys. Status Sol. A*, 1992, **130**: K99.

[216] Uitert L.G. Luminescence of insulating solids for optical masers, in *Luminescence of Inorganic Solids*, Goldberg P, ed. Academic Press, 1966.

[217] Stern W. Colour plasma displays by UV-excited phosphors. A review. *Beitr. Plasmaphys.*, 1985, **25**(1): 57.

[218] Grasser R., Scharmann A., Strack K.R. On the intrinsic nature of the blue luminescence in $CaWO_4$. *J. Lumin.*, 1982, **27**(3): 263.

[219] Zalkin A., Templeton D.H. X-ray diffraction refinement of the calcium tungstate structure. *J. Chem. Phys.*, 1964, **40**(2): 501.

[220] Treadawy M.J., Powell R.C. Luminescence of calcium tungstate crystals. *J. Chem. Phys.*, 1974, **61**: 4003.

[221] Nazarova T.A., Nazarov M.V., Saparin G.V., Obyden S.K., Ivanikov I.P., Popovici E.J., Forgaciu F. Colour cathodoluminescence and photoluminescence of powder samples. *Eur. Microsc. Anal.*, 2001, **15**(2): 19.

[222] Nazarov M.V., Jeon D.Y., Kang J.H., Popovici E.J., Muresan L.E., Zamoryanskaya M.V., Tsukerblat B.S. Luminescence properties of europium-terbium double activated calcium tungstate phosphor. *Solid State Commun.*, 2004, **131**(5): 307.

[223] Popovici E.J., Nemes M., Forgaciu F., Macarovici D., Procedeu de sinteza a luminoforului wolframat de calciu autoactivat, *Romania Patent*, RO114002 B1, 1995.

[224] Popovici E.-J., Nazarov M., Jeon D.Y., Muresan L., Nazarova T.A., Indrea E., Hristea A. Photo- and cathodoluminescence investigation of $CaWO_4$:Eu,Tb phosphors, in *Proc. of Seventh Conference on Optics, ROMOPTO*, Constanta, 2003, Proc. SPIE 2004, **5581**: 754.

[225] Nazarov M.V., Tsukerblat B.S., Popovici E.J., Jeon D.Y. Optical lines in europium-terbium double activated calcium tungstate phosphor. *Phys. Lett. A*, 2004, **330**(3–4): 291.

[226] Nazarov M.V., Tsukerblat B.S., Popovici E.J., Jeon D.Y. Polarization selection rules for the allowed optical transitions in europium-terbium double activated calcium tungstate phosphor. *Solid State Commun.*, 2005, **133**(3): 203.

[227] Mho S.-I., Wright J.C. Site selective spectroscopy of defect chemistry in CdF_2:Eu. *J. Chem. Phys.*, 1982, **77**: 1183.

[228] Judd B.R. Optical absorption intensities of rare-earth ions. *Phys. Rev.*, 1962, **127**(3): 750.

[229] Ofelt G.S. Intensities of crystal spectra of rare-earth ions. *J. Chem. Phys.*, 1962, **37**(3): 511.

[230] Nazarov M.V., Zamoryanskaya M.V., Popovici E.-J., Ungur L., Tsukerblat B.S. Luminescence of calcium tungstate phosphors doped with europium and terbium. *J. Mold. Phys. Sci.*, 2003, **2**(1): 68.

[231] Koster G.F., Dimmok J.O., Wheeler R.G., Statz H. *Properties of the Thirty-Two Point Groups*. M.I.T. Press, Cambridge, 1963.

[232] Judd B.R. Low-lying levels in certain actinide atoms. *Phys. Rev.*, 1962, **125**(2): 613.

[233] Kushida T., Tanaka M. Transition mechanisms and spectral shapes of the 5D-7F line of Eu^{3+} and Sm^{2+} in solids. *Phys. Rev. B*, 2002, **65**(19): 195118.

[234] De Shazer L.G., Dieke G.H. Spectra and energy levels of Eu^{3+} in $LaCl_3$. *J. Chem.Phys*, 1963, **38**: 2190.

[235] Ofelt G.S. Structure of the f^6 configuration with application to rare-earth ions. *J. Chem. Phys.*, 1963, **38**(9): 2171.

[236] Inoue M., Toyazawa Y. Two-photon absorption and energy band structure. *J. Phys. Soc. Jpn.*, 1965, **20**: 363.

[237] Bader T.R., Gold A. Polarization dependence of two-photon absorption in solids. *Phys. Rev.*, 1968, **171**(3): 997.

[238] Tsukerblat B.S., Vitiu E.V. *Optical and Kinetic Effects in Strong Electromagnetic Fields*. Shtiintsa, Kishinev, 1974.

[239] Tsukerblat B.S. *Group Theory in Chemistry and Spectroscopy*. Academic Press, London, 1994.

[240] Chua M., Tanner P.A. Direct calculation of the two-photon line strength of a Γ_{1g}- Γ_{1g} transition in octahedral symmetry. *Phys. Rev. B*, 1996, **54**(16): R11014.

[241] Bouazaoui M., Jacquier B., Linares C., Streck W. Two-photon transitions of Gd^{3+} in cubic $Cs_2NaGdCl_6$. *J. Phys. C: Condens. Matter* 1991, **3**: 921.

[242] Wang D., Ling L., Xia S., Tnner P.A. Theoretical analysis of the two-photon absorption spectrum of Tb^{3+} in $Cs_2NaTbCl_6$. *J. Phys. C: Cond. Matter.*, 2003, **15**: 2681.

[243] Mikhailik V.B., Kraus H., Wahl D., Itoh M., Koike M., Bailiff I.K. One- and two-photon excited luminescence and band-gap assignment in $CaWO_4$. *Phys. Rev. B*, 2004, **69**(20): 205110.

[244] Blaha P., Schwarz K., Madsen G.K.H., Kvasnicka D., Luitz J. In *WIEN2k, An Augmented Plane Wave + Local Orbitals Program for Calculating Crystal Properties*, Schwarz K, ed. Vienna University of Technology, Austria, Vienna, 2001.

[245] Zhang Y., Holzwarth N.A.W., Williams R.T. Electronic band structures of the scheelite materials $CaMoO_4$, $CaWO_4$, $PbMoO_4$, and $PbWO_4$. *Phys. Rev. B*, 1998, **57**(20): 12738.

[246] Issler S.L., Torardi C.C. Solid state chemistry and luminescence of X-ray phosphors. *J. Alloys Compd.*, 1995, **229**(1): 54.

[247] Sonoda M., Takano M., Miyahara J., Kato H. Computed radiography utilizing scanning laser stimulated luminescence. *Radiology*, 1983, **148**(3): 833.

[248] Curry-Iii T.S., J. E. Dowdey, R. C. Murry J. *Christensen's Physics of Diagnostic Radiology*. 4th ed. Lea and Febiger, Philadelphia, 1990.

[249] Brixner L.H. A Study of the calcium molybdate-rare earth niobate systems. *J. Electrochem. Soc.*, 1964, **111**: 690.

[250] Blasse G., Bril A. Luminescence phenomena in compounds with fergusonite structure. *J. Lumin.*, 1970, **3**(2): 109.

[251] Blasse G., Bril A. Photoluminescent efficiency of phosphors with electronic transitions in localized centers. *J. Electrochem. Soc.*, 1968, **115**(10): 1067.

[252] Bril A., Wanmaker W.L., Schuil R.E. In *Proc. of Int. Symp. on Luminescence*. Munich, 1965: 310.

[253] Blasse G., Bril A. Investigation of Tb^{3+} activated phosphors. *Philips Res. Repts.*, 1967, **22**: 481.

[254] Li B., Gu Z.N., Jian J.H., Zeng S.M. *Chem. Chin. Universities* (in Chinese), 2001, **22**: 1.

[255] Chi L.S. *Chin. J. Luminesc.* (in Chinese), 1994, **15**: 15.

[256] Xiao X., Yan B. Synthesis and luminescent properties of novel $RENbO_4$:Ln^{3+} (RE = Y, Gd, Lu; Ln = Eu, Tb) micro-crystalline phosphors. *J. Non-Cryst. Solids*, 2005, **351**(46–48): 3634.

[257] Xiao X., Yan B. Photoluminescence of $Y_{0.6}Gd_{0.4}NbO_4$:Eu^{3+}/Tb^{3+} micrometric phosphors derived from hybrid precursors. *Mater. Lett.*, 2007, **61**(8–9): 1649.

[258] Xiao X., Yan B. Photoluminescent properties of Eu^{3+}(Dy^{3+})-activated YNbx Ta_{1-x} O_4 and $REVTa_2O_9$ (RE = Y, La, Gd) phosphors from the hybrid precursors. *Appl. Phys. A: Mater. Sci. Proc.*, 2007, **88**(2): 333.

[259] Takayama T., Katsumata T., Komuro S., Morikawa T. Growth and characteristics of a new long afterglow phosphorescent yttrium tantalate crystal. *J. Cryst. Growth*, 2005, **275**(1–2): e2013.

[260] Hwang M.H., Kim Y.J. Luminescent properties of Eu^{3+}-doped $YTaO_4$ powders. *Ceram. Int.*, 2008, **34**(4): 1117.

[261] Nazarov M., Tsukerblat B., Byeon C.C., Arellano I., Popovici E.J., Noh D.Y. Polarization selection rules and optical transitions in terbium activated yttrium tantalate phosphor under X-ray, vacuum-ultraviolet, and ultraviolet excitations. *Appl. Optics*, 2009, **48**(1): 17.

[262] Hristea A., Popovici E.-J., Muresan L., Stefan M., Grecu R., Johansson A., Boman M. Morpho-structural and luminescent investigations of niobium activated yttrium tantalate powders. *J. Alloys Compd.*, 2009, **471**(1–2): 524.

[263] Vasiliev V.S., Pinaeva M.M., Shkirman S.F. Subsolidus phase equilibria in the Ta_2O_5-Y_2O_3 system. *J. Inorg. Chem.*, 1979, **24**: 578.

[264] Popovici E.-J., Nazarov M., Muresan L., Noh D.Y., Morar M., Bica E., Indrea E. Studies on terbium activated yttrium based tantalate phosphors. *Radiat. Meas.*, 2010, **45**: 300.

[265] Popovici E.-J., Nazarov M., Muresan L., Noh D.Y., Bica E., Morar M., Arellano I., Indrea E. Studies concerning the properties of some europium activated phosphors based on yttrium tantalate. *Phys. Procedia*, 2009, **2**(2): 185.

[266] Popovici E.-J., Nazarov M., Muresan L., Noh D.Y., Tudoran L.B., Bica E., Indrea E. Synthesis and characterisation of terbium activated yttrium tantalate phosphor. *J. Alloys Compd.*, 2010, **497**: 201.

[267] Blasse G. Vibrational spectra of yttrium niobate and tantalate. *J. Solid State Chem.*, 1973, **7**(2): 169.

[268] Ferguson R.B. The crystallography of synthetic $YTaO_4$ and fused fergusonit. *Can. Mineral*, 1957, **6**: 72.

[269] Wolten G.M. The structure of the M'-phase of $YTaO_4$, a third Fergusonite polymorph. *Acta Crystallogr.*, 1967, **23**: 939.

[270] Blasse G., Bril A. Luminescence in some tantalate host lattices. *J. Solid State Chem.*, 1971, **3**(1): 69.

[271] Reddy V., Method of making $YTaO_4$:Nb phosphor with improved brightness using lithium chloride-lithium sulfate eutectic flux, *US Patent*, US 4938890, 1990.

[272] Hedden D.B., Torardi C.C., Zegarski W. M'-RTaO4 Synthesis: activation of precursor oxides by the reaction flux. *J. Solid State Chem.*, 1995, **118**: 419.

[273] Muresan L., Popovici E.J., Imre-Lucaci F., Grecu R., Indrea E. Studies on Y_2O_3:Eu phosphor with different particle size prepared by wet chemical method *J. Alloys Compd.*, 2009, **483**: 346.

[274] Popovici E.-J., Stefan M., Imre-Lucaci F., Muresan L., Bica E., Indrea E., Barbu-Tudoran L. Studies on the synthesis of cerium activated yttrium aluminate phosphor by wet-chemical route. *Phys. Procedia*, 2009, **2**(2): 603.

[275] Popovici E.J., Muresan L., Hristea A., Indrea E., Vasilescu M., Nazarov M., Jeon D.Y. Synthesis and characterisation of rare earth oxysulphide

phosphors. I. Studies on the preparation of Gd_2O_2S:Tb phosphor by the flux method. *Opt. Mater.*, 2004, **27**(3): 559.

[276] Van Bercum J.G.M., Vermeulen A.C., Delhez R., De Keijser T.H., Mittemeijer E.M. Applicabilities of Warren-Averbach analysis and alternative analysis for separation of size and strain broadening. *J.Appl. Phys.*, 1994, **27**: 345.

[277] Aldea N., Indrea E. XRLINE, a program to evaluate crystallite size of supported metal catalysts by single X-ray profile Fourier analysis. *Comput. Phys. Commun.*, 1990, **60**: 155.

[278] Kraus W., Nolze G. POWDER CELL: a program for representation and manipulation of crystal structures and calculation of the resulting X-ray powder patterns. *J. Appl. Crystallogr.*, 1996, **29**: 301.

[279] Miura N. Phosphors for X-ray and ionizing radiation, in *Practical Applications of Phosphors*, Yen W.M., Shionaya S., Yamamoto H., eds. CRC Press, Boca Raton, 2007.

[280] Massabni A.M.G., Montandon G.J.M., Santos M.A.C.D. Synthesis and luminescence spectroscopy of $YNbO_4$ doped with Eu(III). *Mat. Res.*, 1998, **1**: 1.

[281] Sohn K.S., Zeon W., Chang H., Lee S.K., Park H.D. Combinatorial search for new red phosphors of high efficiency at VUV excitation based on the YRO_4 (R = As, Nb, P, V) System. *Chem. Mater.*, 2002, **14**: 2140.

[282] Nazarov M. Raman spectroscopy of double activated $YNbO_4$:Eu^{3+},Tb^{3+} and $YTaO_4$:Eu^{3+},Tb^{3+}. *J. Mold. Phys. Sci.*, 2010, **9**(1): 5.

[283] Petrov K.I., Gundobin N.V., Kravchenko V.V., Plotkin S.S. *Russ. J. Inorg. Chem.* (in Russian), 1973, **18**: 928.

[284] Jehng J.M., Wachs I.E. Structural chemistry and Raman spectra of niobium oxides. *Chem. Mater.*, 1991, **3**(1): 100.

[285] Rietveld H. Line profiles of neutron powder-diffraction peaks for structure refinement. *Acta Crystallogr.*, 1967, **22**(1): 151.

[286] Lammers M.J.J., Blasse G. Energy transfer phenomena in Tb^{3+}-activated gadolinium tantalate ($GdTaO_4$). *Mater. Res. Bull.*, 1984, **19**(6): 759.

[287] Yang J., Zhang C., Li C., Yu Y., Lin J. Energy transfer and tunable luminescence properties of Eu^{3+} in $TbBO_3$ Microspheres via a facile hydrothermal process. *Inorg. Chem.*, 2008, **47**(16): 7262.

[288] Chen W., Sammynaiken R., Huang Y. Photoluminescence and photostimulated luminescence of Tb^{3+} and Eu^{3+} in zeolite-Y. *J. Appl. Phys.*, 2000, **88**(3): 1424.

[289] Wang H., Yu M., Lin C., Liu X., Lin J. Synthesis and luminescence properties of monodisperse spherical Y_2O_3:Eu^{3+}@SiO_2 particles with core-shell structure. *J. Phys. Chem. C*, 2007, **111**(30): 11223.

[290] Liu X., Lin C., Lin J. White light emission from Eu^{3+} in $CaIn_2O_4$ host lattices. *Appl. Phys. Lett.*, 2007, **90**(8): 081904.

[291] Grisafe D.A., Fritsch C.W. Rare earth niobate systems and the phosphor $GdNbO_4$:Bi. *J. Solid State Chem.*, 1976, **17**(3): 313.

[292] Shin S.H., Jeon D.Y., Suh K.S. Charge-transfer nature in luminescence of $YNbO_4$:Bi blue phosphor. *J. Appl. Phys.*, 2001, **90**(12): 5986.

[293] Blasse G. Some considerations and experiments on concentration quenching of characteristic broad-band fluorescence. *Philips Res. Repts.*, 1968, **23**(4): 344.

[294] Brixner L.H. In *Proc. of Intern. Conf. Luminescence*. Budapest, 1966: 1651.

[295] Boutinand P., Cavalli E., Bettinelli M. Emission quenching induced by intervalence charge transfer in Pr^{3+} -or Tb^{3+}-doped $YNbO_4$ and $CaNb_2O_6$. *J. Phys.: Condens. Matter.*, 2007, **38**: 386230.

[296] Natarajan V., Murthy K.V.R., Jayanth-Kumar M.L. Photoluminescence investigations of Zn_2SiO_4 co-doped with Eu^{3+} and Tb^{3+} ions. *Solid State Commun.*, 2005, **134**(4): 261.

[297] Zhang Z., Wang J., Zhang M., Zhang Q., Su Q. The energy transfer from Eu^{2+} to Tb^{3+} in calcium chlorapatite phosphor and its potential application in LEDs. *Appl. Phys. B: Lasers Opt.*, 2008, **91**(3): 529.

[298] Holloway W.W., Kestigian M., Newman R. Direct evidence for energy transfer between rare earth ions in terbium-europium tungstates. *Phys. Rev. Lett.*, 1963, **11**(10): 458.

[299] Zhang Q.Y., Pita K., Kam C.H. Sol-gel derived zinc silicate phosphor films for full-color display applications. *J. Phys. Chem. Solids*, 2003, **64**(2): 333.

[300] Moadhen A., Elhouichet H., Canut B., Sandu C.S., Oueslati M., Roger J.A. Evidence for energy transfer between Eu^{3+} and Tb^{3+} in porous silicon matrix. *Mat. Sci. Eng.: B*, 2003, **105**(1–3): 157.

[301] Elhouichet H., Daboussi S., Ajlani H., Najar A., Moadhen A., Oueslati M., Tiginyanu I.M., Langa S., Foll H. Strong visible emission from porous GaP doped with Eu and Tb ions. *J. Lumin.*, 2005, **113**(3–4): 329.

[302] Zhang Z.-J., Chen H.-H., Yang X.-X., Zhao J.-T. Preparation and luminescent properties of Eu^{3+} and Tb^{3+} ions in the host of $CaMoO_4$. *Mat. Sci. Eng. B*, 2007, **145**(1–3): 34.

[303] Jorgensen C.K. Electron transfer spectra. *Prog. Inorg. Chem.*, 1970, **12**: 101.

[304] Pauling L. *Nature of the Chemical Bond*. 3rd ed. Cornell University Press, New York, 1960.

[305] Gerard I., Krupa K.C., Simoni E., Martin P. Investigation of charge transfer O^{2-}–Ln^{3+} and F^-–Ln^{3+} in LaF_3:(Ln^{3+}, O^{2-}) and YF_3:(Ln^{3+},O^{2-}) systems. *J. Alloys Compd*, 1994, **207/208**: 120.

[306] Han C.H., Kim H.J., Chang H., Lee S.K., Park H.D. Luminescent properties of YNbO4:Bi phosphors. *J. Electrochem. Soc.*, 2000, **147**(7): 2800.

[307] Lee S.K., Chang H., Han C.H., Kim H.J., Jang H.G., Park H.D. Electronic structures and luminescence properties of $YNbO_4$ and $YNbO_4$:Bi. *J. Solid State Chem.*, 2001, **156**(2): 267.

[308] Perdew J.P., Burke K., Ernzerhof M. Generalized gradient approximation made simple. *Phys. Rev. Lett.*, 1996, **77**: 3865.

[309] Hybertsen M.S., Louie S.G. Electron correlation in semiconductors and insulators: band gaps and quasiparticle energies. *Phys. Rev. B*, 1986, **34**(8): 5390.

[310] Dong X., Bo L., Mu G., Xiaolin L., Shiming H., Chen N. Theoretical calculation of electronic structures and optical properties of $YTaO_4$ and $LuTaO_4$. *Acta Opt. Sin.*, 2009, **29**(2): 448.

[311] Krumpel A.H., Boutinaud P., Van Der Kolk E., Dorenbos P. Charge transfer transitions in the transition metal oxides $ABO_4:Ln^{3+}$ and $APO_4:ln^{3+}$ (A = La, Gd, Y, Lu, Sc; B = V, Nb, Ta; Ln = lanthanide). *J. Lumin.*, 2010, **130**(8): 1357.

[312] Xiao X., Yan B. Hybrid precursors synthesis and optical properties of $LnNbO_4:Bi^{3+}$ blue phosphors and Bi^{3+} sensitizing of on Dy^{3+}'s luminescence in $YNbO_4$ matrix. *J. Alloys Compd.*, 2006, **421**(1–2): 252.

[313] Blasse G., Bril A. Investigations on Bi^{3+} Activated Phosphors. *J. Chem. Phys.*, 1968, **48**: 217.

[314] Ballhausen C.J., Liehr A.D. Intensities in inorganic complexes. Part II. Tetrahedral complexes. *J. Mol. Spectrosc.*, 1958, **2**(1–6): 342.

[315] Ballhausen C.J. A vibrational analysis of the MnO_4^- bands. *Theoretical Chemistry Accounts: Theory, Computation, and Modeling (Theoretica Chimica Acta)*, 1963, **1**(4): 285.

[316] Lee E.Y., Nazarov M., Kim Y.J. Red emission properties of (Y, M) $NbO_4:Eu^{3+}$ (M: Al, Ga) phosphors under near-UV excitation. *J. Electrochem. Soc.*, 2010, **157**(3): J102.

[317] Buth A.H., Blasse G. Luminescence and energy transfer in yttrium niobate ($YNbO_4$). *Phys. Stat. Sol. A*, 1981, **64**(2): 669.

[318] Andrievskaya E.R., Zaitseva Z.A., Shevchenko A.V., Lopato L.M. Recombination luminescence of 88 mol % ZrO_2–Y_2O_3 crystals. *Inorg. Mater.*, 1997, **33**: 390.

[319] Xiao X., Yan B. In-situ chemical co-precipitation composition of hybrid precursors and luminescence of $Y_{1-x}Gd_xNbO_4:RE^{3+}$ (RE = Tb, Eu) micron crystalline phosphors. *J. Alloys Compd.*, 2008, **456**: 447.

[320] Shi S., Liu X., Gao J., Zhou J. Spectroscopic properties and intense red-light emission of (Ca, Eu, M)WO_4 (M = Mg, Zn, Li). *Spectrochim. Acta A*, 2008, **69**: 396.

[321] Kirby A.F., Foster D., Richardson F.S. Comparison of ^7F_J<--$^5D_0$ emission spectra for Eu(III) in crystalline environments of octahedral,

near-octahedral, and trigonal symmetry. *Chem. Phys. Lett.*, 1983, **95**(6): 507.

[322] Nazarov M., Kim Y.J., Lee E.Y., Min K.-I., Jeong M.S., Lee S.W., Noh D.Y. Luminescence and Raman studies of $YNbO_4$ phosphors doped by Eu^{3+}, Ga^{3+}, and Al^{3+}. *J. Appl. Phys.*, 2010, **107**(10): 103104.

[323] Nazarov M.M., Makarova S.A., Shkurinov A.P., Okhotnikov O.G. The use of combination of nonlinear optical materials to control terahertz pulse generation and detection. *Appl. Phys. Lett.*, 2008, **92**: 021114

[324] Nazarov M.M., Shkurinov A.P., Tuchin V.V. Tooth study by terahertz time-domain spectroscopy. *Proc. SPIE Int. Soc. Opt. Eng.*, 2008, **6791**: 679109.

[325] Nazarov M.M., Shkurinov A.P., Tuchin V.V., Zhernovaya O.S. Modification of terahertz pulsed spectrometer to study biological samples. *Proc. SPIE Int. Soc. Opt. Eng.*, 2007, **6535**: 65351J.

[326] Duvillaret L., Garet F., Coutaz J.-L.J. A reliable method for extraction of material parameters in terahertz time-domain spectroscopy. *Sel. Top. Quantum Electron*, 1996, **2**: 739.

[327] Withayachumnankul W., Ferguson B., Rainsford T., Mickan S.P., Abbott D. Material parameter extraction for terahertz time-domain spectroscopy using fixed-point iteration, in *Proc. of Photonic Materials, Devices, and Applications*. Sevilla 2005, Proc. SPIE 2005, **5840**: 221.

[328] Ranson R., Evangelou E., Thomas C. Modeling the fluorescent lifetime of Y_2O_3:Eu. *J. Appl. Phys.*, 1998, **72**: 2663.

[329] Bihari B., Eilers H., Tissue B.M. Spectra and dynamics of monoclinic Eu_2O_3 and Eu^{3+}:Y_2O_3 nanocrystals. *J. Lumin.*, 1997, **75**(1): 1.

[330] Schmechel R., Kennedy M., Von Seggern H., Winkler H., Kolbe M., Fischer R.A., Xaomao L., Benker A., Winterer M., Hahn H. Luminescence properties of nanocrystalline Y_2O_3:Eu^{3+} in different host materials. *J. Appl. Phys.*, 2001, **89**(3): 1679.

[331] Majewski P., Rozumek M., Schluckwerder H., Aldinger F. Phase diagram studies in the systems La_2O_3-SrO-Ga_2O_3 and La_2O_3-MgO-Ga_2O_3 at 1400°C in air. *Int. J. Inorg. Mater.*, 2001, **3**(8): 1343.

[332] Garcнa-Murillo A., Le Luyer C., Garapon C., Dujardin C., Bernstein E., Pedrini C., Mugnier J. Optical properties of europium-doped Gd_2O_3 waveguiding thin films prepared by the sol-gel method. *Opt. Mater.*, 2002, **19**(1): 161.

[333] Jungowska W. Phase equilibria in the system La_2O_3–K_2O–P_2O_5. *J. Thermal Anal. Calorim.*, 2000, **60**: 193.

[334] Nakanishi. Y. In *Proc. Of the III int. meeting on Information Display, IMID*. Daegu, Korea, 2003: 203.

[335] Levshin V.L., Konstantinova M.A., Trapeznikova E.A. On the application of rare-earth elements in the chemistry of phosphors, in *Rare-Earth Elements*, AN Publishing, Moscow, 1959.

[336] Hisamune T. Technical trend of phosphors for plasma display panels, in *Proc. of the 9th International Display Workshops*. San Jose, 2002: 685.

[337] Erdei S., Ainger F.W. In *Proc. of Mater. Res. Soc. Symp.*, 1999: 245.

[338] Database. Powder diffraction file, in *Joint Committee on Powder Diffraction Standards*, 1991.

[339] Lide D.R. *CRC Handbook of Chemistry and Physics*. CRC Press, Boca Raton, 1991.

[340] Newport A., Silver J., Vecht A. The synthesis of fine particle yttrium vanadate phosphors from spherical powder precursors using urea precipitation. *J. Electrochem. Soc.*, 2000, **147**(10): 3944.

[341] Anderson F.G., Weidner H., Summers P.L., Peale R.E., Zhang X.X., Shai B.H.T. Effect of Gd^{3+} substitution for Y^{3+} in $YLiF_4$, $KLiYF_5$, and YVO_4. *J. Lumin.*, 1994, **60&61**: 150.

[342] Wegh R.T., Donker H., Oskam K.D., Meijerink A. Visible quantum cutting in Eu^{3+}-doped gadolinium fluorides via downconversion. *J. Lumin.*, 1999, **82**(2): 93.

[343] Carlos L.D., Videira A.L.L. Emission spectra and local symmetry of the Eu^{3+} ion in polymer electrolytes. *Phys. Rev. B*, 1994, **49**(17): 11721.

[344] Nakamura S., Senoh M., Iwasa N.S., Nagahama S.-I., Yamada T., Mukai T. Superbright green InGaN single-quantum-well-structure light-emitting diodes. *Jpn. J. Appl. Phys.*, 1995, **34**(part 2, No. 10B): L1332.

[345] Nakamura S., Senob M., Iwasa N., Nagahama S. High-power InGaN single-quantum-well-structure blue and violet light-emitting diodes. *Appl. Phys. Lett.*, 1995, **67**: 1868.

[346] Schlotter P., Baur J., Hielscher C., Kunzer M., Obloh H., Schmidt R., Schneider J. Fabrication and characterization of GaN/InGaN/AlGaN double heterostructure LEDs and their application in luminescence conversion LEDs. *Mat. Sci. Eng. B*, 1999, **59**(1–3): 390.

[347] Yum J.-H., Seo S.-Y., Lee S., Sung Y.-E. $Y_3Al_5O_{12}$:$Ce_{0.05}$ Phosphor coatings on gallium nitride for white light emitting diodes. *J. Electrochem. Soc.*, 2003, **150**(2): H47.

[348] Tamura T., Setomoto T., Taguchi T. Illumination characteristics of lighting array using 10 candela-class white LEDs under AC 100 V operation. *J. Lumin.*, 2000, **87–89**: 1180.

[349] Kummer F., Zwaschka F., Ellens A., Debray A., Waitl G., Luminous substance for a light source and light source associated therewith, *US Patent*, US 6669866, 2001

[350] Ellens A., Zwaschka F., Leuchtsto: fuer Lichtquellen und zugehoerige Lichtquelle, *Germany Patent*, DE, 199 51 790 A 1, 2001

[351] Batentschuk M., Osvet A., Schierning G., Klier A., Schneider J., Winnacker A. Simultaneous excitation of Ce^{3+} and Eu^{3+} ions in $Tb_3Al_5O_{12}$. *Radiat. Meas.*, 2004, **38**(4–6): 539.

[352] Guo X., Sakurai K. Formation of yttrium aluminum perovskite and yttrium aluminum garnet by mechanical solid-state reaction. *Jpn. J. Appl. Phys.*, 2000, **39**: 1230.

[353] Inoue M., Otsu H., Kominami H., Inui T. Glycothermal synthesis of rare earth aluminium garnets. *J. Alloys Compd.*, 1995, **226**(1–2): 146.

[354] Kasuya R., Isobe T., Kuma H., Katano J. Photoluminescence enhancement of PEG-modified YAG:Ce^{3+} nanocrystal phosphor prepared by glycothermal method. *J. Phys. Chem. B*, 2005, **109**(47): 22126.

[355] Pereira P.F.S., Matos M.G., Avila L.R., Nassor E.C.O., Cestari A., Ciuffi K.J., Calefi P.S., Nassar E.J. Red, green and blue emission doped YAG phosphors prepared by non-hydrolytic sol-gel route. *J. Lumin.*, 2010, **130**: 488.

[356] Choudhary R., Laishram K., Gupta R.K. Rapid synthesis of Nd:YAG nanopowder by microwave flash combustion. *Mater. Sci. – Poland*, 2009, **27**(4/1): 1025.

[357] Lu C.H., Hsu W.T., Dhanaraj J., Jagannathan R. Sol-gel pyrolysis and photoluminescent characteristics of europium-ion doped yttrium aluminum garnet nanophosphors. *J. Eur. Ceram. Soc.*, 2004, **24**(15–16): 3723.

[358] Popovici E.J., Muresan L., Amalia H., Indrea E., Vasilescu M. Synthesis and characterisation of europium activated yttrium oxide fine powders. *J. Alloys Compd*, 2007, **434–435**: 809.

[359] Muresan L., Popovici E.-J., Grecu R., Tudoran L.B. Studies on the synthesis of europium activated yttrium oxide by wet-chemical method: 1. Influence of precursor quality on phosphor photoluminescence properties. *J. Alloys Compd.*, 2009, **471**(1–2): 421.

[360] Blasse G., Bril A. A new phosphor for flying-spot cathode-ray tubes for color television: yellow-emitting $Y_3Al_5O_{12}$ Single bond Ce^{3+}. *Appl. Phys. Lett.*, 1967, **11**(2): 53.

[361] Blasse G., Bril A. Study of energy transfer from Sb^{3+}, Bi^{3+}, Ce^{3+} to Sm^{3+}, Eu^{3+}, Tb^{3+}, Dy^{3+}. *J. Chem. Phys.*, 1967, **47**(6): 1920.

[362] Kim G.C., Mho S.I., Park H.L. Observation of energy transfer between Ce^{3+} and Eu^{3+} in $YAlO_3$:Ce, Eu. *J. Mater. Sci. Lett.*, 1995, **14**(11): 805.

[363] Miniscalco W.J., Pellegrino J.M., Yen W.M. Measurements of excited-state absorption in Ce^{3+}: YAG. *J. Appl. Phys.*, 1978, **49**(12): 6109.

[364] Park S.H., Mho S.I., Lee K.W. Luminescence characteristics of Y2O2S : Eu^{3+}, Tb^{3+} phosphor. *Bull. Korean Chem. Soc.*, 1996, **17**(5): 487.

[365] Alexander G., Ramakrishnan P., Mukherjee T.K., Nambi K.S.V. Measurement of quantum and cathodoluminescent efficiencies of inorganic powder phosphors using silicon photodiode detector. *Ind. J. Pure Appl. Phys.*, 1993, **31**(8): 531.

[366] Tregellas-Williams J. A review of the measurement of the quantum efficiency of inorganic phosphors. *J. Electrochem. Soc.*, 1958, **105**(3): 173.

[367] Justel T., Krupa J.-C., Wiechert D.U. VUV spectroscopy of luminescent materials for plasma display panels and Xe discharge lamps. *J. Lumin.*, 2001, **93**(3): 179.

[368] Huh Y.-D., Shim J.-H., Kim Y., Do Y.R. Optical properties of three-band white light emitting diodes. *J. Electrochem. Soc*, 2003, **150**(2): H57.

[369] Adams M.J., Higheld J.G., Kirkbright G.F. Determination of the absolute quantum efficiency of sodium salicylate using photoacoustic spectroscopy. *Analyst* 1981, **106**: 850.

[370] Zachau M. Photocalorimetric measurement of the quantum efficiencies of phosphors. *J. Lumin.*, 1997, **72–74**: 792.

[371] Adams M.J., Higheld J.G., Kirkbright G.F. Determination of the absolute quantum efficiency of luminescence of solid materials employing photoacoustic spectroscopy. *Anal. Chem.*, 1980, **52**: 1260.

[372] Peters T.E., Baglio J.A. Luminescence and structural properties of thiogallate phosphors Ce^{3+} and Eu^{2+} activated phosphors. Part I. *J. Electrochem. Soc.*, 1972, **119**(2): 230.

[373] Peters T.E. Luminescent properties of thiogallate phosphors. *J. Electrochem. Soc.*, 1972, **119**(12): 1720.

[374] Loireau-Lozac'h A.M., Guittard M., Flahaut J. Sur une nouvelle famille de combinaisons soufrees, de type "melilite". *Mat. Res. Bull.*, 1973, **8**: 75.

[375] Donohue P.C., Hanlon J.E. The synthesis and photoluminescence of $M^{II}M_2^{III}(S,Se)_4$. *J. Electrochem. Soc.*, 1974, **121**(1): 137.

[376] Eisenmann B., Jakowski M., Schafer H. Die strukturen von $BaGa_2S_4$ und $BaAl_2S_4$. *Mater. Res. Bull.*, 1982, **17**(9): 1169.

[377] Garcia A., Fouassier C., Dougier P. Photo- and cathodoluminescent properties of erbium-doped thiogallates. *J. Electrochem. Soc.*, 1982, **129**(9): 2063.

[378] Davolos M.R., Garcia A., Fouassier C., Hagenmuller P. Luminescence of Eu^{2+} in strontium and barium thiogallates. *J. Solid State Chem.*, 1989, **83**(2): 316.

[379] Georgobiani A.N., Sturov V.V., Tyutyunnikov V.I., Tagiev B.G., Tagiev O.B., Djabbarov R.B. Radical-recombination luminescence ion-luminescence and photoluminescence of $CaGa_2S_4$:Eu. *J. Phys. Chem. Solids*, 2003, **64**: 1519.

[380] Benalloul P., Barthou C., Fouassier C., Georgobiani A.N., Lepnev L.S., Emirov Y.N., Gruzintsev A.N., Tagiev B.G., Tagiev O.B., Jabbarov R.B. Luminescence of Eu^{2+} in calcium thiogallate. *J. Electrochem. Soc.*, 2003, **150**(1): G62.

[381] Jabbarov R.B., Chartier C., Tagiev B.G., Tagiev O.B., Musayeva N.N., Barthou C., Benalloul P. Radiative properties of Eu^{2+} in $BaGa_2S_4$. *J. Phys. Chem. Solids*, 2005, **66**(6): 1049.

[382] Chartier C., Barthou C., Benalloul P., Frigerio J.M. Photoluminescence of Eu^{2+} in $SrGa_2S_4$. *J. Lumin.*, 2005, **111**(3): 147.

[383] Kim J.M., Kim K.N., J.K.Park, Kim C.H., Jang H.G. Synthesis and luminescence characteristics of $SrGa_2S_4$:Eu green phosphor for light emitting diodes by solid-state method. *J. Korean Chem. Soc*, 2004, **48**: 371.

[384] Aidaev F.S. Synthesis and Luminescent Properties of Eu-Activated $CaGa_2S_4$ and $SrGa_2S_4$. *Russ. Inorg. Mater.*, 2003, **39**: 96.

[385] Barthou C., Jabbarov R.B., Benalloul P., Chartier C., Musayeva N.N., Tagiev B.G., Tagiev O.B. Radiative properties of the blue $BaAl_2S_4$:Eu^{2+} phosphor. *J. Electrochem. Soc.*, 2006, **153**(3): G253.

[386] Guo C., Tang Q., Huang D., Zhang C., Su Q. Influence of co-doping different rare earth ions on $CaGa_2S_4$:Eu^{2+}, RE^{3+}(RE = Ln) phosphors. *J. Phys. Chem. Solids*, 2007, **68**(2): 217.

[387] Mueller G.O., Mueller-Mach R. In *Proc. of IDW-2000*. Kobe, Japan, 2000: 821.

[388] Dorenbos P. Energy of the first $4f^7$-->$4f^65d$ transition of Eu^{2+} in inorganic compounds. *J. Lumin.*, 2003, **104**(4): 239.

[389] Kim J.M., Kim K.N., Park J.K., Kim C.H., Jang H.G. Synthesis of the $CaGa_2S_4$:Eu^{2+} phosphors and application in white LEDs, in *Proc.of IMID-2005*. Seoul, 2005: 1623.

[390] Eisenmann B., Jakowski M., Klee W., Schafer H. Luminescence and structural properties of $BaGa_2S_4$ and $BaAl_2S_4$. *Rev. Chim. Miner.*, 1983, **20**: 255.

[391] Fateley W.G., Mcdewitt N.T., Bentley F.F. Infrared and Raman selection rules for lattice vibrations: the correlation method. *Appl. Spectrosc.*, 1971, **25**(2): 155.

[392] Halford R.S. Motions of molecules in condensed systems: I. Selection rules, relative intensities, and orientation effects for Raman and infra-red spectra. *J. Chem. Phys.*, 1946, **14**(1): 8.

[393] Mamedov N., Iida S., Aoki-Matsumoto T., Kato A., Yamazaki M., Yamamoto N., Shishido T., Iwai K. One-phonon spectra and effective phonon density-of-states in $CaGa_2S_4$ *Jpn. J. Appl. Phys.*, 2000, **39**: 307.

[394] Iida S., Mamedov N., Aoki-Matsumoto A., Yamamoto N., Yamakasi M., Kato A., Iwai K. Molecular modes in vibration spectra of $CaGa_2S_4$. *Jpn. J. Appl. Phys.*, 2000, **39** 315.

[395] Chartier C., Jabbarov R., Jouanne M., Morhange J., Benalloul P., Barthou C., Frigerio J., Tagiev B. Raman investigation of polycrystalline $SrGa_2S_4$ compound, in *Proc. of 11th Int. Workshop on Inorg. and org. Electrolumin. 2002. Int. Conf. on the Sci. and Technol. of Emissive Displays and Lighting – EL2002*. Ghent, 2002: 49.

[396] Chartier C., Jabbarov R., Jouanne M., Morhange J., Benalloul P., Barthou C., Frigerio J., Tagiev B., Gambarov E. Raman investigation of orthorhombic $M^{II}Ga_2(S,Se)_4$ compounds. *J. Phys.: Condens. Matter.*, 2002, **14**: 13693.

[397] Schlafer H., Wasgestian H. Ermittlung der F- und G-Matrixelemente für achtfach koordinierte Verbindungen vom Typ des archimedischen Antiprismas. *Theoret. chim. Acta*, 1963, **1**: 369.

[398] Eichenauer L., Jarofke B., Mertins H.C., Dreyhsig J., Busse W., Gumlich H.E., Benalloul P., Barthou C., Benoot J., Fouassier C., Garcia A. Optical characterization of europium and cerium in strontium thiogallate thin films and powders. *Phys. Status Solidi A*, 1996, **153**(2): 515.

[399] Sugar J., Spector N. Spectrum and energy levels of doubly ionized europium (Eu III). *J. Opt. Soc. Am.*, 1974, **64**(11): 1484.

[400] Tagiev B.G., Shakhtakhtinskii M.G., Dzhalilov V.A., Gyul'malyev T.A., Izzatov B.M., Aslanov G.K., Tagiev O.B., Talybov Y.G. Luminescence of the $CaGa_2S_4$ compound doped with rare-earth elements. *Russ. J. Inorg. Mater.* (in Russian), 1993, **29**: 1238.

[401] Nazarov M.V., Kang J.H., Jeon D.Y., Popovici E.J., Muresan L., Tsukerblat B.S. Lattice parameter and luminescence properties of europium activated yttrium oxide. *Solid State Commun.*, 2005, **133**(3): 183.

[402] Iida S., Matsumoto T., Mamedov N.T., An G., Yosuke M., I..Bairamov A., Tagiev B.G., Tagiev O.B., Dzhabbarov R.B. Observation of laser oscillation from $CaGa_2S_4$:Eu^{2+}. *Jpn. J. Appl. Phys.*, 1997, **36**: L857.

[403] Matsumoto T., Iida S., Mamedov N.T., Maruyama Y., An G., Bairamov A.I., Tagiev B.G., Tagiev O.B., Dzhabbarov R.B. Observation of laser oscillation from $CaGa_2S_4$:Eu^{2+}, in *Proc. of 11th Intern.Conf. Ternary and Multinary Compounds, ICTMC-11*, 1998: 1001.

[404] Hidaka C., Yamagishi E., Takizawa T. Preparation of $Ca_{(1-x)}Eu_xGa_2S_4$ crystals and their photoluminescence, absorption and excitation spectra. *J. Phys. Chem. Solids*, 2005, **66**: 2058.

[405] Iida C., Kato A., Tanaka M., Najafov H., Ikuno H. Photoluminescence characterization of rare-earth stoichiometric compounds of $EuGa_2S_4$. *J. Phys. Chem. Solids*, 2003, **64**: 1815.

[406] Henderson B., Imbusch G.F. *Optical Spectroscopy of Inorganic Solids*. Clarendon Press, Oxford, 1989.

[407] Perlin Y.E., Tsukerblat B.S. Optical bands and polarization dichroism of Jahn-Teller centers, in *The Dynamical Jahn-Teller Effect in Localized Systems*, Yu.E. Perlin and Wagner M, eds. Elsevier, Amsterdam, 1984.

[408] Pekar S.I. Theory of F-centers. *JETP (J. Exptl. Theoret. Phys.* (in Russian), 1950, **20:** 267.

[409] Huang K., Rhys A.R. Theory of light absorption and non-radiative transitions in F-centres. *Proc. Royal Soc. Lond. A*, 1950, **204**(1074): 406.

[410] Markham J.J. Interaction of normal modes with electron traps. *Rev. Mod. Phys.*, 1959, **31**: 956.

[411] Perlin Y.E. Modern methods in the theory of many-phonon processes. *Sov. Phys. Usp.*, 1964, **6**(4): 542.

[412] Mannenback M. Computation of the intensities of vibrational spectra of electronic bands in diatomic molecules. *Phys.Status Solidi A*, 1951, **17**(11–12): 1001.

[413] Struck C.W., Fonger W.H. Unified model of the temperature quenching of narrow-line and broad-band emissions. *J. Lumin.*, 1975, **10**(1): 1.

[414] Struck C.W., Fonger W.H. Transition rates in single-h ω models. *J. Lumin.*, 1979, **18–19**(Part 1): 101.

[415] Bartram R.H. Radiationless transitions of point imperfections in solids. *J. Phys. Chem. B*, 1990, **51**(7): 641.

[416] Grinberg M., Jaskylski W., Koepke C., Planelles J., Janowicz M. Spectroscopic manifestation of a confinement-type lattice anharmonicity. *Phys. Rev. B*, 1994, **50**(9): 6504.

[417] Hizhnyakov V., Holmar S., Tehver I. Theory of optical spectra of impurity centres in crystals: general consideration of quadratic vibronic coupling. *J. Lumin.*, 2007, **127**: 7.

[418] Yamamoto H. Fundamentals of luminescence, in *Phosphor Handbook*, S. Shionoya and W.M. Yen, ed. CRC Press, New York, 1999.

[419] Abramowitz M., Stegun I. *Handbook of Mathematical Functions*. Dover, New York, 1972.

[420] Mukai T., Yamada M., Nakamura S. Current and temperature dependences of electroluminescence of InGaN-based UV/blue/green light-emitting diodes. *Jpn. J. Appl. Phys. Part 1*, 1998, **37**: L1358.

[421] Mukai T., Morita D., Nakamura S. High-power UV InGaN/AlGaN double-heterostructure LEDs. *J. Cryst. Growth*, 1998, **189–190**: 778.

[422] Chang S.J., Kuo C.H., Su Y.K., Wu L.W., Sheu J.K., Wen T.C., Lai W.C., Chen J.R., Tsai J.M. 400-nm InGaN-GaN and InGaN-AlGaN multiquantum well light-emitting diodes. *IEEE J. Sel. Top Quantum Electron*, 2002, **8**: 744.

[423] Wang J.G., Jing X.P., Yan C.H., Lin J.H. $Ca_{1-2x}Eu_xLixMoO_4$: a novel red phosphor for solid-state lighting based on a GaN LED. *J. Electrochem. Soc*, 2005, **152**(3): G186.

[424] Hanlon A., Pattison P.M., Kaeding J.F., Sharma R., Sharma R., Fina P., Nakamura S. 292 nm AlGaN Single-quantum well light emitting diodes grown on transparent AlN base. *Jpn. J. Appl. Phys. Part 1*, 2003, **42**: L628.

[425] Sivakumar V., Varadaraju U.V. Intense red-emitting phosphors for white light emitting diodes. *J. Electrochem. Soc.*, 2005, **150**(10): H168.

[426] Nakamura S., Gasol G. The blue laser diode: GaN based light emitters and lasers. Springer, Berlin, 1997.

[427] Neeraj S., Kijima N., Cheetham A.K. Novel red phosphors for solid state lighting; The system $Bi_xLn_{1-x}VO_4$; Eu^{3+}/Sm^{3+} (Ln = Y, Gd). *Solid State Commun.*, 2004, **131**: 65.

[428] Yang W., Luo L., Chen T., Wang N. Luminescence and energy transfer of Eu- and Mn-coactivated $CaAl_2Si_2O_8$ as a potential phosphor for white-light UVLED. *Chem. Mater.*, 2005, **17**: 3883.

[429] Hu Y., Zhuang W., Ye H., Wang D., Zhang S., Huang X. A novel red phosphor for white light emitting diodes. *J. Alloys Compd.*, 2005, **390**(1–2): 226.

[430] Mueller-Mach R., Mueller G., Krames M.R., Hoppe H.A., Stadler F., Schnick W., Juestel T., Schmidt P. Highly efficient all-nitride phosphor-converted white light emitting diode. *Phys. status solidi: A*, 2005, **202**(9): 1727.

[431] Kim J.S., Jeon P.E., Park Y.H., Choi J.C., Park H.L., Kim G.C., Kim T.W. White-light generation through ultraviolet-emitting diode and white-emitting phosphor. *Appl. Phys. Lett.*, 2004, **85**(17): 3696.

[432] Kim J.S., Jeon P.E., Choi J.C., Park H.L., Mho S.I., Kim G.C. Warm-white-light emitting diode utilizing a single-phase full-color $Ba_3MgSi_2O_8:Eu^{2+}$, Mn^{2+} phosphor. *Appl. Phys. Lett.*, 2004, **84**(15): 2931.

[433] Neeraj S., Kijima N., Cheetham A.K. Novel red phosphors for solid-state lighting: the system $NaM(WO_4)_{2-x}(MoO_4)_x:Eu^{3+}$ (M = Gd, Y, Bi). *Chem. Phys. Lett.*, 2004, **387**: 2.

[434] Nazarov M., Yoon. C., White light emitting device, *Korea Patent*, KR10066689(B1), 2006.

[435] Nazarov M., Yoon. C., White light emitting device, KR100665221(B1), *Korea Patent*, KR100665221(B1), 2006.

[436] Nazarov M., Yoon. C., White light emitting device, KR100616812 (B1), *Korea Patent*, KR100616812 (B1), 2006.

[437] Riccardi R., Gout D., Gauthier G., Guillen F., Jobic S., Garcia A., Huguenin D., Macaudiere P., Fouassier C., Brec R. Structural investigation and luminescence properties of the $Ce_3(SiS_4)_2X$ (X = Cl, Br, I) family and the $La_{3-x}Ce_x(SiS_4)_2I$ ($0 \leq x \leq 1$) solid solution. *J. Solid State Chem.*, 1999, **147**: 259.

[438] Li Y.Q., With G., Hintzen H. Synthesis, structure, and luminescence properties of Eu^{2+} and Ce^{3+} activated $BaYSi_4N_7$. *J. Alloys Compd.*, 2004, **385**: 1.

[439] Yamada K., Ohta M., Taguchi T. $Ca(Eu_{1-x}La_x)4Si_3O_{13}$ Red phosphor and its application to tri-chromatic white LEDs. *J. Light Vis. Env.*, 2004, **28**: 73.

[440] Nazarov M., Yoon. C., White light emitting device, *Korea Patent*, KR100616690 (B1), 2006.

[441] Nazarov M., Yoon C. Controlled peak wavelength shift of $Ca_{1-x}Sr_x(S_ySe_{1-y})$: Eu^{2+} phosphor for LED application. *J. Solid State Chem.*, 2006, **179**(8): 2529.

[442] Nazarov M., Kynin A., Yoon C., Biruchinsky S., White light emitting device, *Korea Patent*, KR100714581(B1), 2007.

[443] Nazarov M., Yoon C., Biruchinsky S., White light emitting device *Korea Patent*, KR20070065486(A), 2007.

[444] Nazarov M., Yoon C., Biruchinsky S., White light emitting device, *Korea Patent*, KR100649762(B1), 2006.

[445] Nazarov M. Multiexcited phosphors in UV LED devices. *J. Mold. Phys. Sci.*, 2008, **7**(1): 72.

[446] Schubert E.F., Kim J.K., Luo H., Xi J.Q. Solid-state lighting – a benevolent technology. *Rep. Prog. Phys.*, 2006, **69**: 3069.

[447] Taguchi T., Uchida Y., Kobashi K. Efficient white LED lighting and its application to medical fields. *Phys. Status Solidi A*, 2004, **201**: 2730

[448] Narukawa Y., Narita J., Sakamoto T., Deguchi K., Yamada T., Mukai T. Ultra-high efficiency white light emitting diodes. *Jpn. J. Appl. Phys.*, 2006, **45**: L1084

[449] Yamada M., Naitou T., Izuno K., Tamaki H., Murazaki Y., Kameshima M., Mukai T. Red-enhanced white-light-emitting diode using a new red phosphor. *Jpn. J. Appl. Phys., Part 2*, 2003, **42**: L20.

[450] Sakuma K., Omichi K., Kumura N., Ohashi M., Tanaka D., Hirosaki N., Yamamoto Y., Xie R.J., Suehiro T. Warm-white light-emitting diode with yellowish orange SiAlON ceramic phosphor. *Opt. Lett.*, 2004, **29**: 2001.

[451] Jansen M., Letschert H.P. Inorganic yellow-red pigments without toxic metals. *Nature* 2000, **404**: 980.

[452] Hoppe H.A., Lutz H., Morys P., Schnick W., Seilmeier A. Luminescence in Eu^{2+}-doped $Ba_2Si_5N_8$: fluorescence, thermoluminescence, and upconversion. *J. Phys. Chem. Solids*, 2000, **61**: 2001.

[453] Uheda K., Takizawa H., Endo T., Yamane H., Shimada M., Wang C.M., Mitomo M. Synthesis and luminescent property of Eu^{3+}-doped $LaSi_3N_5$ phosphor. *J. Lumin.*, 2000, **87–89**: 967.

[454] Cao G.Z., Metselaar R. α-Sialon ceramics: a review. *Chem. Mater.*, 1991 **3**(2): 242.

[455] Hampshire S., Park H.K., Thomson D.P. α′-Sialon ceramics. *Nature* 1978, **274**: 880.

[456] Izumi F., Mitomo M., Suzuki J. Structure refinement of yttrium α-sialon from X-ray powder profile data. *J. Mater. Sci. Lett.*, 1982, **1**: 533.

[457] Izumi F., Mitomo M., Bando Y. Rietveld refinements for calcium and yttrium containing α-sialons. *J. Mater. Sci.*, 1984, **19**: 3115.

[458] Van Krevel J.W.H., Rutten J.W.T., Mandal H., Hintzen H.T., Metselaar R. Luminescence properties of terbium-, cerium-, or europium-doped α-sialon materials. *J. Solid State Chem.*, 2002, **165**: 19.

[459] Shannon R.D. Revised effective ionic radii and systematic studies of interatomic distances in halides and chalcogenides. *Acta Crystallogr., Sect. A: Cryst. Phys., Diffr., Theor. Gen. Crystallogr.*, 1976, **32**: 751.

[460] Sakuma K., Hirosaki N., Xie R.J., Yamamoto Y., Suehiro T. Luminescence properties of (Ca,Y)-α-SiAlON:Eu phosphors. *Mater. Lett.*, 2007, **61**(2): 547.

[461] Zhang H., T. Horikawa, H. Hanazawa, A. Hamaguchi, K. Machida. Photoluminescence properties of α-SiAlON:Eu^{2+} prepared by carbothermal reduction and nitridation method. *J. Electrochem. Soc.*, 2007, **154**: J59.

[462] Karunaratne B.S.B., Lumby R.J., Lewis M.H. Rare-earth-doped α'-Sialon ceramics with novel optical properties. *J. Mater. Res.*, 1996, **11**: 2790.

[463] Xie R.J., Hirosaki N., Mitomo M., Yamamoto Y., Suehiro T., Ohashi N. Photoluminescence of cerium-doped α-SiAlON Materials. *J. Am. Ceram. Soc*, 2004, **87**: 1368.

[464] De Graaf D., Hintzen H.T., Hampshire S., With G.D. Long wavelength Eu^{2+} emission in Eu-doped Y–Si–Al–O–N glasses. *J. Eur. Ceram. Soc.*, 2003, **23**: 1094.

[465] Xie R.J., Hirosaki N., Mitomo M., Yamamoto Y., Suehiro T., Sakuma K. Optical Properties of Eu^{2+} in alpha-SiAlON *J. Phys. Chem. B*, 2004, **108**: 12027.

[466] Xie R.J., Hirosaki N., Sakuma K., Yamamoto Y., Mitomo M. Eu^{2+}-doped Ca-α-SiAlON: a yellow phosphor for white light-emitting diodes. *Appl. Phys. Lett.*, 2004, **84**: 5404.

[467] Xie R.J., Mitomo M., Xu F.F., Uheda K., Bando Y. Preparation of Ca-alpha-sialon ceramics with compositions along the Si_3N_4-1/2 Ca_3N_2:3AlN line. *Z. Metallkd.*, 2001, **92**(8): 931.

[468] Peng M., Pei Z., Hong G., Su Q. The reduction of Eu^{3+} to Eu^{2+} in $BaMgSiO_4Eu$ prepared in air and the luminescence of $BaMgSiO_4Eu^{2+}$ phosphor. *J. Mater. Chem.*, 2003, **13**: 1202.

[469] Pires A.M., Davolos M.R. Luminescence of europium(III) and manganese(II) in barium and zinc orthosilicate. *Chem. Mater.*, 2001, **13**: 21.

[470] Xie R.J., Mitomo M., Uheda K., Xu F.F., Akimune Y. Preparation and luminescence spectra of calcium- and rare-earth (R = Eu, Tb, and Pr)-codoped α-SiAlON ceramics. *J. Am. Ceram. Soc.*, 2002, **85**: 1229.

[471] Dorenbos P. Crystal field splitting of lanthanide $4f^{n-1}$- 5d-levels in inorganic compounds. *J. Alloys Compd.*, 2002, **341**: 156.

[472] Kim J.S., Park Y.H., Kim S.M., Choi J.C., Park H.L. Temperature-dependent emission spectra of M_2SiO_4:Eu^{2+} (M = Ca, Sr, Ba) phosphors for green and greenish white LEDs. *Solid State Commun.*, 2005, **133**: 445.

[473] Chen Y., Liu B., Shi C., Ren G., Zimmerer G. The temperature effect of $Lu_2SiO_5:Ce^{3+}$ luminescence. *Nucl. Instrum. Methods Phys. Res. A*, 2005, **537**: 31.

[474] Xie R.J., Hirosaki N., Kimura N., Sakuma K., Mitomo M. 2-phosphor-converted white light-emitting diodes using oxynitride/nitride phosphors. *Appl. Phys. Lett.*, 2007, **90**(19): 191101.

[475] Xiaolin S., Guangyan H., Xinyong D., Dong X., Guilan Z., Guoqing T., Wenju C. Study of energy transfer between rare earth ions in CaS host by photoluminescence spectra. *J. Phys. Chem. Solids*, 2001, **62**(4): 807.

[476] Pham-Thi M. Rare-earth calcium sulfide phosphors for cathode-ray tube displays. *J. Alloys Compd.*, 1995, **225**(1–2): 547.

[477] Summers C.J., Wagner B.K., Tong W., Park W., Chaichimansour M., Xin Y.B. Recent progress in the development of full color SrS-based electroluminescent phosphors. *J. Cryst. Growth*, 2000, **214–215**: 918.

[478] Wu J., Newman D., Viney I.V.F. Study on relationship of luminescence in CaS:Eu,Sm and dopants concentration. *J. Lumin.*, 2002, **99**(3): 237.

[479] Choi H., Kim C.-H., Pyun C.-H., Kim S.-J. Valencies of Eu in CaS:Eu, La. *J. Solid State Chem.*, 1998, **138**(1): 149.

[480] Tian Y., Zaremba D., Hill I., Red photoluminescent phosphors, *US Patent*, US 6783700, 2004.

[481] Van Haecke J.E., Smet P.F., Poelman. D. The influence of source powder composition on the electroluminescence of $Ca_{1-x}Sr_xS$:Eu thin films. *Spectrochim. Acta B*, 2004, **59**(10–11): 1759.

[482] Zhang X., Liang L., Zhang J., Su Q. Luminescence properties of $(Ca_{1-x}Sr_x)Se:Eu^{2+}$ phosphors for white LEDs application. *Mater. Lett.*, 2005, **59**(7): 749.

[483] Poelman D., Vercaemst R., Van Meirhaeghe R.L., Laflнre W.H., Cardon F. The influence of Se-coevaporation on the emission spectra of CaS:Eu and SrS:Ce thin film electroluminescent devices. *J. Lumin.*, 1995, **65**(1): 7.

[484] Jenkins R., Snyder R., eds. *Introduction to X-ray Powder Diffractometry*. John Wiley & Sons, 1996.

[485] Tanaka S. Color electroluminescence in alkaline-earth sulfide thin-films. *J. Lumin.*, 1988, **40–41**: 20

Appendix I

The Luminescence Literature

A. Phosphor Books

Cathodoluminescent Phosphors: Theories and Practical Applications, by Lyuji Ozawa, May 2007

Luminescence and Display Phosphors: Phenomena and Applications, by Arunachalam Lakshmanan, January 2007

Luminescence: From Theory to Applications, by Cees Ronda (Editor), 2007

Fundamentals of Phosphors, by Shigeo Shionoya (Editor), Hajime Yamamoto (Editor), William M. Yen (Editor), December 2006

Measurements of Phosphor Properties, by William M. Yen (Editor), Shigeo Shionoya (Editor), Hajime Yamamoto (Editor), December 2006

Practical Applications of Phosphors, by Shigeo Shionoya (Editor), Hajime Yamamoto (Editor), William M. Yen (Editor), December 2006

Phosphor Handbook, by Shigeo Shionoya (Editor), William M. Yen, Hamime Yamamoto (Editor), December 2006, 2nd edition

Dielectrics in Emerging Technologies and Persistent Phosphors: Joint Proceedings of the International Symposia by K. Worhoff, January 2006

.*Luminescent Materials and Applications*, by A. Kitai (Editor), 2006

Modern Luminescence Spectroscopy of Minerals and Materials, by Michael Gaft, Renata Reisfeld, and Gerard Panczer, 2005

Spectroscopic Properties of Rare Earths in Optical Materials, by Guokui Liu (Editor) and Bernard Jacquier (Editor), 2005

Select Topics In Luminescent Materials, by P. Reineker, V. Stefan (Editor), A. Vitukhnovsky (Editor), December 2004

Inorganic Phosphors: Compositions, Preparation and Optical Properties, by Marvin J. Weber (Editor), W.M. Yen (Editor), William M. Yen, June 2004

Luminescence and the Solid State, Second Edition (Studies in Inorganic Chemistry) by Richard C. Ropp, 2004

Handbook of Luminescence, Display Materials, and Devices, by Hari Singh-Nalwa (Editor), Lauren Shea Rohwer (Editor), July 2003

Advanced Luminescent Materials and Quantum Confinement, by David J. Lockwood (Editor), M. Cahay (Editor), J. P. Leburton (Editor),

S. Bandyopadhyay (Editor), N. Koshida (Editor), M. Zacharias (Editor), December 2002

Polymers, Phosphors, and Voltaics for Radioisotope Microbatteries, by Kenneth E. Bower (Editor), Yuri A. Barbanel (Editor), Yuri G. Shreter (Editor), George W. Bohnert (Editor), June 2002

Luminescence and Luminescent Materials: Symposium Held April 17–19, 2001, San Francisco, California, U.S.A by Peter C. Schmidt (Editor), October 2001.

Luminescent Materials: Symposium Held April 5–8, 1999, San Francisco, California, U.S.A by Baldassare Di Bartolo (Editor), Joanna McKittrick (Editor), Kailash Mishra (Editor), September 1999

Physics and Chemistry of Luminescence, by K. C. Mishra, Baldassare Di Bartolo, C. W. Truck, December 1998

Phosphor Handbook, by Shigeo Shionoya, (Editor), September 1998

Proceedings of the Sixth International Conference on Luminescent Materials, by Electrochemical Society, C. R. Ronda, T. Welker, January 1998

Synthesis and Application of Lanthanide-Doped Materials, by Allan J. Bruce (Editor), B. G. Potter (Editor), Ohio) American Ceramic Society. Meeting (97th: 1995: Cincinnati October 1997

Luminescent Materials, by G. Blasse and B.C. Grabmeyer, 1994

Solid State Luminescence: Theory, Materials and Devices, by A. Kitai, 1993

Cathodoluminescence: Theory and Applications, by Lyuji Ozawa, 1991

Luminescence of Solids, by D. R. Vij, 1988

Fluorescent Lamp Phosphors, by K. H. Butler, 1980

An Introduction to Luminescence of Solids, by Humboldt W. Leverenz, 1950

B. Phosphor Reviews

Inorganic luminescent materials: 100 years of research and application, C. Feldmann, T. Justel, C. R. Ronda, P. J. Schmidt, Advanced Functional Materials 13, 511–516 (2003).

Phosphors for plasma display panels, C.-H. Kim, I.-E. Kwon, C.-H. Park, Y.-J. Hwang, H.-S. Bae, B.-Y. Yu, C.-H. Pyun, and G.-Y. Hong, Journal of Alloys and Compounds 311, 33–39 (2000).

RE-activated lanthanide phosphate phosphors for PDP applications, R. P. Rao and D. J. Devine, Journal of Luminescence 87–89, 1260–1263 (2000).

Recent achievements in research on phosphors for lamps and displays, C. R. Ronda, Journal of Luminescence 72–74, 49–54 (1997).

New X-ray phosphors, by L. H. Brixner, Materials Chemistry and Physics 16, 253–281 (1987).

Scintillation detectors for X-rays, N. Martin, Measurement Science and Technology 17, R37-R54 (2006).

Solid state lighting – a world of expanding opportunities at LED 2002, A. Mills, III-Vs Review 16, 30–33 (2003).

Development of phosphors for LEDS, L. S. Rohwer and A. M. Srivastava, The Electrochemical Society INTERFACE 12, 36–39 (2003).

Phosphors development for LED lighting, A. Mills, III-Vs Review 18, 32–34 (2005).

Phosphors to brighten your future, by A. Mills, III-Vs Review 19, 36–39 (2006).

New phosphors for white LEDs: Material Design Concepts, M. Mikami, H. Watanabe, K. Uheda, S. Shimooka, Y. Shimomura, T. Kurushima, and N. Kijima, IOP Conference Series: Materials Science and Engineering 1, 012002 (2009).

Luminescent materials, C. Fouassier, Current Opinion in Solid State and Materials Science 2, 231–235 (1997).

New developments in the field of luminescent materials for lighting and displays, T. Justel, H. Nikol, and C. Ronda, Angewandte Chemie International Edition 37, 3084–3103 (1998).

Rare earth phosphors: Fundamentals and applications, C. R. Ronda, T. Justel, and H. Nikol, Journal of Alloys and Compounds 275–277, 669–676 (1998).

Recent developments in Rare Earth doped materials for optoelectronics, A. J. Kenyon, Progress in Quantum Electronics 26, 225–284 (2002).

Phosphors, A. M. Srivastava and C. R. Ronda, The Electrochemical Society INTERFACE 12, 48–51 (2003).

Role of the surface in luminescent processes, B. L. Abrams and P. H. Holloway, Chemical Reviews 104, 5783–5802 (2004).

Development of nanophosphors, H. Chander, Materials Science and Engineering: R: Reports 49, 113–155 (2005).

A Review on synthesis of nanophosphors – future luminescent materials, Harish Chander in Proc. of ASID'06, New Delhi, 2006

Taking advantage of luminescent lanthanide ions, Jean-Claude G. Bünzli and Claude Piguet Chem. Soc. Rev 34 (2005) 1048

Synthesis of lamp phosphors: facile combustion approach, S. Ekambaram, K. C. Patil and M. Maaza, Journal of Alloys and Compounds 393 (2005) 81

VUV spectroscopy of luminescent materials for plasma display panels and Xe discharge lamps, Thomas Justel, Jean-Claude Krupa, Detlef U Wiechert, Journal of Luminescence 93 (2001) 179

A review of the measurement of the quantum efficiency of inorganic phosphors, J. Tregellas-Williams, Journal of The Electrochemical Society 105 (1958) 173.

Luminescence spectroscopy with synchrotron radiation: History, highlights, future, Georg Zimmerer, Journal of Luminescence 119–120 (2006) 1

Appendix II

List of Main Abbreviations

AID – activator incorporation degree
CL – cathodoluminescence
CCL – color cathodoluminescence
CIE – International Commission on Illumination (usually abbreviated CIE for its French name, Commission internationale de l'éclairage)
CRI – color rendering index
CRT – cathode ray tube
CTB – charge-transfer band
DFT – density functional theory
DOS – density of states
EDS – energy dispersive spectroscopy
EPMA – electron probe microanalysis
ICSD – inorganic crystallographic structure database
FPD – flat panel display
FWHM – full widths at half maximum
FTIR – Fourier transform infrared
LED – light emitting diode
NIRS – near-infrared spectroscopy
PDP – plasma display panel
PL – photoluminescence
PLE – photoluminescence excitation
RE – rare earth
SEM – scanning electron microscopy
THz – terahertz
TAG – terbium aluminum garnet, $Tb_3Al_5O_{12}$
YAG – yttrium aluminum garnet, $Y_3Al_5O_{12}$
WDS – wavelength-dispersive spectroscopy
XL – X-ray luminescence
XRD – X-ray diffraction
VUV – vacuum ultraviolet

Index